AF675359

Sustainable Recycling Techniques of Pavement Materials

Sustainable Recycling Techniques of Pavement Materials

Editors

Qiang Li
Jiaqing Wang
Kaijian Huang
Dongdong Ge
Fangyuan Gong

MDPI • Basel • Beijing • Wuhan • Barcelona • Belgrade • Manchester • Tokyo • Cluj • Tianjin

Editors

Qiang Li
Nanjing Forestry University
China

Jiaqing Wang
Nanjing Forestry University
China

Kaijian Huang
Nanjing Forestry University
China

Dongdong Ge
Changsha University of Science & Technology
China

Fangyuan Gong
Hebei University of Technology
China

Editorial Office
MDPI
St. Alban-Anlage 66
4052 Basel, Switzerland

This is a reprint of articles from the Special Issue published online in the open access journal *Materials* (ISSN 1996-1944) (available at: https://www.mdpi.com/journal/materials/special_issues/sustainable_recycling_pavement_materials).

For citation purposes, cite each article independently as indicated on the article page online and as indicated below:

LastName, A.A.; LastName, B.B.; LastName, C.C. Article Title. *Journal Name* **Year**, *Volume Number*, Page Range.

ISBN 978-3-0365-6199-8 (Hbk)
ISBN 978-3-0365-6200-1 (PDF)

© 2022 by the authors. Articles in this book are Open Access and distributed under the Creative Commons Attribution (CC BY) license, which allows users to download, copy and build upon published articles, as long as the author and publisher are properly credited, which ensures maximum dissemination and a wider impact of our publications.

The book as a whole is distributed by MDPI under the terms and conditions of the Creative Commons license CC BY-NC-ND.

Contents

materials

Editorial

Sustainable Recycling Techniques of Pavement Materials

Jiaqing Wang [1,*], Qiang Li [1,*], Kaijian Huang [1], Dongdong Ge [2] and Fangyuan Gong [3]

1 College of Civil Engineering, Nanjing Forestry University, Nanjing 210037, China
2 National Engineering Laboratory of Highway Maintenance Technology, Changsha University of Science & Technology, Changsha 410114, China
3 School of Civil and Transportation Engineering, Hebei University of Technology, Tianjin 300401, China
* Correspondence: jiaqingw@njfu.edu.cn (J.W.); liqiang2526@njfu.edu.cn (Q.L.)

Innovative sustainable techniques for transportation infrastructure enhancement have been proposed in recent decades. The recycling of solid wastes or used pavement materials is facing some challenges that are mainly related to reduced performance since the aging effect or other damages accumulated during the service period. Compared with normal pavement materials prepared with natural resources, the mechanical strength, chemical composition, and microstructure are obviously affected. Not only the raw materials but also the energies consumed during the life-cycle could be saved with recycled materials and improved durability by utilizing these innovative sustainable techniques.

This Special Issue attracted innovative and efficient techniques and materials for pavement recycling and reconstruction. The collected studies include pavement recycling techniques, the effective utilization of industrial and construction waste in infrastructure engineering, improvements in sustainable techniques for pavement materials, green low-carbon and durable pavement structures and materials, the smart construction (3D printing) of sustainable infrastructures, the investigation of durability performance enhancement by recycled pavement materials, and the application of life-cycle assessment (LCA) in evaluating infrastructure sustainability.

The innovative green materials, new techniques, and main contributions of these studies are summarized below:

Xue et al. investigated the performance of an asphalt binder and mixture modified by lignin and waste engine oil [1]. Lignin, as a by-product or waste material of biomass materials, is used as a modifier for asphalt materials; similarly, waste motor oil is the non-distillable part collected from the solid–liquid copolymer formed by the consumption of motor oil after oxidation, etc. Waste motor oil and asphalt have relatively similar compositions and good recyclable value and are also considered to be good asphalt modifiers, both of which can be utilized and have great potential for utilization. In this study, two additives, lignin and waste engine oil, were used for matrix asphalt modification (previously, only a single additive was used to improve one aspect of the asphalt properties) and based on the Brookfield viscometer, dynamic shear rheometer, bending beam rheometer, rutting test, and low-temperature bending beam test, to explore the high- and low-temperature lignin-waste-engine-oil-modified asphalt binder and its mixes performance. The lignin increases the rotational viscosity, while waste oil decreases it. The lignin increases the elastic component, while waste motor oil decreases the elastic component. Waste motor oil can improve low-temperature performance. The dynamic stability obtained for asphalt mixtures has a good linear relationship with viscosity.

Plastics are synthetic materials mainly derived from refined crude oil petroleum products; the main sources of waste plastics in the territory are plastic containers, plastic packaging, and other common plastic industrial products. Due to the huge amount of plastic production and the need to reduce environmental concerns, waste plastics are attractive for asphalt modification, and there are great prospects for development [2]. Using waste plastics as an asphalt modifier expands the application field of waste plastics,

Citation: Wang, J.; Li, Q.; Huang, K.; Ge, D.; Gong, F. Sustainable Recycling Techniques of Pavement Materials. *Materials* **2022**, *15*, 8710. https://doi.org/10.3390/ma15248710

Received: 22 November 2022
Accepted: 1 December 2022
Published: 7 December 2022

Publisher's Note: MDPI stays neutral with regard to jurisdictional claims in published maps and institutional affiliations.

Copyright: © 2022 by the authors. Licensee MDPI, Basel, Switzerland. This article is an open access article distributed under the terms and conditions of the Creative Commons Attribution (CC BY) license (https://creativecommons.org/licenses/by/4.0/).

which solves the problem of waste plastic disposal and reduces environmental pollution. According to the source of waste plastics, choose a suitable processing process. When used for asphalt modification, scrap plastics can be made in various forms and sizes depending on the specific application. Asphalt modified with waste plastics alone is a physical modification. LCA methods can effectively quantify the environmental impact of waste plastics on asphalt and asphalt mix modification. The high-temperature performance of waste plastic-modified asphalt is quite effective after the introduction of such plastics. However, the low-temperature performance still needs to be improved in future studies.

Municipal solid waste (MSW) is largely treated using incineration, and the resulting product is divided into fly ash and bottom ash, depending on where it is collected. Fly ash has long been used as a road construction material. Using fly ash in asphalt mixes and concrete structures reduces the requirement to use natural aggregates and mineral fillers [3]. Zhao et al. quantified the properties of asphalt mortars and asphalt mixtures containing municipal solid waste incineration (MSWI) fly ash as a filler, including the rheological properties, high-temperature properties, low-temperature properties, and water stability. The type of filler has little effect on the low-temperature properties of asphalt mortar. With an increased fly ash replacement rate, the high-temperature performance of the mix is enhanced. As the replacement rate of fly ash increases, the tensile strength and stiffness modulus of asphalt mixtures increase, and the low-temperature performance decreases. The addition of fly ash weakened the moisture stability of the asphalt mixture.

The uniformity of natural aggregate quality is difficult to control. The homogeneity of the material (e.g., density, hardness, etc.) and the shape characteristics of the aggregate (e.g., profile shape, angle, surface texture, etc.) control the homogeneity of the coarse aggregate quality. Li et al. [4] propose using 3D-printing technology to generate standard aggregate models, explore the feasibility of a direct preparation of standard aggregates using 3D-printing technology from the technical index and economic perspectives, and clarify the development direction of artificial aggregates. Through a series of indoor material tests, a cement-based material suitable for preparing high-strength artificial aggregates was developed, and a set of preparation processes was determined. Finally, the physical and mechanical properties of the natural aggregates were compared to verify the technical properties of the artificial aggregates and their engineering feasibility. The results demonstrated that photosensitive resin-based materials and ABS resin-based materials have excellent bending properties. However, they have weaker compressive strength and hardness compared to natural aggregates. Preparing resin-based artificial aggregates using 3D-printing technology is technically feasible, but its high cost severely limits the large-scale industrialization of on-site implementation. Cement-based materials are a low-cost and reliable alternative to the preparation of artificial aggregates using 3D printing technology. In this study, the optimal ratio of cementitious materials was proposed. The manufacturing procedures for cement-based artificial aggregates were improved and optimized. According to a comprehensive evaluation, the 3D shape of the prepared cementitious artificial aggregates has good agreement with the 3D morphology of natural aggregates.

Building demolition waste (BDW) is typically generated during the construction, maintenance, repair, rehabilitation, and demolition of buildings and other infrastructure. These wastes are heterogeneous and consist of different materials, such as concrete, brick, ceramics, wood, glass, plastics, soil, and gypsum-based construction materials, with bulky wastes (such as bricks and concrete) accounting for 70–80% of the total weight. Xiao et al. [5] investigated the feasibility and applicability of recycled BDW materials as unbonded granular fillers for pavement subgrade/subbase. Five different gradations were designed to represent the gradation changes caused by granular crushing, and through tests, data were obtained to propose an empirical mechanical prediction model for the modulus of elasticity by combining the gradation changes and stress states. This provided a technical reference and guidance to promote recycled BDW as an unbound granular material in pavement subgrade/subbase applications. In this paper, the authors investigated the effects of crushing-induced gradation changes on the shear strength and the resilient properties of BDW recycled aggregates through a series of monotonic triaxial compression tests and repeated load triaxial (RLT) tests. The authors propose a model for predicting the elastic modulus in combination with particle crushing. The model provides a useful guide for predicting the elastic properties of recycled BDW aggregates with significantly lower strength and friability. In addition, considering the high absorption ratio of recycled BDW aggregates, their water-sensitive properties are being further investigated to provide a more comprehensive understanding of the application of recycled BDW aggregates in pavement granular layers.

Constructing porous asphalt pavements is an effective way to promote sponge city infrastructure construction. The main asphalt binder used in porous asphalt is high-viscosity modified asphalt, which is overexposed to oxygen, temperature, water, and ultraviolet radiation, thus accelerating its aging damage, leading to the deterioration of asphalt pavement performance. In the study proposed by Zhang et al. [6], the aging characteristics of high-viscosity modified asphalt under multi-factor coupled environmental conditions were investigated based on infrared spectroscopy experiments. Finally, the aging kinetic equations of the high-viscosity modified asphalt based on a functional group index were established. The results demonstrated that high-viscosity asphalt is sensitive to complex aging factors through changes in chemical functional groups.

The 3D concrete printing (3DCP) technique uses cementitious composite material deposited in layers of shapes with the help of computer design and large-scale 3D printers. Fly ash (FA), silica fume (SF), and ground granulated blast furnace slag (GGBS) are solid waste by-products from the energy and smelting industries, and cementitious composites using 3DCP technology use a combination of (FA), (SF), and (GGBS) to replace the cement binder. The study conducted by Wang et al. [7] aimed to investigate the effect of different steam-curing conditions on the mechanical properties of concrete using the 3DCP technique. Orthogonal experiments were performed to reduce the experimental workload and to obtain highly sensitive evaluation indices. In addition, coefficients were used to assess the anisotropy of the material quantitatively. Subsequently, the printed interlayer bonding properties were further analyzed. As a result, optimal steam-curing conditions for printed concrete were obtained for various mechanical properties and anisotropy. This study proposes that anisotropy coefficients and orthogonal tests can be used to evaluate the mechanical properties and anisotropy of concrete using the 3DCP technique when various steam-curing conditions are used for printed materials. Subsequently, optimal steam-curing conditions and interlayer bonding were demonstrated.

Wang et al. [8] investigated the mechanisms of water transformation and pore structure evolution of Cement-Stabilized Dredged Sediment (CDS) using the nuclear magnetic resonance (NMR) technique, and quantitative relationships between the macroscopic mechanical properties and the microstructure were established. A series of laboratory samples were prepared for different cement contents and maintenance periods. In addition, stress–strain curves were determined based on unconfined compression tests, and the influence of the water content parameters on strength development was quantitatively assessed from the perspective of stabilization mechanisms. Similarly, a mathematical model of the relationship between the hydraulic conductivity measured with a flexible wall penetrometer and the representative parameters of the pore size distribution was presented. Finally, the reliability of the NMR technique as a new method for studying the microscopic mechanisms of CDS strength and permeability evolution was further validated with scanning electron microscopy (SEM) and mercury-in-pressure porosity (MIP) tests.

Construction and demolition (C&D) waste can be utilized after incineration and activation. The properties of polypropylene fibers prevent cracking and inhibit brittle damage; moreover, polypropylene fibers mitigate sulfate attack. Zhang et al. [9] experimentally investigated the enhancement of cement-stabilized soil (CSS) properties with the introduction of treated C&D waste, polypropylene fibers, and sodium sulfate. For this, 10%, 20% and 30% of C&D waste were used as a substitute for cement. The dosages of polypropylene fiber and sodium sulfate were 1%, 2%, 4%, and 0.2%, and 0.4% and 0.8%, respectively. Uniaxial compressive strength (UCS) tests, flexural strength (FS) tests, and direct shear tests were conducted to examine the effect of coupling reinforcement on the mechanical properties of CSS. A backpropagation neural network (BPNN) and FR-tuned hyperparameters with the BAS algorithm were used to predict the UCS and FS properties of CSS based on 84 experimental results, which can be used as design guidelines for the future applications of such sustainable materials.

Zheng et al. [10] investigated the effect of carbonation on cementitious materials under waterlogged and saturated humidity environments by increasing CO_2 concentration using an accelerated CO_2 erosion tester that simulates the environment used in inflow tunnels. The results of the study can provide some theoretical and technical support for the design, construction, and safe operation of the diversion tunnel lining for the better sustainability of cementitious materials under carbonation environments.

Steel slag is a significant solid waste from steelmaking. The application of recycled steel slag in asphalt mixes is mainly hindered by the volume expansion after water absorption. Proper treatment methods must be proposed to solve the problem of water-induced volume expansion of steel slag and evaluate the submersion stability of the treated steel slag and its asphalt mixture. The mechanisms of oxalic acid and water erosion on the performance of steel slag and its asphalt mixture were analyzed by Huang et al. [11], and the suitability of steel slag and its mixture treated with oxalic acid was evaluated, providing engineering suggestions and theoretical guidance for the practical application of recycled steel slag in asphalt mixtures. In this study, steel slag was treated with a modified oxalic acid solution to inhibit the swelling and water erosion of steel slag and its asphalt mixture at different times before and after the treatment. Then, the basic properties of the steel slag and its asphalt mixture were tested by comparison. The mechanisms of oxalic acid and water erosion on the mixture properties of steel slag and its asphalt mixture were investigated. The potential of oxalic acid treatment of steel slag in asphalt mixes was examined.

Bio-oil can be recycled from plant and animal remains, restaurant waste oil, and discarded handicrafts. The use of biomass materials to modify asphalt is sustainable, and bio-oil can be added to asphalt/water emulsions as a penetrating agent, which is effective in increasing the penetration depth. Shi et al. [12] investigated the effect of emulsifier content on the stability of biomass asphalt/water emulsions based on molecular dynamics (MD) simulation techniques combined with macroscopic and microscopic experiments. The article investigated the effect of emulsifier dosage on the stability of biomass anionic bitumen/water emulsions from three perspectives, including interaction energy, interfacial layer thickness, and radial distribution function. The optimal emulsifier dosage was determined and verified by storage stability, particle size analysis and macroscopic and microscopic tests were conducted by infrared spectroscopy. Some waste shells have also been studied as biomass asphalt modifiers [13]. The road performance of modified asphalt with Trigonella mussel shell powder has been studied. The authors applied Trigonella mussel shell powder to modify asphalt in order to improve the road performance of asphalt pavement. The potential of shell powder as an asphalt modifier was found, and the high- and low-temperature properties and water stability of asphalt mixtures can be enhanced with the waste shell power modifier.

Recycling materials and most solid wastes can be treated and applied in infrastructure constructions for performance modification and enhancement. Natural resources and energy consumption can be significantly saved through these proposed green materials and innovative techniques, which would contribute to the development of infrastructure sustainability.

Author Contributions: Conceptualization, J.W.; methodology, J.W.; formal analysis, J.W. and Q.L.; investigation, Q.L.; resources, J.W. and Q.L.; data curation, D.G. and K.H.; writing—original draft preparation, J.W.; writing—review and editing, J.W., Q.L., D.G., K.H. and F.G.; supervision, J.W. and Q.L. All authors have read and agreed to the published version of the manuscript.

Funding: This research was funded by the National Natural Science Foundation of China (grant No. 52108408) and the Natural Science Foundation of Jiangsu Province (grant No. BK20210617).

Conflicts of Interest: The authors declare no conflict of interest.

References

1. Xue, X.; Gao, J.; Wang, J.; Chen, Y. Evaluation of High-Temperature and Low-Temperature Performances of Lignin–Waste Engine Oil Modified Asphalt Binder and Its Mixture. *Materials* **2021**, *15*, 52. [CrossRef] [PubMed]
2. Xu, F.; Zhao, Y.; Li, K. Using Waste Plastics as Asphalt Modifier: A Review. *Materials* **2021**, *15*, 110. [CrossRef] [PubMed]
3. Zhao, X.; Ge, D.; Wang, J.; Wu, D.; Liu, J. The Performance Evaluation of Asphalt Mortar and Asphalt Mixture Containing Municipal Solid Waste Incineration Fly Ash. *Materials* **2022**, *15*, 1387. [CrossRef] [PubMed]
4. Li, W.; Wang, D.; Chen, B.; Hua, K.; Huang, Z.; Xiong, C.; Yu, H. Preparation of Artificial Pavement Coarse Aggregate Using 3D Printing Technology. *Materials* **2022**, *15*, 1575. [CrossRef] [PubMed]
5. Xiao, Y.; Kong, K.; Aminu, U.F.; Li, Z.; Li, Q.; Zhu, H.; Cai, D. Characterizing and Predicting the Resilient Modulus of Recycled Aggregates from Building Demolition Waste with Breakage-Induced Gradation Variation. *Materials* **2022**, *15*, 2670. [CrossRef] [PubMed]
6. Zhang, W.; Li, Q.; Wang, J.; Meng, Y.; Zhou, Z. Aging Behavior of High-Viscosity Modified Asphalt Binder Based on Infrared Spectrum Test. *Materials* **2022**, *15*, 2778. [CrossRef] [PubMed]
7. Wang, B.; Yao, X.; Yang, M.; Zhang, R.; Huang, J.; Wang, X.; Dong, Z.; Zhao, H. Mechanical Performance of 3D Printed Concrete in Steam Curing Conditions. *Materials* **2022**, *15*, 2864. [CrossRef] [PubMed]
8. Wang, S.; He, X.; Cai, G.; Lang, L.; Ma, H.; Gong, S.; Niu, Z. Investigation on Water Transformation and Pore Structure of Cement-Stabilized Dredged Sediment Based on NMR Technology. *Materials* **2022**, *15*, 3178. [CrossRef] [PubMed]
9. Zhang, G.; Ding, Z.; Wang, Y.; Fu, G.; Wang, Y.; Xie, C.; Zhang, Y.; Zhao, X.; Lu, X.; Wang, X. Performance Prediction of Cement Stabilized Soil Incorporating Solid Waste and Propylene Fiber. *Materials* **2022**, *15*, 4250. [CrossRef] [PubMed]
10. Zheng, J.; Zeng, G.; Zhou, H.; Cai, G. Experimental Study on Carbonation of Cement-Based Materials in Underground Engineering. *Materials* **2022**, *15*, 5238. [CrossRef] [PubMed]
11. Huang, X.; Yan, F.; Guo, R.; He, H. Study on the Performance of Steel Slag and Its Asphalt Mixture with Oxalic Acid and Water Erosion. *Materials* **2022**, *15*, 6642. [CrossRef] [PubMed]
12. Shi, S.; Lin, L.; Hu, Z.; Gu, L.; Zhang, Y. Study on the Stability of Bio-Oil Modified Prime Coat Oil Based on Molecular Dynamics. *Materials* **2022**, *15*, 6737. [CrossRef] [PubMed]
13. Fan, G.; Liu, H.; Liu, C.; Xue, Y.; Ju, Z.; Ding, S.; Zhang, Y.; Li, Y. Analysis of the Influence of Waste Seashell as Modified Materials on Asphalt Pavement Performance. *Materials* **2022**, *15*, 6788. [CrossRef] [PubMed]

materials

MDPI

Article

Evaluation of High-Temperature and Low-Temperature Performances of Lignin–Waste Engine Oil Modified Asphalt Binder and Its Mixture

Xue Xue [1,2], Junfeng Gao [1,3,*], Jiaqing Wang [4,*] and Yujing Chen [1]

1 Key Laboratory of Transport Industry of Road Structure and Materials (Xi'an), Chang'an University, Xi'an 710064, China; xxue@chd.edu.cn (X.X.); yujingchen90@163.com (Y.C.)
2 Xi'an Municipal Engineering Design & Research Institute Co., Ltd., Xi'an 710065, China
3 National & Local Joint Engineering Laboratory of Transportation and Civil Engineering Materials, Chongqing Jiaotong University, Chongqing 400074, China
4 College of Civil Engineering, Nanjing Forestry University, Nanjing 210037, China
* Correspondence: jfgao@cqjtu.edu.cn (J.G.); jiaqingw@njfu.edu.cn (J.W.)

Abstract: This research aims to explore the high-temperature and low-temperature performances of lignin–waste engine oil-modified asphalt binder and its mixture. For this research, the lignin with two contents (4%, 6%) and waste engine oil with two contents (3%, 5%) were adopted to modify the control asphalt binder (PG 58-28). The high-temperature rheological properties of the lignin–waste engine oil-modified asphalt binder were investigated by the viscosity obtained by the Brookfield viscometer and the temperature sweep test by the dynamic shear rheometer. The low-temperature rheological property of the lignin–waste engine oil-modified asphalt binder was evaluated by the stiffness and m-value at two different temperatures (−18 °C, −12 °C) obtained by the bending beam rheometer. The high-temperature and the low-temperature performances of the lignin–waste engine oil-modified asphalt mixture were explored by the rutting test and low-temperature bending beam test. The results displayed that the rotational viscosity and rutting factor improved with the addition of lignin and decreased with the incorporation of waste engine oil. Adding the lignin into the control asphalt binder enhanced the elastic component while adding the waste engine oil lowered the elastic component of the asphalt binder. The stiffness of asphalt binder LO60 could not meet the requirement in the specification, but the waste engine oil made it reach the requirement based on the bending beam rheometer test. The waste engine oil could enhance the low-temperature performance. The dynamic stabilities of LO40- and LO60-modified asphalt mixture increased by about 9.05% and 17.41%, compared to the control mixture, respectively. The maximum tensile strain of LO45 and LO65 increased by 16.39% and 25.28% compared to that of LO40 and LO60, respectively. The high- and low-temperature performances of the lignin–waste engine oil-modified asphalt LO65 was higher than that of the control asphalt. The dynamic stability had a good linear relationship with viscosity, the rutting factor of the unaged at 58 °C, and the rutting factor of the aged at 58 °C, while the maximum tensile strain had a good linear relationship with m-value at −18 °C. This research provides a theoretical basis for the further applications of lignin–waste engine oil-modified asphalt.

Keywords: lignin; waste engine oil; modified asphalt; high-temperature performance; low-temperature performance

Citation: Xue, X.; Gao, J.; Wang, J.; Chen, Y. Evaluation of High-Temperature and Low-Temperature Performances of Lignin–Waste Engine Oil Modified Asphalt Binder and Its Mixture. *Materials* **2022**, *15*, 52. https://doi.org/10.3390/ma15010052

Academic Editor: Francesco Canestrari

Received: 4 November 2021
Accepted: 8 December 2021
Published: 22 December 2021

Publisher's Note: MDPI stays neutral with regard to jurisdictional claims in published maps and institutional affiliations.

Copyright: © 2021 by the authors. Licensee MDPI, Basel, Switzerland. This article is an open access article distributed under the terms and conditions of the Creative Commons Attribution (CC BY) license (https://creativecommons.org/licenses/by/4.0/).

1. Introduction

The vigorous development of road construction and maintenance, the continuous increase in traffic volume, and the continuous changes of the local climate and environment in China have necessitated higher requirements for the performance of various aspects of asphalt pavement materials [1,2]. In order to improve the overall performance of asphalt materials, different chemical or physical modifiers are used to add to the base asphalt [3–6]. In recent years, some waste recyclable materials [7,8], especially the biomass materials [9–11], have been introduced into the road engineering field as a good modifier, and they have received more and more attention. The use of waste and recyclable materials can improve the performance of asphalt materials, promote the recycling of waste, and reduce the pollution caused by its burial and disposal, which has good application prospects [12].

Among the waste and recyclable materials, lignin has attracted many researchers. As a by-product or waste material of biomass materials, lignin is used as a modifier of asphalt materials and has great potential performance. It is a type of complex organic polymer, which forms important structural materials in the supporting tissues of vascular plants and some algae [13]. Among woody plants, lignin accounts for 25%, making it the second most abundant organic matter in the world. Lignin is a biopolymer with a three-dimensional network structure formed by connecting three kinds of phenylpropane units through ether bonds and carbon-carbon bonds [14–16]. It is rich in aromatic ring structures, aliphatic and aromatic hydroxyl groups, and active groups (such as quinone groups). In the field of papermaking, lots of lignin by-products are (about 50 million tons) produced, which are directly discharged into natural water in the form of pollutant "black liquor" [17]. Meanwhile, less than 20% of the lignin can be effectively used because of the concentration and burning of the byproducts [18]. Thus, less lignin can be used for resource utilization, resulting in a waste of organic resources.

In recent years, according to the properties and application status of lignin, researchers in the road field have carried out research on the preparation of modified petroleum-based asphalt by combining lignin and asphalt binder. Gao et al. introduced lignin as a biological additive to modify virgin asphalt, and studied the high-temperature rheological properties of lignin-modified asphalt and base asphalt [19]. The results displayed that the activation energy had an increasing trend with the increase in the lignin compared to the virgin asphalt. When the content of lignin was less than 8%, the reduction in fatigue life of lignin modified asphalt binder was small. Pan took biopolymers, namely coniferol lignin, as an example, and studied the chemical and physical basis of general chemical antioxidants [18]. He also used the X-ray photoelectron spectroscopy (XPS) technology to prove the validity of the lignin-modified and unmodified asphalt models, and the XPS results were highly consistent with the model predictions. Batista et al. investigated the physical and chemical properties of lignin-modified asphalt binder, and explored the feasibility for the applying of the lignin from the pulp and paper industry to the asphalt pavements [20]. The results revealed that the lignin-modified asphalt had low carbonyl index, which was beneficial to the weather aging resistance. Xie et al. proved the potential of lignin as an asphalt binder modifier by deriving the lignin fractions through enzyme-mediator-based biological processing and formic acid-based chemical processing [21]. Arafat et al. used three different types of lignin (kraft paper lignin, black liquor precipitation lignin, laboratory lignin produced from rice husks using a deep eutectic solvent) to replace a small portion (up to 6%) of asphalt to determine the effect of lignin on the aged and unaged asphalt binders and mixtures [22]. It was concluded that 6% of base asphalt binder can be replaced by the lignin from agricultural waste.

At the same time, waste engine oil is the non-distillable part collected from the solid-liquid copolymer formed from engine oil after being consumed by oxidation, etc. [23]. Waste engine oil and asphalt have relatively similar components and good recyclability value. In recent years, waste engine oil and its extracts have been regarded as good asphalt modifiers. Some researchers in the field of road pavement have delved into these aspects. Liu et al. used 4 wt% and 8 wt% waste engine oil to modify the asphalt binder from three different sources, and analyzed the chemical composition, molecular weight, and rheological properties [24]. They concluded that waste engine oil significantly reduced the viscosity of the adhesive and lowered the application temperature of asphalt binder. They also evaluated the aging behavior of waste engine oil-modified asphalt binder through testing the conventional performance, rheological properties, and micro characteristics of the aging residue [25]. Hesp et al. recorded the waste oil residues found on the pavement in Ontario and Canada, and inferred that the use of waste oil residues in asphalt was common based on zinc dialkyldithiophosphate as a general additive in engine oil. They also found that typical modification levels are in the range of 5–20% based on the XRF analysis of pure waste oil residues [26]. Li et al. evaluated a limited but well-controlled set of asphalt binders with the same performance level containing the recycled engine oil residues by the content of 0%, 2.5%, 6%, and 15% [27]. It observed that the fatigue cracking performance of the recycled engine oil residues at moderate temperature depended on the aging conditions and a stress/strain control model.

Based on the previous literature review, it can be observed that most of the researches focused on the lignin or waste engine oil, and the single additive adding into the asphalt binder could improve one aspect of the property of asphalt and decreased other properties. Meanwhile, there is rarely information about the performances of lignin–waste engine oil-modified asphalt from the perspective of binder and mixture. The relationship between the indexes of binder and indicators of lignin–waste engine oil-modified asphalt mixture was also rarely investigated. Thus, this research aims to add both the lignin and waste engine oil into the base asphalt to modify it, and explore the high-temperature and low-temperature performance of lignin–waste engine oil-modified asphalt binder and its mixture based on the Brookfield viscometer, dynamic shear rheometer, bending beam rheometer, the rutting test, and the low-temperature bending beam test.

2. Materials and Methods

2.1. Materials

2.1.1. Base Asphalt

The base asphalt utilized was PG 58-28, which was regarded as the control in the experiments. The specific gravity was 1.03. The properties of this base asphalt met the requirements in the specification.

2.1.2. Lignin

The lignin used in this study was provided by Shanghai Jack Gump Industrial Co. Ltd (Shanghai, China). It was obtained from the waste wood chips. The carbon, hydrogen, and oxygen of this kind of lignin were about 64.2%, 5.8%, and 29.2%, respectively.

2.1.3. Waste Engine Oil

The waste engine oil was replaced from the engine oil of the car by the car repairing shop. It was black and brown. The viscosity was 50 mPa·s, while the specific gravity was 0.91.

2.1.4. Aggregate and Filler

The aggregates were limestone and the filler was powdered from the limestone. The properties of the coarse aggregate and fine aggregate met the requirements in the specification [28].

2.2. Preparation of Lignin–Waste Engine Oil-Modified Asphalt

To prepare the lignin–waste engine oil-modified asphalt, the base asphalt binder contained by a container was firstly heated to flow in the heating oven at about 160 °C. Then, the weighted lignin was added into the base asphalt by different dosages and they were placed for the shearing by the high-speed shear mixer. The shear speed, shear time, and shear temperature were 6000 rpm, 40 min, and 160 °C, respectively. After that, the prepared waste engine oil with different contents were slowly added into the mixed asphalt binders and stirred for about 30 min to make the mixture evenly mixed. The dosages of lignin used in this research were 4% and 6% while the contents of waste engine oil utilized were 3% and 5% based on the previous studies [19,23]. To conduct experiments clearly and efficiently, the modified asphalt binder with m% lignin and n% waste engine oil was denoted as LOmn. For instance, LO43 refers to the modified asphalt with 4% content of lignin and 3% of waste engine oil, while LO60 refers to the modified asphalt with 6% content of lignin and 0% of waste engine oil. In addition, PG 58-28 (Shell Bitumen China Solutions Centre, Shanghai, China) was used as the control for comparing.

2.3. Mixture Gradation and Preparation of Lignin–Waste Engine Oil-Modified Asphalt Mixture

To investigate the high-temperature and low-temperature performances of the lignin–waste engine oil-modified asphalt mixture, the gradation of AC-16 was selected for the asphalt mixture, as shown in Table 1. The optimum asphalt was 4.5%, which was determined according to the Marshall test in the specification [28]. The slab specimen was formed to conduct the rutting test; it was also cut further for the low-temperature bending beam test.

Table 1. The gradation of AC-16 in this research.

Sieve Size/mm	**The Mass Percentage Passing through the Following Sieves/%**										
	19	**16**	**13.2**	**9.5**	**4.75**	**2.36**	**1.18**	**0.6**	**0.3**	**0.15**	**0.075**
Composite gradation	100	95	88	75	47	32	23	17	12	8.5	6
Upper limit	100	100	92	80	62	48	36	26	18	14	8
Lower limit	100	90	76	60	34	20	13	9	7	5	4

2.4. Test Methods

2.4.1. Viscosity Test

The viscosity was tested by the Brookfield viscometer. Four temperatures (110, 135, 150, and 165 °C) and three rotational speeds (10, 20, and 50 rpm) were selected to test the viscosities of the lignin and waste engine oil-modified asphalt.

2.4.2. Temperature Sweep Test

The temperature sweep test of the lignin–waste engine oil-modified asphalt binder was conducted by the dynamic shear rheometer. The unaged and rolling thin film oven (RTFO) aged-modified asphalt binders were used for the temperature sweep test. The diameter and spacing for the test sample were 25 mm and 1 mm, respectively. In this study, the temperatures were 52, 58, 64, 70, 76, and 82 °C, while the test frequency was 1.59 Hz. The phase angle and rutting factors obtained from the dynamic shear rheometer were used to analyze the high properties of the lignin–waste engine oil-modified asphalt binder.

2.4.3. Bending Beam Rheometer Test

The low-temperature property of asphalt binder was tested with the bending beam rheometer (BBR) test at different temperatures. The bending beam rheometer test provides a measure of low-temperature stiffness and relaxation properties of asphalt binders. These parameters give an indication of an asphalt binder's ability to resist low-temperature cracking. The asphalt for this test was firstly aged by the rolling thin film oven (RTFO) and pressure aging vessel (PAV) test. The bending creep stiffness modulus (S) and the slope of the creep curve (m) obtained from this test were utilized to evaluate the flexibility of asphalt and the ability to resist the crack under the low temperature. In this research, two temperatures, −18 °C and −12 °C, were selected based on the performance grade of the control asphalt.

2.4.4. Rutting Test for Mixture

To evaluate the high-temperature performance of the mixture, a rutting test was conducted. The formed slab specimen was placed in the chamber for wheel tracking at 60 °C. The dynamic stability (DS) and rutting depth at 60 min obtained by the wheel tracker were analyzed to investigate the high-temperature performance of the lignin–waste engine oil-modified asphalt mixture. The detailed process was shown in the specification Test Methods of Bitumen and Bituminous Mixture for Highway Engineering (JTG E20-2011) [29].

2.4.5. Low-Temperature Bending Beam Test

The low-temperature bending beam test was used to investigate the low-temperature performance of mixture. The size of the beam was 250 mm × 30 mm × 35 mm while the temperature was −10 °C ± 0.5 °C. The maximum tensile strain and strength obtained was analyzed to evaluate the low-temperature performance of the lignin–waste engine oil-modified asphalt mixture. The detailed process was shown in the specification Test Methods of Bitumen and Bituminous Mixture for Highway Engineering (JTG E20-2011) [29]. The map of this study is shown in Figure 1.

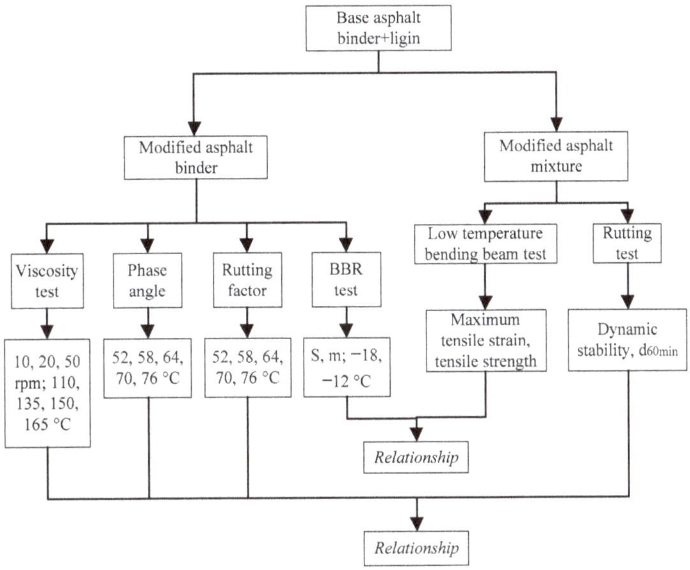

Figure 1. The map of this study.

3. Results and Discussion

3.1. Viscosity

Viscosity is related to the resistance to the deformation under different actions from the external. Figure 2a–c shows the viscosity of control (PG 58-28) asphalt binder and lignin–waste engine oil-modified asphalt binders under different rotational speeds and temperatures.

It can be found from Figure 2 that the viscosities of the lignin–waste engine oil-modified asphalt binders at 135 °C were less than the requirement value, 3.00 Pa·s, in the specification. Increased temperature led to the decreased viscosities of different asphalt binder types. The decrease in viscosity between 150 °C and 165 °C was lower than that of the viscosity between 135 °C and 150 °C, regardless of the modified asphalt binder types and the rotational speeds. For instance, at 50 rpm, the viscosities of LO45 at 135 °C, 150 °C, and 165 °C were 0.41 Pa·s, 0.21 Pa·s, and 0.12 Pa·s, respectively. Furthermore, the decrease in viscosity from 150 °C to 165 °C was 0.09 Pa·s, while the decrease from 135 °C to 150 °C was 0.20 Pa·s. This meant that the temperature had a significant influence on the viscosity. Take the asphalt binder control, LO40 and LO60 to analyze, it can be concluded that the addition of lignin increased the viscosity of asphalt binder, regardless of the rotational speeds. However, compared to LO40, the viscosities of LO43 and LO45 decreased, and the viscosity of LO45 decreased more than that of LO43. The viscosities of LO63 and LO65 showed the same change rule with that of LO43 and LO45. This indicated that the incorporation of waste engine oil decreased the viscosity of base asphalt binder and modified asphalt binder. This may be caused by the light component existing in the waste engine oil.

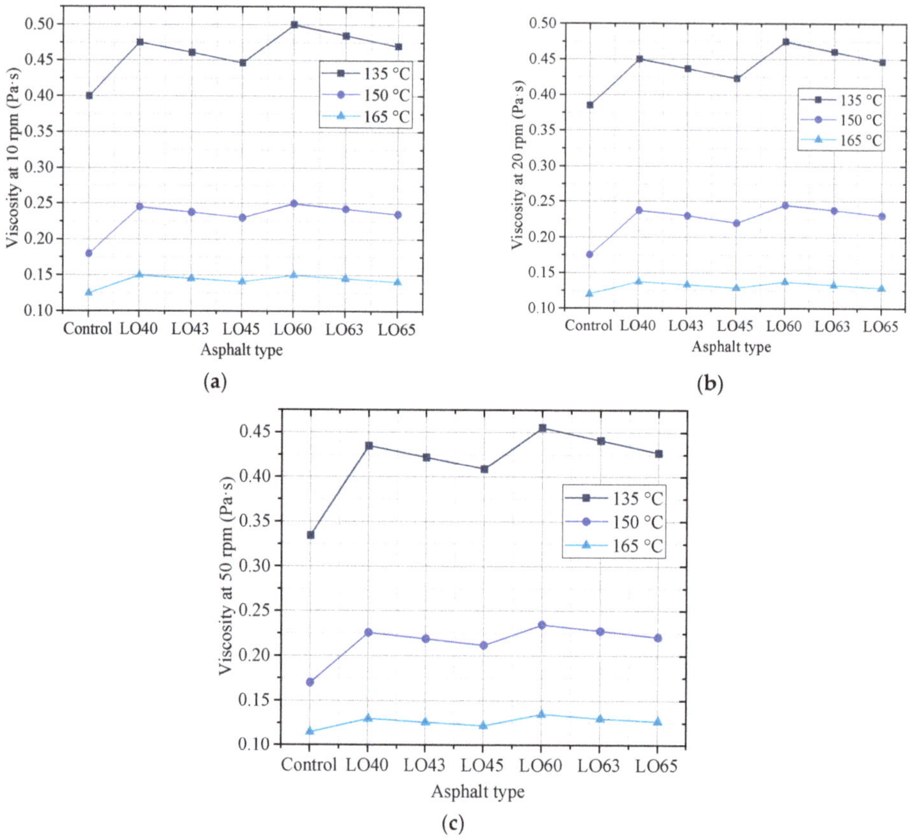

Figure 2. The viscosity of different asphalt types at different rotational speeds. (**a**) 10 rpm; (**b**) 20 rpm; (**c**) 50 rpm.

Figure 3 displays the viscosity of the control asphalt binder and the lignin–waste engine oil-modified asphalt binder at 110 °C. As shown in Figure 3, the addition of lignin increased the viscosity of asphalt binder, while the incorporation of waste engine oil decreased the viscosity of the lignin-modified asphalt binder. In addition, with the increase in the rotational speed, the viscosity decreased, regardless of the asphalt types at 110 °C. This indicated that the lignin–waste engine oil asphalt binders were non-Newtonian fluid, following the previous findings on the asphalt binder.

3.2. Temperature Sweep

3.2.1. Phase Angle

Figure 4a,b shows the phase angle of the unaged and RTFO-aged control asphalt binder and the lignin–waste engine oil-modified asphalt binder.

The phase angle reflects the viscoelasticity of the asphalt binder. The smaller the phase angle, the higher the elastic part. It can be seen from Figure 4 that the phase angles of the control asphalt binder and lignin–waste engine oil asphalt binder increased with the increase in the temperature. This indicates that high temperatures give the asphalt binder less elastic portion. For the unaged asphalt binder, compared to the control, the phase angles of asphalt binder LO40 and LO60 decreased regardless of the test temperature, illustrating the improvement in the elastic portion by the addition of lignin. The phase angles of asphalt binder LO43 and LO45 increased compared to LO40, while the phase angles of LO63 and LO65 also enhanced compared to LO60. This meant that the incorporation of waste engine oil could improve the viscous component of the asphalt. Take the phase angles at 58 °C as an example, for the unaged asphalt, the phase angles of control, LO40, LO43, LO45, LO60, LO63, an dLO65 were 86.34°, 85.22°, 86.07°, 86.50°, 84.19°, 85.03°, and 85.45°, respectively. For the RTFO-aged asphalt binders, the phase angle of the control was higher than that of other asphalt binders, while the phase angle of asphalt binder LO40 and LO60 was lower than that of other asphalt binders. This was because the RTFO aging of lignin made the asphalt harder, and the more content of lignin, the obvious of this results. On the contrary, the waste engine oil could make the asphalt binder softer, leading to the difference of phase angle to the lignin-modified asphalt without waste engine oil.

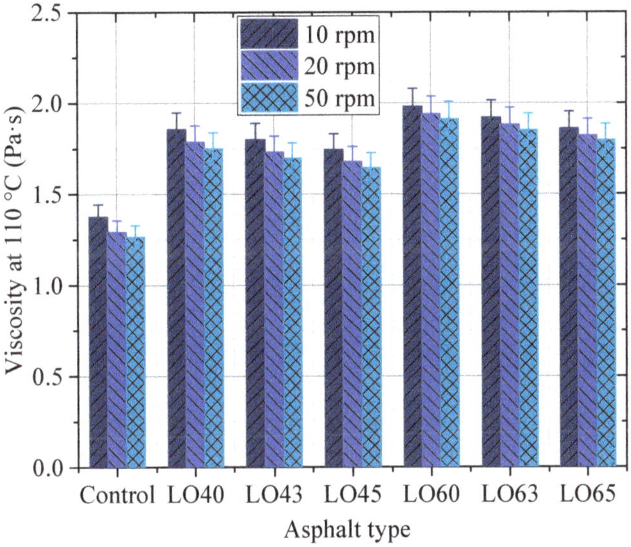

Figure 3. The viscosity of different asphalt types at 110 °C.

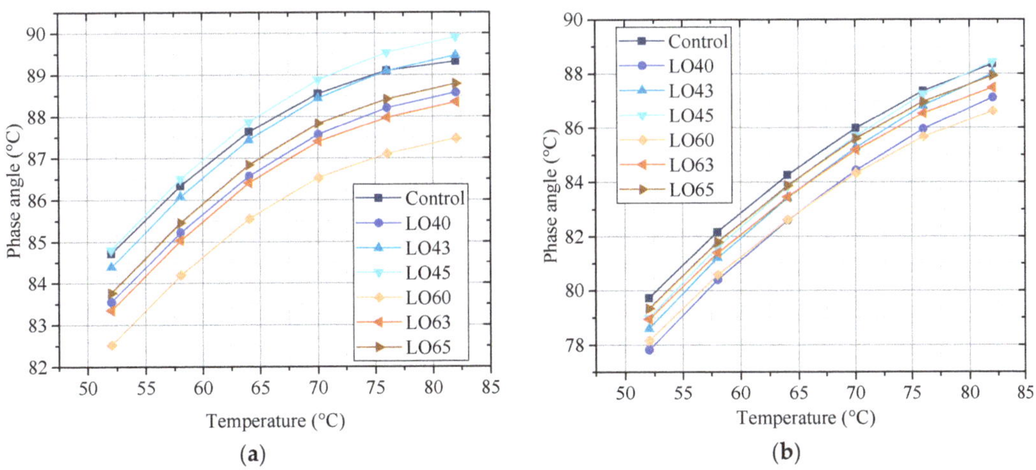

Figure 4. Phase angle of the unaged and the RTFO-aged asphalt binders. (**a**) The unaged; (**b**) The RTFO-aged.

3.2.2. Rutting Factor

The rutting factor of the unaged and RTFO-aged control asphalt binder and lignin–waste engine oil-modified asphalt binder is shown in Figure 5a,b.

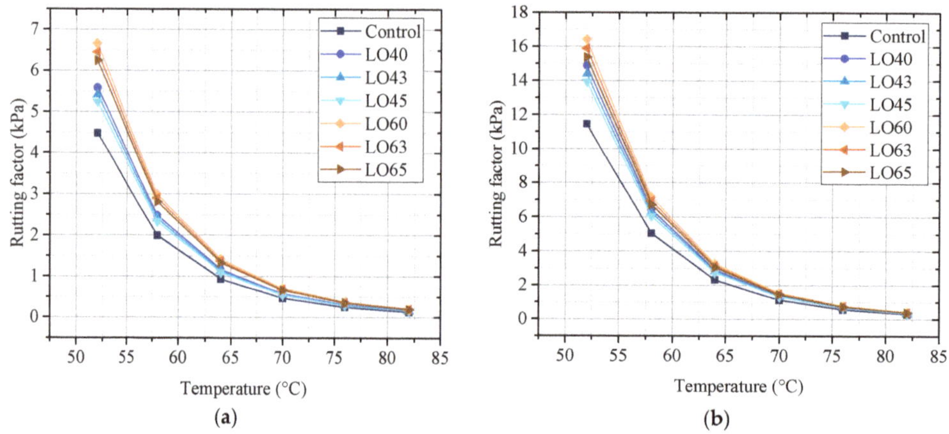

Figure 5. Rutting factor of the unaged and the RTFO-aged asphalt binders. (**a**) The unaged; (**b**) The RTFO-aged.

Rutting factor is used to evaluate the high property of the asphalt binder. As shown in Figure 5, the rutting factors of lignin–waste engine oil-modified asphalt dramatically decreased with the increase in temperature from 52 °C to 64 °C, while it slowly decreased with the changes in temperature from 64 °C to 82 °C. The temperature had a significant influence on rutting factor, and the higher temperature lead to a lower rutting factor, indicating lower high property of asphalt binder. For the unaged, the rutting factor of asphalt binder could be divided to three groups: the control, the LO4 group (LO40, LO43, LO45), and the LO6 group (LO60, LO63, LO65). Furthermore, it can be seen as obvious that the rutting factor of group LO6 was higher than that of group LO4, and that the rutting factor of the control was the lowest. This demonstrated that the incorporation of lignin could improve the high property of asphalt binder. Within the group of LO4, the rutting factor of LO40 was the highest, followed by LO43 and LO45, indicating the influence of waste engine oil on the lignin modified asphalt binder. The rutting factor within the group of LO6 had the same changes. For instance, when the temperature was 58 °C, the rutting factors of LO60, LO63, and LO65 were 2.995 kPa, 2.901 kPa, and 2.809 kPa, respectively. For the RTFO-aged, the rutting factors of different lignin–waste engine oil-modified asphalt binder displayed the same change as the unaged, though the group was not divided.

3.3. Bending Beam Rheometer Test

The stiffness and m-value of the control and lignin–waste engine oil-modified asphalt binder obtained by the bending beam rheometer test are shown in Figures 6 and 7.

It can be observed from Figures 6 and 7 that a higher test temperature led to a lower stiffness. For example, the stiffness of asphalt binder LO63 at −18 °C and −12 °C were 247 Pa and 232 Pa, respectively. This could provide evidence for the utilized temperature condition of the modified asphalt binder. Besides, take the stiffness of the lignin–waste engine oil-modified asphalt binder at −18 °C as an example, all the results could meet the requirement for the using of asphalt binder in the specification, which is 300 Pa. The stiffness of asphalt binder LO40 and LO60 were 249 Pa and 254 Pa, respectively, which increased by about 5.1% and 7.2% compared to that of the control asphalt binder, respectively. Moreover, the stiffness of asphalt binder LO45 and LO65 decreased by 2.8% and 3.9% compared to that of LO40 and LO60, respectively. This indicated that the addition of lignin made the stiffness higher, while the waste engine oil had the opposite influence. Thus, the waste engine oil was beneficial to the low-temperature resistance of asphalt binder. In addition, the m-value of different kinds of asphalt binders, except for LO60 at −18 °C, was greater than 0.3. This meant that the stiffness of asphalt binder LO60 could not meet the requirement in the specification, but the waste engine oil could make it reach the requirement. The stiffness and m-value of lignin–waste engine oil-modified asphalt binder at −12 °C showed the same change rule as them at −18 °C. This can be attributed to the alkylbenzenes, as well as linear alkanes contained in the waste engine oil, which could dissolve the asphalt binder [30].

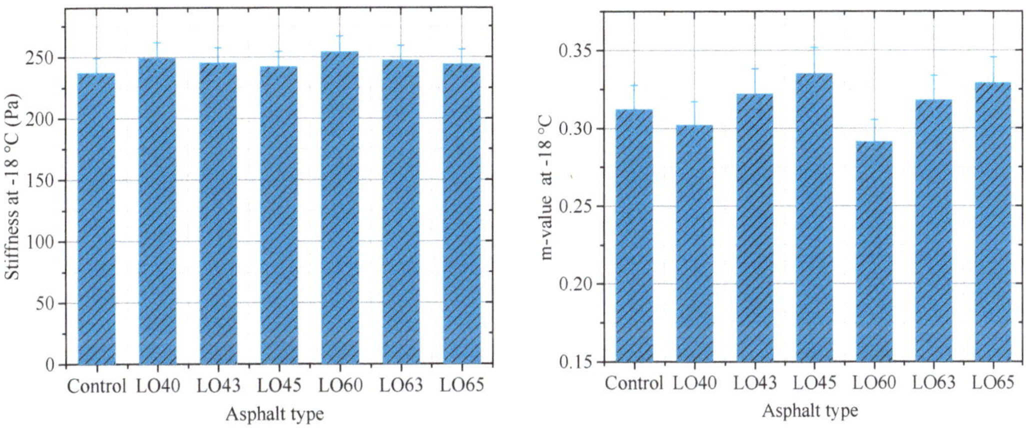

(a) (b)

Figure 6. The BBR results at −18 °C. (**a**) Stiffness; (**b**) m-value.

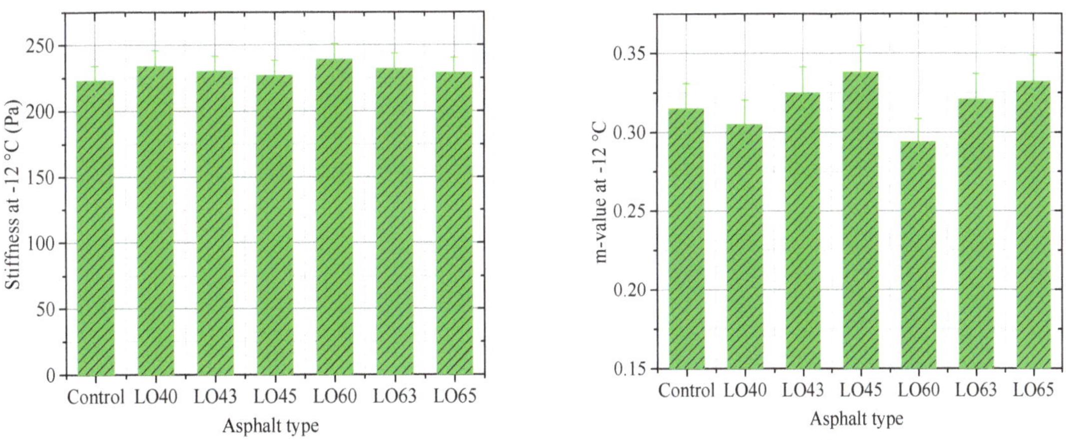

(a) (b)

Figure 7. The BBR results at −12 °C. (**a**) Stiffness; (**b**) m-value.

3.4. Rutting Test

The dynamic stability (DS) is the number of wheel passing on the surface of the slab specimen when the rutting depth reaches 1 mm. In this study, to investigate the high-temperature performance of lignin–waste engine oil-modified asphalt mixture, the dynamic stability (DS) and rutting depth at 60 min were obtained and analyzed.

Figure 8 illustrates the dynamic stability and rutting depth at 60 min of different lignin–waste engine oil-modified asphalt mixtures. As shown in Figure 8a, the dynamic stability of the control asphalt mixture was 2598.32 times/mm. The dynamic stabilities of LO40- and LO60-modified asphalt mixtures were 2833.56 times/mm and 3050.69 times/mm, which increased by about 9.05% and 17.41% compared to the control mixture, respectively. With the incorporation of waste engine oil, the dynamic stability decreased. For instance, the dynamic stability of LO43 was lower than that of LO40, followed by LO45. In addition, the dynamic stability of LO63 was lower than that of LO60, followed by LO65. However, the dynamic stability of the LO45 asphalt mixture was 2699.98 times/mm, which was still higher than that of the control. This illustrated that the addition of lignin could balance the attenuation of the high performance of asphalt mixture affected by the incorporation of waste engine. Moreover, the rutting depths at 60 min of different kind of asphalt mixtures from Figure 7b displayed opposite changes compared to the dynamic stability, illustrating the same change law as the dynamic stability Figure 8a. All the rutting depths at 60 min of the lignin–waste engine oil-modified asphalt mixture were higher than that of the control asphalt mixture. In all, the high-temperature performance of the lignin–waste engine oil-modified asphalt LO65 was higher than that of the control asphalt. This was in accordance with the findings by the rutting factor from the temperature sweep.

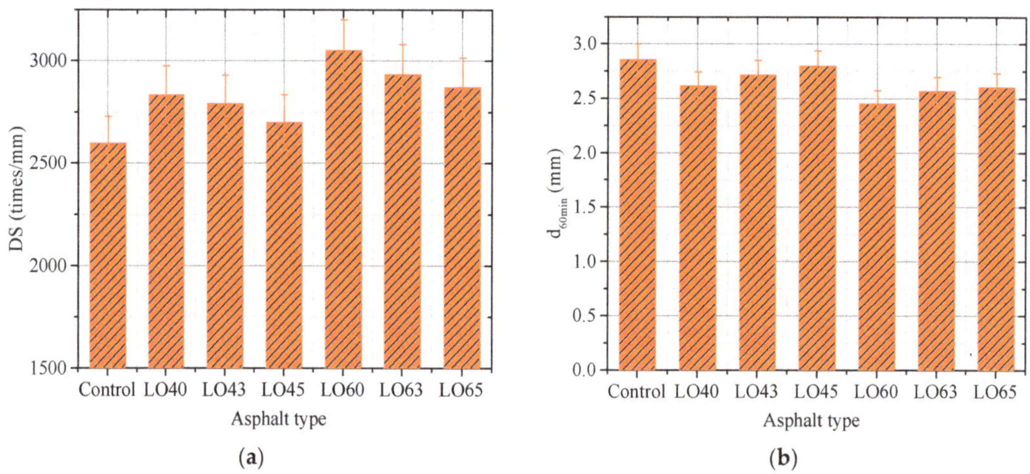

Figure 8. The dynamic stability and rutting depth at 60 min. (**a**) Dynamic stability (DS); (**b**) d_{60min}.

3.5. Low-Temperature Bending Beam Test

The maximum tensile strain and tensile strength obtained and calculated by the low-temperature bending beam test were used evaluate the low-temperature performance crack resistance of the asphalt mixture.

Figure 9 describes the maximum tensile strain and tensile strength of different kinds of lignin–waste engine oil-modified asphalt mixture. It can be investigated that the maximum tensile strain and tensile strength of LO60 asphalt mixture were 2155.61 με and 8.511 MPa, respectively, which were the lowest among the lignin–waste engine oil-modified asphalt mixture. The maximum tensile strain and tensile strength of LO40 asphalt mixture were higher than that of LO60, but they were also lower than other asphalt mixture. This indicated that the addition of lignin decreased the low-temperature resistance of asphalt mixture. Meanwhile, the waste engine oil increased the maximum tensile strain and tensile strength of LO40 and LO60. Take the maximum tensile strain as an example, the maximum tensile strain of LO45 and LO65 were 2903.55 and 2700.62, which increased by 16.39% and 25.28% compared to that of LO40 and LO60, respectively. Ultimately, the low-temperature performance of the lignin–waste engine oil-modified asphalt LO65 was higher than that of the control asphalt. This was consistent with the results of asphalt binder from the bending beam rheometer.

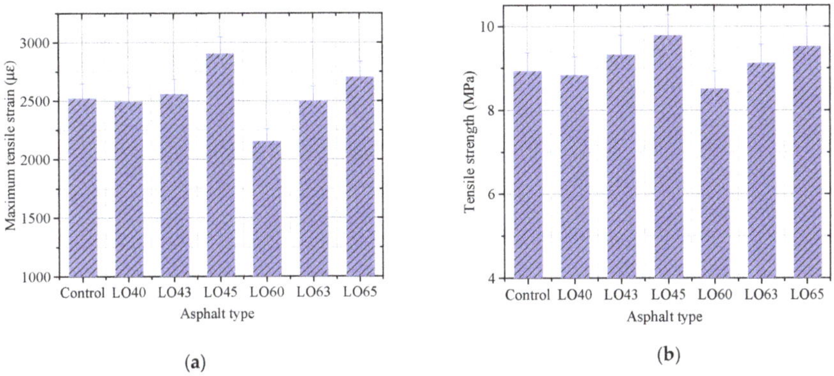

Figure 9. The maximum tensile strain and tensile strength. (**a**) Maximum tensile strain; (**b**) Tensile strength.

3.6. The Linear Fit of the Binder and Mixture

To analyze the relationship between the indexes of binder and indicators of the lignin–waste engine oil-modified asphalt mixture, the linear fit was conducted.

Figure 10a,b shows the linear fit of dynamic stability (DS) from the asphalt mixture, viscosity, and rutting factor, which indicated the high property of the lignin–waste engine oil-modified asphalt binder. The viscosity of different kinds of asphalt binders with 20 rpm at 135 °C was selected, and the rutting factors of the unaged and RTFO-aged asphalt binder at 58 °C were selected. It can be observed from Figure 10a,b that the R-square of dynamic stability and viscosity, the rutting factor of the unaged at 58 °C, and the rutting factor of the aged at 58 °C were 0.9352, 0.9284, and 0.9256, respectively. This indicated that the dynamic stability obtained by the asphalt mixture had a good linear relationship with viscosity, the rutting factor of the unaged at 58 °C, and the rutting factor of the aged at 58 °C, from the perspective of the high-temperature performance. The linear fit of maximum tensile strain and m-value is displayed in Figure 11. The m-value at −18 °C obtained by the bending beam rheometer test was selected. As shown in Figure 11, the R-square of maximum tensile strain and m-value was 0.8646, illustrating a good linear relationship between maximum tensile strain and m-value at −18 °C from the perspective of the low-temperature performance.

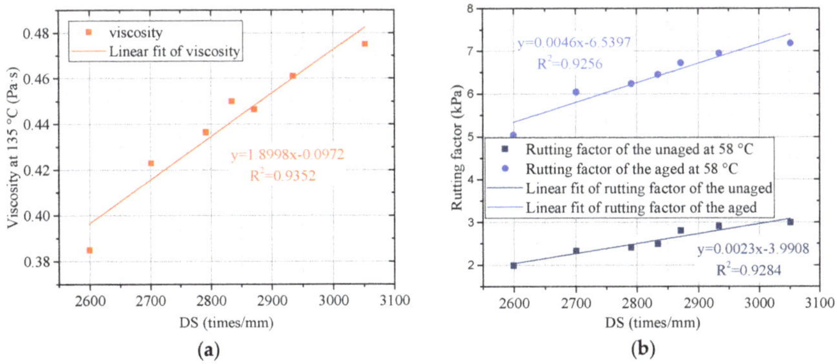

Figure 10. The linear fit of DS, viscosity, rutting factor. (**a**) DS with viscosity; (**b**) DS with rutting factor.

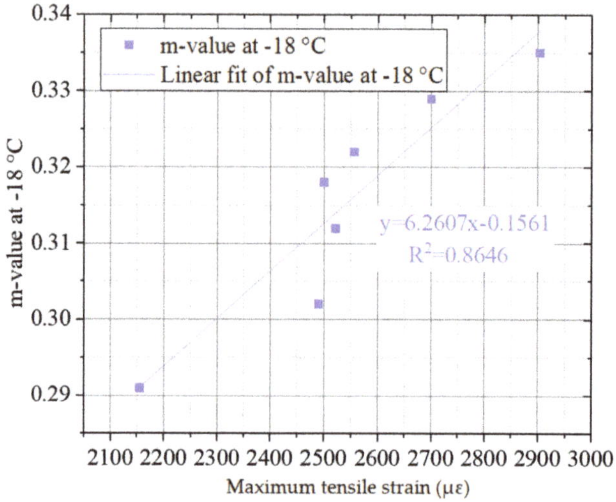

Figure 11. The linear fit of maximum tensile strain and m-value.

4. Conclusions

The high-temperature and low-temperature performances of lignin–waste engine oil-modified asphalt binder and its mixture were evaluated. The Brookfield viscometer, dynamic shear rheometer, and bending beam rheometer were utilized to test the properties of the asphalt binder, while the rutting test and low-temperature bending beam test were conducted to investigate the performances of the asphalt mixture. Based on the test results, the following conclusions can be drawn.

(1) The rotational viscosity improved with the addition of lignin and decreased with the incorporation of waste engine oil. Meanwhile, the viscosity of lignin–waste engine oil-modified asphalt can still meet the requirement for mixing and construction of asphalt binder in the specification.
(2) The rutting factors of lignin–waste engine oil-modified asphalt dramatically decreased with the increase in temperature from 52 °C to 64 °C, while it slowly decreased with the changes in temperature from 64 °C to 82 °C. Adding the lignin into the control asphalt binder enhanced the elastic component, while adding the waste engine oil lowered the elastic component of the asphalt binder.
(3) The stiffness of asphalt binder LO60 could not meet the requirement in the specification, but the waste engine oil helped it reach the requirement based on the bending beam rheometer test. The waste engine oil could enhance the low-temperature performance.
(4) The dynamic stabilities of LO40- and LO60-modified asphalt mixture increased compared to the control mixture, respectively. The maximum tensile strain of LO45 and LO65 increased compared to that of LO40 and LO60, respectively. The high- and low-temperature performances of the lignin–waste engine oil-modified asphalt LO65 was higher than that of the control asphalt.
(5) The dynamic stability obtained by the asphalt mixture had a good linear relationship with viscosity, a rutting factor of the unaged at 58 °C, and a rutting factor of the aged at 58 °C, from the perspective of a high-temperature performance. The maximum tensile strain and m-value at −18 °C had a good linear relationship from the perspective of a low-temperature performance.

Due to the limitation of experimental conditions of this study, the durability of the lignin–waste engine oil-modified asphalt was not studied. In the future, the comprehensive performances during the application in the field and the durability will be investigated.

Author Contributions: Conceptualization and methodology, J.G. and X.X.; writing—original draft preparation, X.X.; writing—review and editing, J.W., Y.C.; supervision, J.G. All authors have read and agreed to the published version of the manuscript.

Funding: This research was funded by the National Natural Science Foundation of China (NSFC), grant number 51778062, 52108399.

Institutional Review Board Statement: Not applicable.

Informed Consent Statement: Not applicable.

Data Availability Statement: All data generated or analysed during this study are included in this published article.

Conflicts of Interest: The authors declare no conflict of interest.

References

1. Qian, C.; Fan, W.; Liang, M.; He, Y.; Ren, S.; Lv, X.; Nan, G.; Luo, H. Rheological properties, storage stability and morphology of CR/SBS composite modified asphalt by high-cured method. *Constr. Build. Mater.* **2018**, *193*, 312–322. [CrossRef]
2. Li, Y.; Zhang, S.; Wang, R.; Dang, F. Potential use of waste tire rubber as aggregate in cement concrete—A comprehensive review. *Constr. Build. Mater.* **2019**, *225*, 1183–1201. [CrossRef]
3. Gao, J.; Wang, H.; Chen, J.; Meng, X.; You, Z. Laboratory evaluation on comprehensive performance of polyurethane rubber particle mixture. *Constr. Build. Mater.* **2019**, *224*, 29–39. [CrossRef]
4. Lei, Y.; Wang, H.; Fini, E.H.; You, Z.; Yang, X.; Gao, J.; Dong, S.; Jiang, G. Evaluation of the effect of bio-oil on the high-temperature performance of rubber modified asphalt. *Constr. Build. Mater.* **2018**, *191*, 692–701. [CrossRef]
5. Intari, D.E.; Fathonah, W.; Saputro, B. Performance of asphalt concrete mixture (AC-WC) using asphalt added with the waste of rice husk ash. *IOP Conf. Ser. Mater. Sci. Eng.* **2019**, *673*, 012037. [CrossRef]
6. Gheni, A.A.; Alghazali, H.H.; ElGawady, M.A.; Myers, J.J.; Feys, D. Durability properties of cleaner cement mortar with by-products of tire recycling. *J. Clean. Prod.* **2019**, *213*, 1135–1146. [CrossRef]
7. Lei, Y.; Wang, H.; You, Z.; Wang, Y.; Jiang, X.; Gao, J. Effect of mixing sequence on compaction property of hot recycled asphalt mixtures. *Eng. J. Wuhan Univ.* **2019**, *52*, 317–323.
8. Arabani, M.; Tahami, S.A.; Taghipoor, M. Laboratory investigation of hot mix asphalt containing waste materials. *Road Mater. Pavement Des.* **2017**, *18*, 713–729. [CrossRef]
9. Gao, J.; Wang, H.; You, Z.; Hasan, M.R.M. Research on properties of bio-asphalt binders based on time and frequency sweep test. *Constr. Build. Mater.* **2018**, *160*, 786–793. [CrossRef]
10. Gao, J.; Wang, H.; You, Z.; Hasan, M.R.M.; Lei, Y.; Irfan, M. Rheological Behavior and Sensitivity of Wood-Derived Bio-Oil Modified Asphalt Binders. *Appl. Sci.* **2018**, *8*, 919. [CrossRef]
11. Cuong Manh, V.; Quang-Vu, B. Effects of DOPO-Grafted Epoxidized Soybean Oil on Fracture Toughness and Flame Retardant of Epoxy Resin/Rice Husk Silica Hybrid. *Macromol. Res.* **2020**, *28*, 826–834.
12. Shadmani, A.; Tahmouresi, B.; Saradar, A.; Mohseni, E. Durability and microstructure properties of SBR-modified concrete containing recycled asphalt pavement. *Constr. Build. Mater.* **2018**, *185*, 380–390. [CrossRef]
13. Sundstrom, D.W.; Klei, H.E.; Daubenspeck, T.H. Use of byproduct lignins as extenders in asphalt. *Ind. Eng. Chem. Prod. Res. Dev.* **1983**, *22*, 496–500. [CrossRef]
14. Li, C.; Zhao, X.; Wang, A.; Huber, G.W.; Zhang, T. Catalytic Transformation of Lignin for the Production of Chemicals and Fuels. *Chem. Rev.* **2015**, *115*, 11559–11624. [CrossRef]
15. Chio, C.; Sain, M.; Qin, W. Lignin utilization: A review of lignin depolymerization from various aspects. *Renew. Sustain. Energy Rev.* **2019**, *107*, 232–249. [CrossRef]
16. Mai, C.; Milstein, O.; Hüttermann, A. Chemoenzymatical grafting of acrylamide onto lignin. *J. Biotechnol.* **2000**, *79*, 173–183. [CrossRef]
17. Ben-Iwo, J.; Manovic, V.; Longhurst, P. Biomass resources and biofuels potential for the production of transportation fuels in Nigeria. *Renew. Sustain. Energy Rev.* **2016**, *63*, 172–192. [CrossRef]
18. Pan, T. A first-principles based chemophysical environment for studying lignins as an asphalt antioxidant. *Constr. Build. Mater.* **2012**, *36*, 654–664. [CrossRef]
19. Gao, J.; Wang, H.; Liu, C.; Ge, D.; You, Z.; Yu, M. High-temperature rheological behavior and fatigue performance of lignin modified asphalt binder. *Constr. Build. Mater.* **2020**, *230*, 117063. [CrossRef]
20. Batista, K.B.; Padilha, R.P.L.; Castro, T.O.; Silva, C.F.S.C.; Araújo, M.F.A.S.; Leite, L.F.M.; Pasa, V.M.D.; Lins, V.F.C. High-temperature, low-temperature and weathering aging performance of lignin modified asphalt binders. *Ind. Crop. Prod.* **2018**, *111*, 107–116. [CrossRef]
21. Xie, S.; Li, Q.; Karki, P.; Zhou, F.; Yuan, J.S. Lignin as Renewable and Superior Asphalt Binder Modifier. *ACS Sustain. Chem. Eng.* **2017**, *5*, 2817–2823. [CrossRef]

22. Arafat, S.; Kumar, N.; Wasiuddin, N.M.; Owhe, E.O.; Lynam, J.G. Sustainable lignin to enhance asphalt binder oxidative aging properties and mix properties. *J. Clean. Prod.* **2019**, *217*, 456–468. [CrossRef]
23. Wang, W.; Jia, M.; Jiang, W.; Lou, B.; Jiao, W.; Yuan, D.; Li, X.; Liu, Z. High temperature property and modification mechanism of asphalt containing waste engine oil bottom. *Constr. Build. Mater.* **2020**, *261*, 119977. [CrossRef]
24. Liu, S.; Peng, A.; Wu, J.; Zhou, S.B. Waste engine oil influences on chemical and rheological properties of different asphalt binders. *Constr. Build. Mater.* **2018**, *191*, 1210–1220. [CrossRef]
25. Liu, S.; Peng, A.; Zhou, S.; Wu, J.; Xuan, W.; Liu, W. Evaluation of the ageing behaviour of waste engine oil-modified asphalt binders. *Constr. Build. Mater.* **2019**, *223*, 394–408. [CrossRef]
26. Hesp, S.A.M.; Shurvell, H.F. X-ray fluorescence detection of waste engine oil residue in asphalt and its effect on cracking in service. *Int. J. Pavement Eng.* **2010**, *11*, 541–553. [CrossRef]
27. Li, X.; Gibson, N.; Andriescu, A.; Arnold, T.S. Performance evaluation of REOB-modified asphalt binders and mixtures. *Road Mater. Pavement Des.* **2017**, *18*, 128–153. [CrossRef]
28. China, M.o.C.o.t.P.s.R.o. *Technical Specifications for Construction of Highway Asphalt Pavements*; China Communications Press Co., Ltd.: Beijing, China, 2004.
29. China, M.o.C.o.t.P.s.R.o. *Standard Test Methods of Bitumen and Bituminous Mixture for Highway Engineering*; China Communications Press Co., Ltd.: Beijing, China, 2011.
30. Ahmed, R.B.; Hossain, K. Waste cooking oil as an asphalt rejuvenator: A state-of-the-art review. *Constr. Build. Mater.* **2020**, *230*, 116985. [CrossRef]

materials

MDPI

Review

Using Waste Plastics as Asphalt Modifier: A Review

Fengchi Xu, Yao Zhao * and Kangjian Li

College of Civil Engineering, Nanjing Forestry University, Nanjing 210037, China; echo23336@163.com (F.X.); kangjian0119@163.com (K.L.)
* Correspondence: zhaoyaonfu@163.com; Tel.: +86-25-85427758

Abstract: The use of waste products in the production of asphalt binders and asphalt mixtures has become widespread due to economic and environmental benefits. In particular, the use of recycled waste plastic in asphalt binders and mixtures is gaining more attention. This review presents analyses and comparisons of various forms of waste plastic used in asphalt modification, and approaches to incorporating waste plastic into asphalt mixtures, both for single and composite modifications. It focuses on the properties of waste plastics, asphalt binders, and asphalt mixtures. Overall, the incorporation of plastic waste into asphalt mixtures can significantly improve high-temperature performance and has potential economic and environmental benefits. The performance of modified asphalt is highly dependent on multiple factors, such as waste sources, waste plastic dosages, blending conditions, and the pretreatment methods for waste plastic. There are different ways to apply waste plastics to blend into a mixture. In addition, this paper discusses the current challenges for waste plastic-modified asphalt, including the stability, low-temperature performance, modification mechanism, and laboratory problems of the blends. The use of chemical methods, such as additives and functionalization, is considered an effective way to achieve better interactions between waste plastics and the binder, as well as achieving a higher sufficiency utilization rate of waste plastics. Although both methods provide alternative options to produce waste plastic-modified asphalt with stability and high performance, the optimal proportion of materials used in the blends and the microcosmic mechanism of composite modified asphalt are not clear, and should be explored further.

Keywords: recycled waste plastic; asphalt binder and mixture; composite modification; performance; compatibility

Citation: Xu, F.; Zhao, Y.; Li, K. Using Waste Plastics as Asphalt Modifier: A Review. *Materials* **2022**, *15*, 110. https://doi.org/10.3390/ma15010110

Academic Editors: Francesco Canestrari, Krzysztof Formela and Simon Hesp

Received: 26 October 2021
Accepted: 20 December 2021
Published: 24 December 2021

Publisher's Note: MDPI stays neutral with regard to jurisdictional claims in published maps and institutional affiliations.

Copyright: © 2021 by the authors. Licensee MDPI, Basel, Switzerland. This article is an open access article distributed under the terms and conditions of the Creative Commons Attribution (CC BY) license (https://creativecommons.org/licenses/by/4.0/).

1. Introduction

1.1. Environmental Problems Caused by Plastic Waste

Plastics have been widely used in diverse fields for their substantial benefits in terms of cost-effectiveness, light weight, durability, and ease of processing relative to many other materials [1,2]. The increasing demand in various fields promotes the rapid growth of the productivity of plastics. In 2019, the global production of plastics reached 368 million metric tons. China is one of the largest producers, accounting for around 30% [3]. China's plastics industry is in a stage of rapid development in this context. For primary plastics production, it has increased by an average annual rate of 9.1% during the past 10 years, as shown in Figure 1a [4].

The widespread use of plastics brings great convenience to everyday life and promotes economic and social development. However, approximately 50% of plastic materials are single-use materials, such as packaging, agricultural films, and disposable consumer items. Between 20 and 25% of plastics are used for long-term infrastructures such as pipes, cable coatings, and structural materials [1]. At the same time, the huge amount of plastic consumption has led to stress on plastic waste management and eco-environmental protection. Currently, many countries have various management strategies for the total municipal solid waste (MSW) stream. Unfortunately, most of the world's postconsumer plastic materials are treated as waste. Plastic waste in the US, China, and other countries has long been disposed of simply by landfill and incineration (for energy recovery). In 2018, landfills in the US received 27 million tons of plastic, accounting for 18.5% of all MSW landfilled [5]. Landfill, as a traditional plastic waste disposal method, has a high demand for land resources, which has also become a key issue for plastic waste disposal in many countries [2]. The long-term risk of chemicals leaching from the plastics into soils and waters is a serious environmental threat [6]. Moreover, the most severe issue is that this leaching of chemicals may take at least several decades, and probably centuries, as most plastics are not biodegradable. On the other hand, incineration effectively reduces both the volume and mass of plastic waste, but air pollution occurs during the heat treatment process in the incinerator, releasing carbon monoxide, dioxin, and other toxic emissions [7]. The waste plastics also contain heavy metals such as cadmium (Cd) and lead (Pb), which discharge from smoke dust and residues produced during the process of incineration [8].

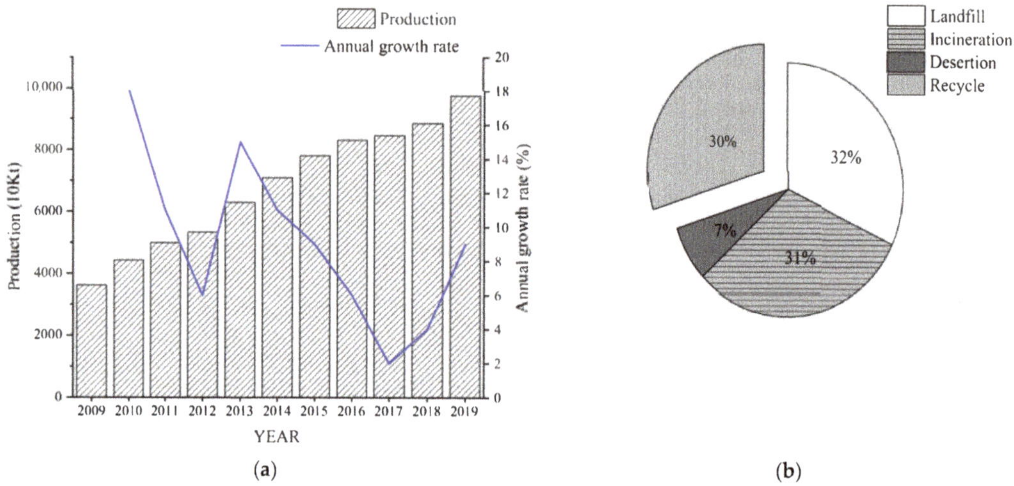

Figure 1. The production and disposal of plastics in China: (**a**) primary plastics production (2009–2019); (**b**) disposal of plastic waste in 2019.

Additionally, the improper handling of plastic waste leads to a high content of plastics in MSW incineration-bottom ash (MSWI-BA). The disposal of MSWI-BA results in an increase in contaminants that pollute water, soil, and oceans [9–11]. Microplastics derived from the fragmentation and degradation of plastics pose an even more serious concern for public health, as they are small enough to pass through waste filtration systems and are difficult to remove [12,13]. Consequently, "plastic pollution" or "white pollution" has become one of the most pressing environmental issues of the modern world [14].

The recycling and utilization of plastic waste have been considered a golden management strategy for reducing environmental impact and natural resource depletion [1,15]. With the increasing awareness of the dangers of improper plastic treatment, this issue has also attracted wide international attention. Countries around the world have introduced policies to ban the use of plastics. By 2020, most countries had begun to prohibit the use of disposable plastic products, as shown in Table 1. From 2008 to 2016, the consumption of plastic shopping bags in supermarkets decreased by more than 2/3, a cumulative reduction of about 1.4 million tons of plastic shopping bags, equivalent to a reduction of nearly 30 million tons of carbon dioxide in China. At the same time, some plastic waste enterprises quickly adjusted to establish and run a waste plastic industry recycling system to support the government strategy, which was shown to work effectively in China. In 2019, approximately 30% of plastic waste was recycled in China, as shown in Figure 1b [16].

Table 1. Laws and main measures in different countries or areas [17].

Countries or Areas	Laws and Main Measures
USA	In 2019, California Conference Bill No. 1080 phased out disposable plastics by 2030.
Canada	It was announced that the use of disposable plastic products would be banned from 2021.
Japan	Enactment of the 'Plastic Resource Recycling Promotion Act' in 2019 to reduce disposable plastic products by 25% by 2030.
Africa	South Africa introduced a plastic bag levy policy in 2003; plastic shopping bags are banned in 16 countries in West Africa.
India	From 2 October, 2019, the use of plastic bags, bottles, plates, and straws was prohibited nationwide.
UK	The ban on the use of plastic straws, plastic cotton swabs, etc., began on 1 October 2020.
Europe	The European Commission's proposal to prohibit the use of disposable plastic products, covering ten kinds of disposable plastic products, was enacted on 3 July 2021 with a view to reducing disposable plastic containers and packaging in Europe by 2030.
Norway	From 3 July 2021, the use of disposable plastic products such as plastic straws and tableware is prohibited.
Iceland	From 3 July 2021, it is prohibited to put commonly used disposable plastic products on the market.
China	In 2007, the General Office of the State Council of China published a Notice for limiting the use of free plastic bags. In 2020, the National Development and Reform Commission and the Ministry of Ecological Environment of China published a Notice for a ban on the use of disposable plastic products.

1.2. The Benefits of Using Waste Plastic in Asphalt

China's road network has developed an ascending trend since 2008, and as of 2020, consists of over 5.20 million kilometers of roads due to rapid urbanization and economic growth [18]. Asphalt is a thermoplastic material that demonstrates viscoelastic properties under most pavement operative conditions, thus playing an important role in pavement performance [19].

In an attempt to improve the durability and reliability of asphalt pavements to meet the climatic, traffic, and other requirements, the use of modified asphalt instead of raw asphalt has long been recommended as an effective approach [20]. It has been known for some time that virgin polymers can improve asphalt performance, especially of the high-temperature stability [21]. But virgin polymer materials are difficult to find and are uneconomical when used as modifiers [22]. High construction costs, when combined with awareness regarding environmental stewardship, have encouraged the use of waste plastics in asphalt modification.

Various studies and research projects have been conducted to find appropriate applications of using waste plastics in asphalt production, discussing the properties of waste plastic-modified asphalt, modified mechanisms, and environmental concerns [23,24]. In general, there is a desire to improve the utilization of waste plastic materials in asphalt, as long as performance is not adversely impacted [25]. According to evidence from previous literature, utilizing waste plastic as a modifier in asphalt production provides asphalt with a similar property to virgin polymers, substantially reducing the construction cost, and protecting the environment from additional contamination [26–28].

This paper is a literature review that critically presents the recent progress, developments, and challenges in the application of waste plastic-modified asphalt binder and mixture technologies. This paper is not the first review of this topic, but it renews the latest developments in the field. It focuses on the approaches for waste plastic-modified asphalt and mixture production, the influences of the main factors including the types and dosages of waste plastic, blending conditions and pretreatment methods on the properties of modified asphalt binder, and the discussion of the performance of waste plastic-modified asphalt mixtures, life cycle assessment (LCA) and practical engineering applications. In addition, compared with other reviews, this paper discusses the current challenge for waste plastic-modified asphalt including the stability, low-temperature performance, the modification mechanism, and laboratory problems of the asphalt blends, as well as provides potential ways to improve the properties of waste plastic asphalt and mixtures.

2. Waste Plastic and Sources

Plastics are synthetic materials derived primarily from refined crude oil petroleum products [25,29]. Table 2 provides a summary of common waste plastic products that can be recycled in accordance with ASTM D7611 (ASTM 2019) and GB/T 37547-2019. In general, the main sources of waste plastic in the environment are plastic containers, plastic packaging, and other common plastic industrial products, as shown in Table 2. Specifically, most single-use plastic products, such as bottles, packaging and disposable products, are manufactured from low-density polyethylene (LDPE) [30], high-density polyethylene (HDPE) [31,32], and polystyrene (PS) [33], while long-term plastic items are made from polyethylene terephthalate (PET) [28], polypropylene (PP) [34], polyvinyl chloride (PVC) [35,36], ethylene-vinyl acetate copolymer (EVA) [37,38] and others. Additionally, researchers have investigated other waste plastic types (e.g., polyurethane (PU) [39] and acrylonitrile butadiene styrene (ABS) [40,41]).

Table 2. Melting point and main sources of waste plastics.

Type	Melting Point (°C)	Sources [21,42]
LDPE	110–120 [43]	Soft drink and mineral water bottles
HDPE	130 [44]	Plastic bottles and packaging
PP	145–165 [45]	Straw, furniture, and wrapping industries
PVC	160–210 [46]	Fittings and plumbing pipes
PET	260 [47]	Soft drink and water bottles
PS	210–249 [48]	Disposable plates and cups, carry-out containers, and compact disc cases
EVA	65–80 [49]	Soles, thin films, and wire cables
ABS	No true melting point [42]	Electronic devices
PU	No true melting point [42]	Upholstered furniture and mattresses, shoes, cars, medical devices, buildings, and technical equipment

Different types of waste plastic have drastically different characteristics, which are mainly affected by chemical composition, chemical structure, and average molecular weight [26]. For example, the characteristics of LDPE are mainly affected by a large number of branched chains, and the crystallinity is only 55–65% [50]. By contrast, HDPE only has a few short branched chains, but the crystallinity is 80–90%—much higher than that in LDPE—so it is difficult to immerse in asphalt [51]. The melting point is suggested as one of the primary characteristics used to determine whether a waste plastic can be used as an asphalt modifier. There is an obvious difference among the melting points of these waste plastics, as listed in Table 2. PET has a melting point of around 260 °C, which is much higher than the temperatures for typical asphalt binder production and storage. PVC also does not meet the necessary criteria, as its melting point is 160–210 °C. If waste plastics with significantly different melting points are heated together, some will melt while others do not. Some plastics may not have melted yet, while others may be about to reach their decomposition temperature. Moreover, the performance of the mixed waste plastics may vary, with some plastics degrading when several types of plastic are heated together. Thus, it is necessary to classify and recover waste plastics before utilization. It is recommended that waste plastics such as PET, PVC, and PS are more suitable for dry process modification, because their melting points are higher than the heating temperature for preparing modified asphalt [42]. Waste plastics such as LDPE, HDPE, PP, and EVA, which have melting temperatures below the production temperatures of typical asphalt, are preferred for use in wet processes as modifiers (or potential modifiers) in asphalt production [52].

3. The Use of Waste Plastics in Asphalt

3.1. Forms of Waste Plastics Used in Asphalt

Waste plastics can be used as asphalt modifiers in a variety of forms through further processes. In the early days, waste plastics were processed into pellet form (Figure 2a) and were intended to be incorporated directly into the asphalt production plant [29,53]. These pellets were produced from 100% waste plastics, with sizes measuring between 0.3 mm and 0.5 mm [9]. In recent years, waste plastics have begun to be processed into shredding form, as shown in Figure 2b. However, the waste plastics in both pellet and shredding forms can only be processed through the complicated industrial system. Researchers have recently produced waste plastics as a modifier in more accepted forms by using common methods such as scissors and crushers in the lab. For example, Modarres and Hamedi [54] cut waste PET bottles and cans into small pieces (larger than shredding) and crushed them into flake form using a special crusher, as shown in Figure 2c. Kumar and Garg [55] and Lin et al. [56] made modified asphalt with waste plastics in thin strip form (20×3 mm^2) and fiber form (less than 2 mm). Furthermore, waste HDPE powder and electronic-plastic (e-waste) powder (as shown in Figure 2d) have been used as asphalt modifiers [32,40].

(a) (b) (c) (d)

Figure 2. Various forms of waste plastics used as modifiers: (**a**) pellet [57]; (**b**) shredding [54]; (**c**) flake [57]; (**d**) powder [58]. Reprinted with permission from Refs. [54,57,58]. Copyright 2014 Elsiver publisher.

3.2. Approaches to Incorporating Waste Plastics into Asphalt

There are two main approaches used to incorporate waste plastics into the asphalt: the wet process and the dry process [9]. In the wet process, waste plastics are added directly into the asphalt binder at high temperatures, where mechanical mixing is required to achieve a homogeneous plastic-modified binder blend. The mixing temperature and mixing time depend on the nature of the waste plastic source and asphalt binder. In the dry process, waste plastics are added directly to the asphalt mixture, either as a partial aggregate replacement or a mixture modifier [59]. When the addition of waste plastic is carried out by a wet process, the waste plastics are added to the asphalt binder to modify their properties before coming into contact with the aggregates [60]. When the plastics are added using a dry process, the waste plastics are mixed with aggregates so that they actually act as reinforcement materials [9].

Both the wet and dry methods have advantages as well as drawbacks, as shown in Table 3. The wet process is a conventional way of adding waste plastics, whereby they are mixed with the asphalt in a high-shear mill. Thus, the wet process requires specialized mixing and storage facilities [61], and it is better for controlling the properties of the modified asphalt binder [62]. This is likely the reason why the wet method is currently the most widely used in asphalt modification. By contrast, the dry process does not require professional equipment. It can be applied in any asphalt plant without major modifications [9]. The results from previous research have shown that the modified asphalt binder produced by the wet process has a higher viscosity, which allows a better coating of the aggregate particles, without exudation or drainage problems [63]. The modified asphalt mixture produced by the dry method has relatively poor water stability [62]. In terms of cost, the AC-16 mixture production materials using the dry method costs around CNY 290,000 per kilometer, which is lower than the wet method [64].

Table 3. The advantages and drawbacks of different processes.

Method	Production Cost	Technological Problem		Performance of Mixture	
		Advantage	Drawback	Advantage	Drawback
Wet process	Expensive (AC-16)	Normative guidance and engineering experience	Complex production process (specialized mixing and storage facilities)	Higher viscosity	Poor storage stability
Dry process	Cheap (AC-16)	Lack of normative guidance	Simple production process (no need of professional facility)	-	Poor water stability

The performances of asphalt mixtures containing different waste plastics are significantly different because of the wet or dry process. Overall, the asphalt mixtures containing waste HDPE and EVA show similar properties for both the wet and dry processes. However, the waste HDPE mixture produced using the dry process exhibits poor water sensitivity [62].

Most previous studies focused on the wet process. The wet process is currently the most widely used for polymer asphalt modification because of its enhanced thermal behavior. However, the dry process is more cost-effective and has a simpler production process, meaning it is more convenient for waste plastic-modified asphalt production. Thus, further research is needed due to the lack of normative guidance and engineering experience for the dry process.

3.3. *Single and Composite Modification*

3.3.1. Single Modification

Each type of waste plastic has its own chemical composition, unique structure, molecular weight, etc., all of which affect the performance of modified asphalt [26] when such plastics are used as a modifier independently in asphalt production.

(1) Waste LDPE

LDPE has a lower specific gravity, strength, and temperature resistance than HDPE because of its long, flexible, and linear polyethylene chain [65]. Due to the irregular structure of the multimolecular chain arrangement of LDPE, the branched chains in asphalt combine with each other to form reticular three-dimensional structures, which can better improve the properties of modified asphalt. Thus, LDPE is widely used as a modified material for asphalt. Since the 1990s, several studies in China, Europe, the US, and the UK have reported the use of modifiers made by recycled LDPE independently [66,67]. Khan et al. [68] studied waste LDPE, HDPE, and crumb rubber (CR) as an addition to base bitumen, and showed that modified asphalt binder with 10% LDPE offers the best resistance against rutting compared to HDPE and CR. Ho et al. [69] investigated combinations of three types of recycled LDPE as asphalt modifiers. The results have shown that the molecular weight and molecular weight distribution of waste LDPE have significant effects on the asphalt's low-temperature performance, thermal storage stability, and polymer phase distribution. The recycled LDPE with lower molecular weight and wider molecular weight distribution is more suitable for asphalt modification, compared with high molecular weight LDPE with very narrow molecular weight distribution.

(2) Waste HDPE

As discussed above, the high crystallinity of HDPE makes it difficult to immerse in asphalt, which also affects the compatibility of modified asphalt. It is agreed that the waste HDPE-modified asphalt has higher stiffness and viscosity, and better moisture resistance [70]. Costa et al. [37] indicated that waste HDPE-modified asphalt has higher stiffness and lower penetration, but worse resilience and creep recovery, compared with SBS-modified asphalt.

(3) Waste PP

Recycled PP-modified asphalt has the common characteristics of thermoplastic polyester modified asphalt, especially the superiority of high-temperature performance. However, the addition of waste PP reduces the ductility of modified asphalt and decreases the fatigue cracking performance [71]. Specifically, the reduction of the ductility is around 20% when 5% of waste PP is added to the asphalt [71,72]. Thus, it is recommended that waste PP-modified asphalt is suitable for high-temperature and high-humidity areas, but the viscosity needs to be improved [45].

(4) Waste PVC

Recent studies have shown that the addition of waste PVC increases the viscosity and stiffness of base asphalt so that the modified asphalt has better rutting resistance. One possible reason is that the chloride and carbon bond dipole in PVC provides a greater stiffness [73]. A study by Ziari [74] indicated that waste PVC improves the fatigue resistance, but the thermal cracking resistance is poor. It is noteworthy that hydrogen chloride (HCL) can be formed and discharged into the atmosphere when PVC is heated to a high temperature [26]. Thus, measures should be taken to avoid air pollution.

(5) Waste PET

According to the Wellness Recovery Action Program (WRAP), PET is one of the most recycled plastic wastes [47]. Because of the high melting point, most researchers tend to use waste PET for dry modification [75]. Results have shown that waste PET-modified mixtures developed using the dry method have an improved high-temperature performance and reduced fracture resistance when the dosage of PET is 30% and 50% [76].

(6) Waste PS

Waste PS-modified asphalt mixtures developed using the dry process are found to have higher rigidity, but this could be a problem in colder areas in terms of cracking resistance [33]. Specifically, waste PS exhibits the lowest elastic behavior in the modified asphalt mixture compared with waste PE, PP, and rubber asphalt mixture using the dry process [42]. Fang et al. [77] successfully used a very low-density PS waste to expand the stiffness of asphalt and improve its rutting resistance. Hasan et al. [71] indicated that the addition of waste high impact PS (HIPS) in asphalt improves the asphalt's stiffness, but decreases the low-temperature properties. Furthermore, much more attention should be paid to the fact that harmful substances are released when PS is heated above 70 °C.

(7) Waste EVA

Waste EVA has good compatibility with asphalt, so it has been widely studied and applied. The results have shown that the large volume of the vinyl acetate group becomes a non-crystalline area or amorphous area, which plays a role similar to rubber when EVA is mixed with asphalt. The crystalline area of EVA has high stiffness, which acts as a reinforcing bar, and greatly improves the high-temperature stability, low-temperature cracking resistance, and viscosity of modified asphalt [78]. It also exhibits certain improvements in low-temperature performance when small amounts of waste EVA are added (2–4%) [79].

(8) Waste ABS

The most common e-plastics used in the manufacture of electronic devices are ABS. Evidence from recent studies indicated that the use of e-plastic powders for asphalt modification helped improve the asphalt's viscosity and blending and mixing temperatures, meanwhile decreasing rutting susceptibility compared to virgin asphalt [80]. The low-temperature performance of ABS-modified asphalt is equivalent to that of virgin asphalt binders [81]. Compared with waste EVA- and PE-modified asphalt, waste ABS has poor performance as an asphalt modifier, but it seems to have better storage stability [26]. Moreover, the pavement performance of waste ABS-modified asphalt is better than unmodified asphalt [82]. According to the Mechanistic-Empirical Pavement Design Guide (M-E PDG) [83], using e-waste materials as modifiers for asphalt mixtures using the dry method would decrease the design thickness of the asphalt layers [84].

(9) Waste PU

Bazmara et al. [85] used thermoplastic PU and synthetic PU as modifier additives in asphalt production. The results showed that the addition of synthetic PU increased the asphalt's viscosity and stiffness. Both types of PU improved the performance of base asphalt at high temperatures, including high rutting resistance and performance grade; however, they had no notable effects on asphalt performance at low temperatures. A similar result was reported by Cong [86], who noted that waste PU-modified asphalt had good deformation resistance, aging resistance, fatigue resistance, and high-temperature storage performance. Waste PU-modified asphalt mixture developed using the wet method also had excellent water stability and deformation resistance [87]. Hot-mix asphalts with PU-modified bitumen yielded improvements in stability and lower deformation [88]. With regard to the PU-modified mixture, Salas et al. showed that, compared with the virgin sample, the PU-modified MA from the wet method exhibited lower indentation, and thus the modified mastic asphalts (MA) can be used for heavy-traffic roads [89].

Table 4 presents a summary of the effects of waste plastics on asphalt performance with respect to compatibility, high- and low-temperature performance, and viscosity, based on the most recent literature reviewed in this paper. It is evident that the addition of waste plastics can most likely increase the high-temperature stability and viscosity as well as decrease the low-temperature flexibility. Waste LDPE, PP, EVA, ABS, and PU have good compatibility with asphalt compared with other types of waste plastics, which can be seen from the summary of the rheological results of various waste plastic-modified asphalt in Figure 3. The high-temperature rheological property of PP-modified asphalt was the best, followed by PE- and PVC-modified asphalt, and PS-modified asphalt was the least effective. However, further research is needed due to the difference in dosage and asphalt.

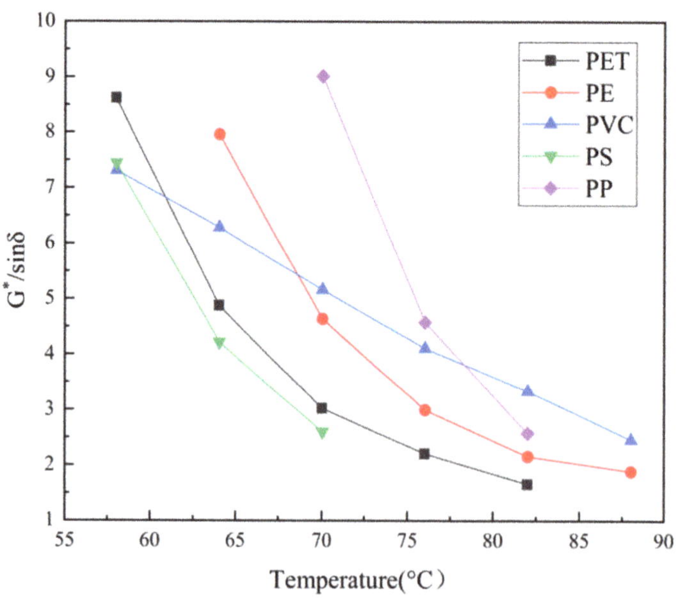

Figure 3. Rheological results of various waste plastic-modified asphalts [36,91–94].

Table 4. Characteristics of common waste plastics.

Type	Characteristics of Waste Plastic				Reference
	Compatibility	High-Temperature Stability	Low-Temperature Flexibility	Viscosity	
LDPE	√	√	-	√	[30]
HDPE		√	-		[31,32]
PP	√	√	-	√	[34,45,90]
PVC	-	√	-	√	[35,36]
PET	-	√	-	√	[28]
PS	-	√	-	√	[33]
EVA	√	√	√	√	[32,37,38,78]
ABS	√	√	-	-	[26]
PU	√	-	-	√	[85,87]

3.3.2. Composite Modification

The application of waste plastic as an independent modifier in asphalt is rare in current studies and engineering practices. This is because the key properties of asphalt cannot be improved by using only one type of waste plastic. In order to enhance and optimize the properties of waste plastic-modified asphalt binder to meet the needs of increased traffic demands, there has been growing interest in composite modification. Recently, more and more studies have investigated the properties of modified asphalt binders containing waste plastic and various materials [95]. This interesting trend means that the application of waste plastic as an asphalt modifier has been accepted by researchers and engineering practice. Some researchers investigated modified blends containing two or more types of waste plastic. For instance, Brovelli et al. [96] and García-Morales et al. [97] assessed the high-temperature stability of base asphalt modified by combining LDPE and EVA. Lai et al. [98] studied the compatibility and performance of waste HDPE/LDPE/PP-modified asphalt. Other researchers focused on the application of various modified blends of waste plastics and common polymers. Nasr and Pakshir [99] tested three melt-compounding combinations of waste PET and crumb rubber to improve the rutting and fatigue damage resistance of two base asphalt binders. A study reported by Al-Abdul Wahhab et al. [44] suggested that waste LDPE/HDPE-modified asphalt, in combination with an elastomeric SBS, can obtain higher recovery and strain resistance, which are better than using the same amount of SBS alone. Additionally, Krzysztof et al. [100] improved the conventional and thermal properties of asphalt by blending waste LDPE, ground tire rubber (GTR), and elastomer. Other studies have reported that waste plastics can be mixed with some common materials such as sulfur [101,102], carbon black [103], and polyphosphoric acid [104,105] as asphalt modifiers.

4. Factors Affecting Properties of Waste Plastic-Modified Asphalt

4.1. Waste Plastic Properties

The characteristics of waste plastics, such as type, chemical composition and structure, and molecular weight, affect the time required for blending, as plastics with higher molecular weight require more time to blend homogeneously with the asphalt binder. Additionally, waste plastics are produced in smaller sizes to help disperse and dissolve into the asphalt binder [94].

PE is one of the most popular thermos-plastics, and it is one of the earliest waste plastics to be used as an asphalt modifier in the world. PE has the simplest polymer structure, with each carbon atom connected to two hydrogen atoms. PE is categorized based on density into HDPE, LDPE, and Linear LDPE (LLDPE) [106]. Evidence from the literature shows that, compared with HDPE, the intermolecular force in LDPE is weaker, which is beneficial to the compatibility between asphalt and LDPE [31,32,50]. It is consequently suggested that both HDPE and LDPE are appropriate for asphalt modification, but LDPE is better.

In recent years, various types of waste plastics have been used as an asphalt modifier. Hu et al. [92] reported that the rheological properties of asphalt binder are enhanced when using waste packaging tape PP as the modifier. Gürü et al. [107] confirmed that thin liquid polyol PET (TLPP) and viscous polyol PET (VPP) made from waste PET bottles can improve the low-temperature performance and fatigue resistance of the modified asphalt. Furthermore, Köfteci et al. [108] found a significant difference in the performances of asphalt binders modified by different waste PVC sources (window, blinds, and cable wastes).

4.2. Asphalt Binder Properties

Asphalts are composed of two main phases: (1) an oily phase consisting of saturated hydrocarbons, aromatic cyclic products, and resins, and (2) a non-oily phase formed by asphaltenes and carbenes; however, the chemical composition and structure are different [109]. Elemental analyses indicate that most asphalts contain 79–88 weight percent (wt%) carbon, 7–13 wt% hydrogen, 2–8 wt% oxygen, traces to 8 wt% sulfur, and traces to 3 wt% nitrogen. Asphaltene contents provide a basis for the classification of asphalts into sol- or gel-types. In general, an asphalt low in asphaltene content (5–10 wt%) has properties characteristic of sol-type asphalt, and has high-temperature susceptibility, high ductility, and a low oxidative hardening rate. In contrast, asphalt with high asphaltene content (20–30 wt%) is gel-type, and has low-temperature susceptibility, low ductility, and a susceptibility to oxidative age hardening. Certainly, asphalt with intermediate asphaltene contents has properties intermediate between the sol- and gel-type behavior [110–112]. A study reported by Lesueur [113] shows that high asphaltene content decreases the compatibility between polymer and asphalt. Furthermore, Giavarini et al. [114] found that modified asphalt's properties depend not only on the difference in density and viscosity between asphalt and polymer, but also on asphalt structure.

4.3. Waste Plastic Dosage

The utilization of suitable waste plastics in asphalt modification shows an improvement in asphalt properties; however, there is an optimum point between the waste dosage and asphalt properties. Many studies available investigated the effects of modified asphalt containing waste plastics at various dosages on its properties and pavement performance. The dosage range of common polymer-modified asphalt is between 2.5 wt% and 3.5 wt%, with a higher dose range greater than 7 wt% being referred to as highly modified asphalt [115]. Mashaan et al. [28] reported that the ideal content of waste plastic is 6–8 wt% to improve the rutting and aging resistances of modified asphalt. Naskar et al. [90] found that modified asphalt with 5 wt% waste plastic has the highest thermal stability compared to the other binders investigated. However, the penetration, softening point, and elasticity of the binder are negatively affected when the waste plastic content is up to 7 wt%. Fuentes-Audén et al. [116] also concluded that only low waste plastic concentrations (0–5 wt%) can be used for road paving whereas high waste plastic concentrations (10–15 wt%) are suitable for roofing. A similar study by Fernandes et al. [117] reported that increasing the waste plastic content in base asphalt improves the softening point temperature, resilience, and viscosity. Moreover, Karmakar and Roy [118] and Ameri et al. [79] indicated that an increased modifier content negatively affects the compatibility between modifier and asphalt, and low-temperature performance. Therefore, the authors recommended that a waste plastic concentration of approximately 5 wt% is better for asphalt properties [21]. Specifically, the optimal percentages of waste PET, PVC, and PP are 3.47 wt%, 6.25 wt%, and 4.64 wt%, respectively [95]. Briefly, the optimal dosage of waste plastics in modified asphalt is not only highly dependent on the properties of the waste plastic and base binder, but also the specific requirements of the asphalt binder.

4.4. Blending Conditions

The results from recent studies indicated that the blending conditions (blending temperature, blending time, and blending speed) used for asphalt modification strongly affect the asphalt properties [78,112,119]. Many studies have been conducted to find the optimum blending conditions for various waste plastic-modified asphalt production. García et al. [120] found that the stirring speed for asphalt preparation determines the size of the polymer particles and thus the rheological properties of modified asphalt. Babalghaith et al. [121] suggested that the optimal blending time for common modified asphalt to achieve the best rheological properties is about 30 min. Fang et al. [57] reported that the optimum parameters for waste PE-modified asphalt preparation were a shear rate of 3750 revolutions per minute (rpm), a temperature of 150 °C, and a shear time of 1.5 h. Another study found a longer blending time of 6 h [97].

The optimum blending conditions used for various types of waste plastic-modified asphalt are summarized in Table 5. The selection of mixing methods mostly depends on the specific type of waste plastic and its content use—the optimum content is about 3–6 wt%. For these seven types of waste plastics, the blending temperature is about 150 to 180 °C, blending time is between 1 and 3 h, and blending speed is from 1200 to 5000 rpm.

Table 5. Blending conditions commonly used for waste plastic-modified asphalt.

Waste Plastic	Optimum Content (wt.%)	Blending Temperature (°C)	Blending Time (min)	Blending Speed (rpm)	Reference
PE	3–6	145–190	60–150	1750–4000	[31,32,106,122]
PP	3–6	160–180	45–90	1800–4000	[45,72,90,123]
PVC	4–8	160–180	60–180	1300–2000	[36,108]
PET	2–8	180	60	13,000	[37,38,56,78]
EVA	3–5	140–180	80–120	1800–3000	[37,38,78]
PS	4–6	150–190	90–120	3000	[33,77]

4.5. Pretreatment Methods for Waste Plastic

It is necessary to carefully consider the appropriate pretreatment method for waste plastic before it is added into the asphalt, as it affects the compatibility between the asphalt and modifier. Generally, pretreatment methods can be divided into two main methods: the physical method and the chemical method. It seems that most researchers (approximately 85%) prefer using the physical method to pretreat waste plastics, according to all of the related research papers referenced in this review.

4.5.1. Physical Method

Waste plastics can be processed into smaller particles through physical methods such as grinding, shredding, pulverization, and extrusion.

Table 6 shows a comparison of the common physical methods used for pretreating waste plastics. Waste plastics can be pretreated using a suitable physical method into various forms of smaller size, based on their sources and purpose of use. For example, waste PET bottles can be processed into particles of larger or smaller sizes by crushing or grinding.

Table 6. Comparison of physical methods for pretreating waste plastics.

Type	Source	Physical Method	Form	Size	Reference
Waste plastic bag waste plastic pipe	-	Shredding	Strip Fiber	1–2 cm 20 × 3 mm^2	[35,55,90]
Waste plastic bottle	PET	Cutting and crushing	Particle	0.45–1.18 mm	[54]
Waste milk bag	-	Extruding	Pellet	-	[57]
Waste plastic bottle waste express bag	PET	Grinding	Particle Piece	0.45–0.5 mm 2–5 mm	[56,124]
Waste window blind and cable	PVC	Pulverization	Powder	-	[108]

4.5.2. Chemical Method

The use of the chemical method for the preparation of waste plastic is commonly carried out through grafting or irradiation to provide better compatibility and properties with the modifier, so that the performance of modified asphalt is improved. Li et al. [125] successfully increased the reaction of LDPE and asphalt through grafting LDPE with glycidyl methacrylate (GMA) containing carbon double bonds, epoxy groups, and epoxy functional groups, resulting in improved performance of the modified asphalt. Vargas et al. [43] improved the asphalt performance at a higher temperature using a similar method of grafting HDPE with glycidyl methacrylate. Yeh et al. [126] reported a similar result by grafting maleated PP.

In recent studies, a novel method named irradiation was used for pretreating waste plastic. Ahmedzade et al. [127] found that the chemical interaction between the waste HDPE and asphalt could be caused by irradiated waste HDPE, consequently enhancing the physical properties of the asphalt. Thus, the irradiated HDPE used as a modifier is recommended. A similar result was achieved by replacing electron beams with gamma rays [128].

5. Engineering Properties, Environmental Concerns and Practical Engineering Applications of Waste Plastic Asphalt Mixtures

5.1. Engineering Properties

Most studies have demonstrated that the addition of waste plastics significantly increases the stiffness and rutting resistance of base asphalt and mixture, and thus, has the potential to extend the service lives of asphalt pavements [31,68,99,129]. Asphalt mixtures modified with waste PE, PP, and rubber have a similar performance of increased rutting resistance [130]. Specifically, the rut depth values are reduced by more than half for waste EVA-modified asphalt mixtures, and by up to a third for waste HDPE-modified asphalt mixtures compared with conventional asphalt mixtures [62]. The performance of waste HDPE-modified mixture in terms of fatigue resistance is similar to or slightly worse than that of the conventional mixture, while the performance of the waste EVA mixture shows significant improvement. Both of the waste PE- and PP-modified mixtures have superior durability and properties with higher resistance to permanent deformation and moisture damage [131]. The waste PVC mixture shows a stronger rutting resistance by improving the rheological properties of the asphalt binder [132].

Based on the results of laboratory tests on asphalt binders, it can be expected that highly modified asphalt mixtures will have better functional properties, durability, and anti-rutting resistance [133]. Laboratory results indicate that asphalt mixtures containing 20% PP/rubber have better low- and high-temperature performances and water sensitivity than those with SBS modified asphalt mixtures [134].

Currently, the main types of waste plastic-modified asphalt mixture are stone matrix asphalt (SMA) and asphalt concrete (AC). Plastic gives an increase in stability, split tensile strength, and compressive strength compared to the conventional SMA mixture [135]. But the waste plastic might increase the cracking potential of the mixture [136]. Laboratory tests indicated that SMA mixtures containing waste plastic increased resistance, anti-rutting performance, fatigue resistance [137], and moisture damage [138]. However, the effect of PET on the moisture susceptibility of SMA mixtures was found to be negligible [139]. Some researchers studied the performance of AC mixtures by adding waste plastic and found that the mixture exhibited an increase in rutting resistance [130,140,141] and Marshall stability and stiffness [130,140,141], and a reduction in the moisture damage resistance, workability, fatigue resistance [141,142], and thermal susceptibility [140].

Some scholars have tried to use waste plastics in the reclaimed asphalt pavement (RAP) mixture, which also gives us an alternative way to alleviate the environmental problems caused by plastics. The results have shown that the addition of waste plastic materials into the RAP mixture potentially improves the durability of the pavement [143]. By contrast, the study of Yamin [144] showed that the addition of RAP has a negative effect on durability compared with the conventional mixture. The result of Leng's [145] research indicated that the samples containing RAP- and PET-derived additives offered better overall performance than conventional mixtures, increasing the rutting resistance by at least 15% and fatigue cracking resistance by more than 60%. Therefore, the rutting resistance and fatigue cracking resistance performance of the RAP mixture with waste plastics have been improved, and the durability remains to be further studied.

5.2. Environmental Concerns

The environmental benefit is obvious, even if the total amount of waste plastic is a small part of the total asphalt mixture. Studies have shown that the CO_2 equivalent emissions can be decreased by 10.2% when 8% of virgin PP is replaced with the same amount of waste PP, and the emission decreases by 15.6% when waste PP is used to replace SBS. More evidence is needed to show that the implementation of these waste plastics into the pavement is eco-friendly [146].

The LCA method is typically applied to quantify the environmental impacts of using waste plastics in asphalt and asphalt mixtures throughout their entire life cycle—from raw material extraction, to transport, manufacturing, and use. Poulikakos et al. [147] assessed four hypothetical roads using the LCA method, with considerable savings in cost, CO_2, and energy compared to conventional asphalt mixtures using all of the virgin components. Another study showed that the energy consumption can be reduced by 2.23% when an asphalt mixture with waste plastic instead of a conventional asphalt mixture is used as a surface course material [148]. Santos [149] also obtained similar conclusions by using waste PP instead of virgin PP. Bart evaluated the environmental behavior of asphalt mixtures with waste PET using the LCA method. The results showed that the use of waste PET in asphalt pavements not only benefits in terms of energy saving and reduced greenhouse gas emissions, but also improves the resistance of cracks, thereby decreasing maintenance requirements [147]. The cradle-to-gate LCA modeling and sensitive analysis suggest that highly modified asphalt mixtures (containing 20% PP/rubber) are more eco-friendly in terms of energy consumption and greenhouse gas emissions [134].

Multi-attribute analysis methods, including environmental factors, costs, and engineering properties, have been conducted to investigate the overall problem of plastic recycling on the road for mixture sustainable factors [149]. The research combined the laboratory experimental performance with the environmental LCA results using the multi-attribute grey relational analysis (GRA) method, and comprehensively sorted the scheme, providing an innovative perspective for the study of recycled materials for road and pavement engineering. Yu et al. [134] evaluated the waste PP asphalt mixture and SBS asphalt mixture from environmental concerns, using cradle-to-gate LCA modeling. The results indicate that the waste PP asphalt mixture is more eco-friendly compared with the SBS asphalt mixture. Two highly modified asphalt mixtures, replacing 25% of bitumen with two types of plastic waste, show environmental and economic advantages; specifically, a 17% reduction in environmental impact and an 11% reduction in economic impact can be achieved [150].

5.3. Practical Engineering Applications

Some researchers have taken a step in evaluating the performance of waste plastic-modified asphalt mixtures by assessing actual field performance, not only at the laboratory level.

(1) India

Waste plastics have been widely used in road construction in India. Since 2002, more than 2500 km of asphalt concrete pavements have been modified with polymer waste using the dry method. In the following years, six sites were selected to investigate the status of waste plastic asphalt pavement. The results showed that these roads worked well, without potholes, raveling, and rutting [151].

The Indian Highway Congress published guidance for the use of waste plastics in the bituminous mixture in 2013, while the National Rural Highway Development Agency provided guidelines on the use of waste plastics in rural road construction [151].

(2) UK

MacRebur's recycled waste plastic was incorporated into the asphalt instead of traditional bitumen and used by Durham County Council in the UK for resurfacing a section of A689 near Sedgefield, and for resurfacing runways and taxiways at Carlisle Airport in the UK.

The Department for Transport provided £1.6 million to extend Cumbria's existing road that is built from recycled plastic mixed with asphalt. The trial will also produce a guidance document on the use of plastic asphalt.

(3) USA

UC San Diego built the first asphalt road made with recycled plastic binder as opposed to a petroleum-based bitumen binder in the USA.

(4) Africa

The local government is building the first plastic road in the city of Kouga municipality, South Africa. It is reported that a 1-kilometer 'plastic road' could consume nearly 700,000 plastic bottles or 1.8 million plastic bags.

(5) The Netherlands

In June 2021, the PlasticRoad Company built the first parking spaces and residential street using waste plastic in the municipality of Almere; in the same year, waste plastic was used to construct a car-sharing location in the province of Overijssel and the municipality of Hardenberg, and to build a bike path at Delft University of Technology.

(6) Mexico

The modular design of Mexico City's PlasticRoad offers temporary water storage and drainage features.

The use of waste plastics on actual roads has just started, and most plastic roads are piloted in bicycle lanes, runways, and rural roads. Only the waste plastic roads in Africa have follow-up reports, and the long-term outcomes of many pilot roads are not yet clear.

6. Challenges of Using Waste Plastics with Asphalt

6.1. *Low-Temperature Performance*

There is no doubt that the utilization of waste plastics with asphalt significantly improves the asphalt's high-temperature performance [9,31,129]. However, it increases the creep stiffness of asphalt at low temperatures, resulting in the degradation of the low temperature cracking resistance of asphalt at the same time [50]. Rubber is an elastomer and can help improve the low-temperature crack and fatigue resistance of asphalt. As a consequence, the invention of a modifier prepared by mixing waste plastic and rubber has been proven successful to solve this problem.

Studies reported by Yan et al. [152] and Kalantar et al. [21] show a great improvement of both the high- and low-temperature performances of asphalt modified with waste PE and waste rubber blending. The researchers explained this phenomenon through laboratory experiments. Ma et al. [153] found a significant enhancement of the compatibility of asphalt modified with waste plastic/rubber blends compared with adding plastic and rubber independently or with additional compatibilizer. Other researchers give an explanation that the elastic component of rubber powder compensates for the side effects of plastic due to its continuous elastic network formed by the rubber/plastic composite modified asphalt [154].

However, the evaluation index of waste plastic-modified asphalt on low-temperature performance is controversial. The assessment of asphalt's low-temperature performance in China is mostly based on the Bending Beam Rheometer test and the ductility test. Yang [50] suggested that the use of ductility, as an index for evaluating the asphalt's low-temperature performance, has a certain limitation in the evaluation of modified asphalt. The results from Yu [155] lead to a similar conclusion: the network structure of plastic-modified asphalt samples prepared and stored at low temperatures can reduce the crack extension and improve the toughness of asphalt, which is contradictory to the result of the ductility test. This means that the ductility is not enough to evaluate the low-temperature performance of asphalt. Further studies should be performed to develop an assessment system for the performance of waste plastic-modified asphalt.

6.2. Storage Stability

Poor storage stability is one of the main drawbacks of polymer-modified asphalt [156]. It is particularly prominent in asphalt with composite modification. The compatibility between modifier and asphalt binder is one factor that strongly influences the storage stability of a modified asphalt binder [153].

Composite modified asphalt is a multiphase system. The composite modifier is an intimate blend that comprises at least two incompatible polymers forming a two- or multi-phase blend including waste plastic and other polymers. Phase separation could occur caused by the asymmetric molecular dynamics among the components in the composite modifier [157]. The modifier and asphalt binder are also incompatible due to the great differences in the density, structure, and molecular weight, from the point of view of thermodynamics [158]. Only when the compatibility among the polymers and asphalt is high enough to avoid phase separation in the binder can a proper, good quality pavement be achieved [21]. Experimental results confirmed that the addition of sulfur, carbon black, hydrophobic clay minerals, nano clay, and montmorillonite along with waste plastic into a modifier can significantly improve the compatibility of asphalt. Functionalization is another method to solve the compatibility problem. As shown in Table 7, functionalization refers to the addition of specific functional groups, free radicals, or grafting in the polymer to make chemical reactions occur in the blend to improve the specific functions of asphalt. For example, grafting LDPE with glycidyl methacrylate (GMA) and epoxy functional groups can successfully increase the reaction between LDPE and asphalt, thus improving the compatibility with the asphalt [125]. In addition, the preparation of waste LDPE and SBS blends by Gao et al. [2] improved the storage stability by forming a small amount of copolymer, which was similar to the effect of a compatibilizer obtained through a chemical reaction between the functional groups.

Table 7. The influences of additives and functionalization on the improvement of compatibility.

	Method	Modifying Influence	Reference
Additive	Phosphoric acid	Improve rheological behaviors of modified asphalt and increase the storage stability at the storage temperature.	[159]
	Montmorillonite	Improve the storage stability of modified asphalt and does not compromise its excellent high temperature rheological properties.	[155]
	Nano clay	Improves the stability of modified asphalt.	[160]
	Hydrophobic clay minerals	The storage stability of modified asphalt is improved by reducing the density difference between polymer modifiers and asphalt.	[161]
	Carbon black	Reduces the density difference between polymer and asphalt, thus improving the storage stability of modified asphalt.	[103,162]
	Sulfur	In this process, the loss of unsaturation, the shift of the double bonds and a molecular isomerization occur.	[101,102,163]

Table 7. *Cont.*

	Method	Modifying Influence	Reference
Functionalization	Copolymer	Improves the compatibility of modified asphalt.	[2,164,165]
	Cross-linking agent	Makes the polymer react with asphalt, so as to provide a chemical connection between the two and form a three-dimensional network structure and improve the compatibility of asphalt.	[38,93]
	Radical initiator	Promotes direct covalent molecular bonding between e-waste plastic powders and the modified asphalt.	[41]
	Antioxidants	Believed to play a role by scavenging free radicals and decomposing the hydroperoxides generated during oxidation.	[166]
	Functional groups	It is generally expected that the added functional groups will interact with some components of asphalt in various ways, such as forming hydrogen bonds or chemical bonds, which may improve compatibility to some extent.	[24]
	Grafting	Maleic anhydride (MAH), methacrylic acid (MAA) and glycidyl methacrylate (GMA) were used to graft some currently used polymer modifiers and were found to improve the storage stability of asphalt.	[24]

6.3. Microcosmic Modification Mechanism

The modification mechanism of polymer on asphalt is complex. The interaction principle of the waste polymer has not been comprehensively and systematically investigated. Most scholars have successfully explained the modification mechanism of waste polymers and base asphalt by the "swelling mechanism". As can be seen from Figure 4, the waste polymer absorbs the light components in the base asphalt, which subsequently swells and agglomerates together, forming a thick gel layer around it. This is the third phase interface between the polymer phase and the base asphalt phase, forming a partially compatible system, during the mixing process of the waste polymer and base asphalt [167]. It has been demonstrated that the swelling of the polymer is the key point for asphalt modification, and sufficient swelling ensures the formation of the interface layer [168]. The interface layer is a third phase between the two phases in terms of composition, structure, and performance, which makes the blend system stable and provides special performance [168].

Fang et al. [169] found that PVC absorbs the light components of asphalt, and swells in asphalt and forms the network structure, affecting the physical properties of the modified asphalt. A further conclusion is that the swelling of waste LDPE and the network structure of the blend helps improve the properties of the base asphalt [65]. Similarly, the swelling of waste PS and waste PE in asphalt is caused by the absorption of the low molecular composition of asphalt, forming a network structure, thereby improving the viscoelasticity and high-temperature performance of the asphalt [129].

Microscopic technology and infrared spectroscopy are effective tools for helping scholars to investigate the modification mechanism. The modifier and binder in the system may go through physical blending or chemical reaction, and the process is difficult to define. Fourier transform infrared spectroscopy (FTIR) can be used to qualitatively and quantitatively analyze the composition of functional groups of materials according to the principle that specific functional groups and chemical bonds selectively absorb infrared light at specific wavelengths, to clarify the modification mechanism of asphalt [170].

The use of FTIR can quantify the absorption bands. Some researchers found that there was no change in the main functional groups of waste PVC asphalt binder, indicating the modification is a physical effect [169]. Other researchers reported that there are slight increases in absorption peaks at 2925 cm^{-1} and 2856 cm^{-1} because the modified asphalt contains a large number of methyl and methylene groups, which coincide with the absorption peaks of the asphalt before modification. However, other functional groups do not change before and after modification, indicating that the waste LDPE modification of asphalt is also a physical process [46].

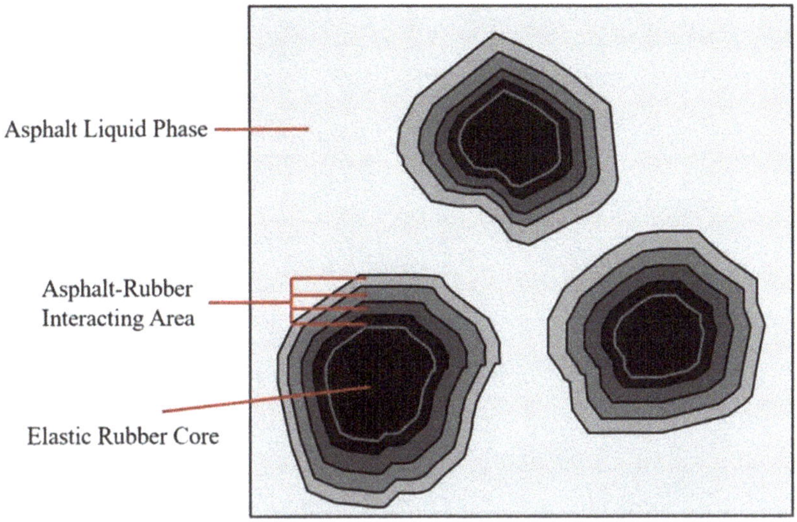

Figure 4. Swelling mechanism of rubber-modified asphalt [168]. Reprinted with permission from Ref. [168]. Copyright 2020 Elsevier publisher.

Microtechnologies, including scanning electron microscopy (SEM) [126,155], atomic force microscopy (AFM) [116,118], and confocal laser scanning microscopy (CLSM) [43,111], are also used to characterize the morphology of modifiers and modified asphalt. It is observed by CLSM that the morphology of the polymer/asphalt system moves from a continuous asphalt-rich phase, with a dispersion of polymer molecules, to a continuous asphalt-rich phase, with a dispersion of asphalt globules, with an increase in the amount of blended polymer [111].

The dispersion of modifiers and the structure form can be seen clearly in the microscopic image. This shows the impacts of modifiers on asphalt and reveals the modification mechanism. This paper summarizes the fluorescence microscopic images of different types of plastics, as shown in Figure 5. The morphology of asphalt/LDPE is shown in Figure 5a. The LDPE phases are dispersed in the asphalt phase in the form of nearly spherical particles, and the edges of the particles become blurred by swelling. In the HDPE/asphalt blend, HDPE is less compatible with asphalt than LDPE, the HDPE phase is limited in extent, and there is no visible connection between the particles (Figure 5b). In the PP/asphalt blend, the yellow PP phases become the dominant phases covering most areas, and the brown asphalt (background) begins to fade and the boundary is blurry (Figure 5c). The morphology of PVC in asphalt is long stripe shape (Figure 5d), while the EVA and PS are branched network structures accompanied by some dispersed granulated particles, which is beneficial to improve the properties of asphalt (Figure 5e,f).

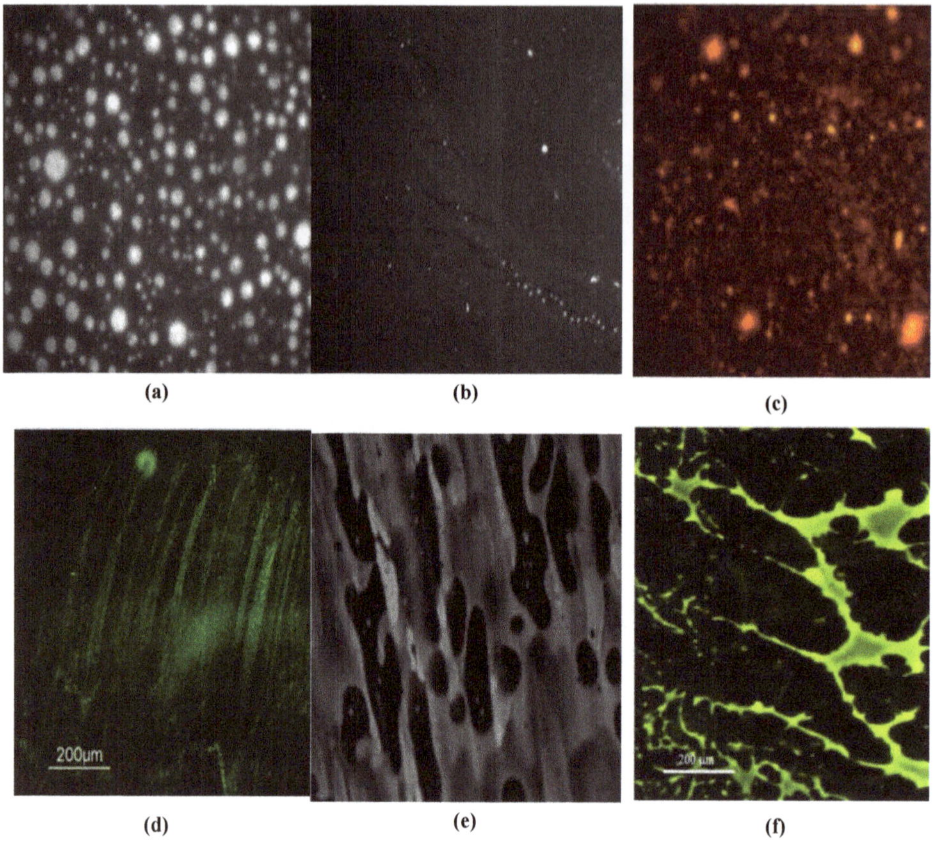

Figure 5. Fluorescence microscopic images of blends of asphalt with different types of waste plastics: (**a**) LDPE/asphalt [43]; (**b**) HDPE/asphalt [43]; (**c**) PP/asphalt [45]; (**d**) PVC/asphalt [46]; (**e**) EVA/asphalt [49]; (**f**) PS/asphalt [77]. Reprinted with permission from Refs. [43,45,46,49,77]. Copyright 2013 Elsevier publisher.

6.4. Laboratory Operational Problems

The uneven dispersion and aggregation of modifiers may be encountered in the laboratory during the modification process of waste plastics. Since asphalt binder is colloidal material, the agglomeration phenomenon naturally appears [171]. The dispersion and aggregation of modifiers in asphalt significantly affect the performance of waste plastic-modified asphalt and its further application [172].

The aggregation of modifiers can be improved by improving the mixing conditions to a certain extent. As described in Section 4.4, 'The Influence of Blending Conditions' in this review, the mixing conditions of different types of waste plastics are different, and the appropriate mixing time, temperature, and speed need to be determined. Similarly, changing the size of the modifier can improve the problem of modifier aggregation. Smaller particle sizes have a relatively larger surface area per unit mass of the modifier and can help the modifier to achieve a homogeneous mix, thus overcoming phase separation and improving the stability of the modified asphalt [173,174].

7. Conclusions

Waste plastic binders and mixtures are gaining increasing attention due to their engineering performance and economic and eco-friendly benefits. This paper presented a review of the waste plastics most commonly used in asphalt binders and mixtures, analyzed and compared various approaches for waste plastics-modified asphalt and mixture production, and discussed the influence of the main factors on the properties of modified asphalt and mixtures. The paper discussed the current challenges for waste plastic-modified asphalt, such as the stability, low-temperature performance, the modification mechanism, and laboratory problems. Based on this review, the following points can be drawn:

(1) The use of waste plastic as an asphalt modifier expands the application field of waste plastic and avoids the waste of resources. It is also an effective way to solve the waste plastic disposal problem and reduce environmental pollution. However, more attention should be paid to PS and PVC, as these plastics produce harmful emissions when heated at high temperatures.

(2) The source of waste plastics is one of the main factors that affect the performance of the modified asphalt due to the differences in chemical composition and structure, resulting in different basic characteristics. High melting point plastics, such as PVC, PS, and PVC, are more suitable for the dry processes; LDPE, HDPE, PP, and EVA, with low melting points, are more appropriate for the wet processes.

(3) Waste plastic can be made into various forms and sizes depending on the specific purpose when used for asphalt modification. Waste plastic with a smaller size and larger surface area, combined with composite modification, is recommended to achieve better high- and low-temperature performance and storage stability.

(4) A physical or a chemical modification can be accurately distinguished upon comparing the FTIR spectra before and after the asphalt modification. The microstructure of the modified asphalt, modification process, and mechanism can be further characterized using microcosmic technologies such as SEM, AFM, and CLSM. Previous studies have provided evidence that asphalt modified with waste plastic independently is a physical modification.

(5) The use of the LCA method can effectively quantify the environmental impacts of using waste plastics for asphalt and asphalt mixture modifications. Waste plastic-modified asphalt mixtures indeed reduce energy consumption and emissions, and the implementation of these waste plastics into the pavement is eco-friendly. The use of waste plastics in actual roads has just started, therefore the long-term outcomes require further assessment.

(6) There is no doubt that the high-temperature performance of waste plastic-modified asphalt is quite effective. However, there are still limitations for waste plastic-modified asphalt applications due to the separation between the waste plastic modifier and the asphalt, and the poor low-temperature performance of waste plastic-modified asphalt. Pretreatment methods (grinding, grafting, irradiating) and composite modification (additives, functionalization) can improve the disadvantages of waste plastic-modified asphalt. However, due to the cost and application limitations, waste plastic asphalt modification requires suitable pretreatment and modification methods.

8. Future Recommendations

As discussed above, waste plastic-modified asphalt is an important approach currently available to reduce environmental impact and natural resource depletion, and it deserves more attention to improve the interaction between the modifier and the asphalt to improve the asphalt properties and pavement performance.

(1) Most of the studies presented in this review focus on assessing the properties of waste plastic-modified asphalt, but the microcosmic mechanism of asphalt modification has not received enough attention. Several studies report results that are characterized by micro-experiments, but the microcosmic mechanism of waste plastic modification is rarely involved. The microcosmic mechanism of waste plastic-modified asphalt is still not well understood, and more attention should be paid to address this knowledge gap.

(2) There is a need to better understand and control the effects of using different types of waste plastic and other polymers as modifiers on the properties of the asphalt, and find

an optimum proportion for the modifier. This will promote the application of recycled waste plastics in modified asphalt.

Author Contributions: Investigation, F.X.; resources, K.L.; data curation, F.X.; writing—original draft preparation, F.X.; writing—review and editing, Y.Z.; funding acquisition, Y.Z. and F.X. All authors have read and agreed to the published version of the manuscript.

Funding: This research was funded by the Postgraduate Research and Practice Innovation Program of Jiangsu Province, grant number [SJCX21_0344]; the Science and Technology Plan Program of the Ministry of Housing and Urban–Rural Development of the People's Republic of China, grant number [2018-K9–074; 2019-K-140]; and the Youth Science and Technology Innovation Fund of Nanjing Forestry University, grant number [CX2019031].

Institutional Review Board Statement: Not applicable.

Informed Consent Statement: Not applicable.

Data Availability Statement: Not applicable.

Conflicts of Interest: The authors declare no conflict of interest.

References

1. Hopewell, J.; Dvorak, R.; Kosior, E. Plastics recycling: Challenges and opportunities. *Philos. Trans. R. Soc. B Biol. Sci.* **2009**, *364*, 2115–2126. [CrossRef] [PubMed]
2. Gao, G.; Zhu, Y.; Zhang, Y.; Zhang, Y. Preparation of storage stable LDPE/SBS Blend-modified Asphalt. *J. Highw. Transp. Res. Dev.* **2002**, *19*, 28–31. (In Chinese)
3. Plastics Europe. Available online: https://www.plasticseurope.org/en/resources/market-data (accessed on 28 July 2021).
4. National Bureau of Statistics. Available online: https://data.stats.gov.cn/easyquery.htm?cn=C01 (accessed on 28 July 2021).
5. Environmental Protection Agency of the United States. Plastic: Material-Specific Data. Available online: https://www.epa.gov/facts-and-figures-about-materials-waste-and-recycling/plastics-material-specific-data (accessed on 28 July 2021).
6. Biber, N.F.A.; Foggo, A.; Thompson, R.C. Characterising the deterioration of different plastics in air and seawater. *Mar. Pollut. Bull.* **2019**, *141*, 595–602. [CrossRef] [PubMed]
7. Wang, X.; Zhang, Y.; Huang, B.; Chen, Z.; Zhong, M.; Wang, W.; Liu, X.; Fan, Y.N.; Hu, W. Atmospheric phthalate pollution in plastic agricultural greenhouses in Shaanxi Province, China. *Environ. Pollut.* **2021**, *269*, 116096. [CrossRef]
8. Brasileiro, L.; Moreno-Navarro, F.; Tauste-Martínez, R.; Matos, J.; Rubio-Gámez, M. Reclaimed Polymers as Asphalt Binder Modifiers for More Sustainable Roads: A Review. *Sustainability* **2019**, *11*, 646. [CrossRef]
9. Li, M.; Zhang, M.; Rong, H.; Zhang, X.; He, L.; Han, P.; Tong, M. Transport and deposition of plastic particles in porous media during seawater intrusion and groundwater-seawater displacement processes. *Sci. Total Environ.* **2021**, *781*, 146752. [CrossRef]
10. Litchfield, S.G.; Schulz, K.G.; Kelaher, B.P. The influence of plastic pollution and ocean change on detrital decomposition. *Mar. Pollut. Bull.* **2020**, *158*, 111354. [CrossRef]
11. Qiu, Z.; Zhao, S.; Feng, X.; He, Y. Transfer learning method for plastic pollution evaluation in soil using NIR sensor. *Sci. Total Environ.* **2020**, *740*, 140118. [CrossRef]
12. Ding, Y.; Zou, X.; Wang, C.; Feng, Z.; Wang, Y.; Fan, Q.; Chen, H. The abundance and characteristics of atmospheric microplastic deposition in the northwestern South China Sea in the fall. *Atmos. Environ.* **2021**, *253*, 118389. [CrossRef]
13. Prata, J.C.; da Costa, J.P.; Lopes, I.; Andrady, A.L.; Duarte, A.C.; Rocha-Santos, T. A One Health perspective of the impacts of microplastics on animal, human and environmental health. *Sci. Total Environ.* **2021**, *777*, 146094. [CrossRef]
14. Wang, M.H.; He, Y.; Sen, B. Research and management of plastic pollution in coastal environments of China. *Environ. Pollut.* **2019**, *248*, 898–905. [CrossRef]
15. Okan, M.; Aydin, H.M.; Barsbay, M. Current approaches to waste polymer utilization and minimization: A review. *J. Chem. Technol. Biotechnol.* **2019**, *94*, 8–21. [CrossRef]
16. Shanghai Resource Recycling Trade Association. Available online: http://www.sh-recycle.org/articledetail.asp?id=4074 (accessed on 29 July 2021).
17. Chen, Y.; Awasthi, A.K.; Wei, F.; Tan, Q.; Li, J. Single-use plastics: Production, usage, disposal, and adverse impacts. *Sci. Total Environ.* **2021**, *752*, 141772. [CrossRef]
18. Ministry of Transport of China. *2019 Statistical Bulletin on the Development of the Transport Industry*; National Bereau of Statistics of China: Beijing, China, 2020. Available online: http://www.stats.gov.cn/tjsj/ndsj/2019/indexeh.htm (accessed on 22 December 2021).
19. Chen, S.; Che, T.; Mohseni, A.; Azari, H.; Heiden, P.A.; You, Z. Preliminary study of modified asphalt binders with thermoplastics: The Rheology properties and interfacial adhesion between thermoplastics and asphalt binder. *Constr. Build. Mater.* **2021**, *301*, 124373. [CrossRef]

20. Ragab, A.A.; Farag, R.K.; Kandil, U.F.; El-Shafie, M.; Saleh, A.M.M.; El-Kafrawy, A.F. Thermo-mechanical properties improvement of asphalt binder by using methylmethacrylate/ethylene glycol dimethacrylate. *Egypt. J. Pet.* **2016**, *25*, 397–407. [CrossRef]
21. Kalantar, Z.N.; Karim, M.R.; Mahrez, A. A review of using waste and virgin polymer in pavement. *Constr. Build. Mater.* **2012**, *33*, 55–62. [CrossRef]
22. Gao, J.; Wang, H.; Chen, J.; Meng, X.; You, Z. Laboratory evaluation on comprehensive performance of polyurethane rubber particle mixture. *Constr. Build. Mater.* **2019**, *224*, 29–39. [CrossRef]
23. Mashaan, N.; Chegenizadeh, A.; Nikraz, H. Laboratory Properties of Waste PET Plastic-Modified Asphalt Mixes. *Recycling* **2021**, *6*, 49. [CrossRef]
24. Zhu, J.; Birgisson, B.; Kringos, N. Polymer modification of bitumen: Advances and challenges. *Eur. Polym. J.* **2014**, *54*, 18–38. [CrossRef]
25. Jamshidi, A.; White, G. *Use of Recyclable Materials in Pavement Construction for Environmental Sustaionability, Proceedings of the Eighteenth Annual International Conference on Pavement Engineering, Liverpool, UK, 27–28 February 2019*; Asphalt Reinforcement Services: Liverpool, UK, 2019.
26. Costa, L.M.B.; Silva, H.M.R.D.; Oliveira, J.R.M.; Fernandes, S.R.M. Incorporation of Waste Plastic in Asphalt Binders to Improve their Performance in the Pavement. *Int. J. Pavement Res. Technol.* **2013**, *6*, 457–464.
27. Joohari, I.B.; Giustozzi, F. Chemical and high-temperature rheological properties of recycled plastics-polymer modified hybrid bitumen. *J. Clean. Prod.* **2020**, *276*, 123064. [CrossRef]
28. Mashaan, N.S.; Chegenizadeh, A.; Nikraz, H.; Rezagholilou, A. Investigating the engineering properties of asphalt binder modified with waste plastic polymer. *Eng. J.* **2021**, *12*, 1569–1574. [CrossRef]
29. Nikolaides, A.F.; Manthos, E. *Bituminous Mixtures and Pavements VII*, 1st ed.; CRC Press: Thessaloniki, Greece, 2019.
30. Hariadi, D.; Saleh, S.M.; Anwar Yamin, R.; Aprilia, S. Utilization of LDPE plastic waste on the quality of pyrolysis oil as an asphalt solvent alternative. *Therm. Sci. Eng. Prog.* **2021**, *23*, 100872. [CrossRef]
31. Appiah, J.K.; Berko-Boateng, V.N.; Tagbor, T.A. Use of waste plastic materials for road construction in Ghana. *Case Stud. Constr. Mater.* **2017**, *6*, 1–7. [CrossRef]
32. Hinislioglu, S.; Agar, E. Use of waste high density polyethylene as bitumen modifier in asphalt concrete mix. *Mater. Lett.* **2004**, *58*, 267–271. [CrossRef]
33. Vila-Cortavitarte, M.; Lastra-González, P.; Calzada-Pérez, M.Á.; Indacoechea-Vega, I. Analysis of the influence of using recycled polystyrene as a substitute for bitumen in the behaviour of asphalt concrete mixtures. *J. Clean. Prod.* **2018**, *170*, 1279–1287. [CrossRef]
34. Razali, M.N.; Aziz, M.A.A.; Jamin, N.F.M.; Salehan, N.A.M. Modification of bitumen using polyacrylic wig waste. *AIP Conf. Proc.* **2018**, *1930*, 020051.
35. Abdel-Goad, M.A.H. Waste polyvinyl chloride-modified bitumen. *J. Appl. Polym. Sci.* **2006**, *101*, 1501–1505. [CrossRef]
36. Behl, A.; Sharma, G.; Kumar, G. A sustainable approach: Utilization of waste PVC in asphalting of roads. *Constr. Build. Mater.* **2014**, *54*, 113–117. [CrossRef]
37. Costa, L.M.B.; Silva, H.M.R.D.; Peralta, J.; Oliveira, J.R.M. Using waste polymers as a reliable alternative for asphalt binder modification—Performance and morphological assessment. *Constr. Build. Mater.* **2019**, *198*, 237–244. [CrossRef]
38. Fang, C.; Zhou, S.; Zhang, M.; Zhao, S.; Wang, X.; Zheng, C. Optimization of the modification technologies of asphalt by using waste EVA from packaging. *J. Vinyl Addit. Technol.* **2009**, *15*, 199–203. [CrossRef]
39. Lakusic, S. Mastic asphalt modified with recycled polyurethane foam. *J. Croat. Assoc. Civ. Eng.* **2018**, *70*, 403–412. [CrossRef]
40. Shahane, H.A.; Bhosale, S.S. E-Waste plastic powder modified bitumen: Rheological properties and performance study of bituminous concrete. *Road Mater. Pavement Des.* **2019**, *22*, 682–702. [CrossRef]
41. Mohd Hasan, M.R.; Colbert, B.; You, Z.; Jamshidi, A.; Heiden, P.A.; Hamzah, M.O. A simple treatment of electronic-waste plastics to produce asphalt binder additives with improved properties. *Constr. Build. Mater.* **2016**, *110*, 79–88. [CrossRef]
42. Wu, S.; Montalvo, L. Repurposing waste plastics into cleaner asphalt pavement materials: A critical literature review. *J. Clean. Prod.* **2021**, *280*, 124355. [CrossRef]
43. Vargas, M.A.; Vargas, M.A.; Sánchez-Sólis, A.; Manero, O. Asphalt/polyethylene blends: Rheological properties, microstructure and viscosity modeling. *Constr. Build. Mater.* **2013**, *45*, 243–250. [CrossRef]
44. Al-Abdul Wahhab, H.I.; Dalhat, M.A.; Habib, M.A. Storage stability and high-temperature performance of asphalt binder modified with recycled plastic. *Road Mater. Pavement Des.* **2016**, *18*, 1117–1134. [CrossRef]
45. Ahmedzade, P.; Demirelli, K.; Günay, T.; Biryan, F.; Alqudah, O. Effects of Waste Polypropylene Additive on the Properties of Bituminous Binder. *Procedia Manuf.* **2015**, *2*, 165–170. [CrossRef]
46. Fang, C.; Liu, X.; Yu, R.; Liu, P.; Lei, W. Preparation and Properties of Asphalt Modified with a Composite Composed of Waste Package Poly(vinyl chloride) and Organic Montmorillonite. *J. Mater. Sci. Technol.* **2014**, *30*, 1304–1310. [CrossRef]
47. Cuadri, A.A.; Partal, P.; Ahmad, N.; Grenfell, J.; Airey, G. Chemically modified bitumens with enhanced rheology and adhesion properties to siliceous aggregates. *Constr. Build. Mater.* **2015**, *93*, 766–774. [CrossRef]
48. Rogers, T. Everything You Need to Know about Polystyrene (PS). Available online: https://www.creativemechanisms.com/blog/polystyrene-ps-plastic/ (accessed on 27 August 2021).
49. Sengoz, B.; Isikyakar, G. Evaluation of the properties and microstructure of SBS and EVA polymer modified bitumen. *Constr. Build. Mater.* **2008**, *22*, 1897–1905. [CrossRef]

50. Yang, X.; Liu, K.; Yang, D. Some Problems of PE Modified Asphalt. *J. China Foreign Highw.* **2008**, *28*, 203–207. (In Chinese)
51. Munaro, M.; Akcelrud, L. Correlations between composition and crystallinity of LDPE/HDPE blends. *J. Polym. Res.* **2007**, *15*, 83–88. [CrossRef]
52. White, G.; Hall, F. *Laboratory Comparison of Wet-Mixing and Dry-mixing of Recycled Waste Plastic for Binder and Asphalt Modification, Proceedings of the 100th Annual Meeting of the Transportation Research Board, Washington, DC, USA, 5–29 January 2021*; National Academy of Medicine: Washington, DC, USA, 2021.
53. White, G.; Reid, G. *Recycled Waste Plastic for Extending and Modifing Asphalt Binders, Proceedings of the 8th Symposium on Pavement Surface Characteristics: SURF 2018, Brisbane, Australia, 2–4 May 2018*; Australian Road Research Board: Brisbane, Australia, 2018.
54. Modarres, A.; Hamedi, H. Effect of waste plastic bottles on the stiffness and fatigue properties of modified asphalt mixes. *Mater. Des.* **2014**, *61*, 8–15. [CrossRef]
55. Kumar, P.; Garg, R. Rheology of waste plastic fibre-modified bitumen. *Int. J. Pavement Eng.* **2011**, *12*, 449–459. [CrossRef]
56. Lin, Y.; Hu, C.; Adhikari, S.; Wu, C.; Yu, M. Evaluation of Waste Express Bag as a Novel Bitumen Modifier. *Appl. Sci.* **2019**, *9*, 1242. [CrossRef]
57. Fang, C.; Liu, P.; Yu, R.; Liu, X. Preparation process to affect stability in waste polyethylene-modified bitumen. *Constr. Build. Mater.* **2014**, *54*, 320–325. [CrossRef]
58. Balasubramanian, B.; Gopala Krishna, G.V.T.; Saraswathy, V.; Srinivasan, K. Experimental investigation on concrete partially replaced with waste glass powder and waste E-plastic. *Constr. Build. Mater.* **2021**, *278*, 122400. [CrossRef]
59. Yue, H.; Bird, R.N.; Heidrich, O. A review of the use of recycled solid waste materials in asphalt pavements. *Resour. Conserv. Recycl.* **2008**, *52*, 58–73.
60. Movilla-Quesada, D.; Raposeiras, A.C.; Silva-Klein, L.T.; Lastra-González, P.; Castro-Fresno, D. Use of plastic scrap in asphalt mixtures added by dry method as a partial substitute for bitumen. *Waste Manag.* **2019**, *87*, 751–760. [CrossRef]
61. White, G.; Magee, C. Laboratory Evaluation of Asphalt Containing Recycled Plastic as a Bitumen Extender and Modifier. *J. Traffic Transp. Eng.* **2019**, *7*, 218–235.
62. Ranieri, M.; Costa, L.; Oliveira, J.R.M.; Silva, H.M.R.D.; Celauro, C. Asphalt Surface Mixtures with Improved Performance Using Waste Polymers via Dry and Wet Processes. *J. Mater. Civ. Eng.* **2017**, *29*, 04017169. [CrossRef]
63. Chavez, F.; Marcobal, J.; Gallego, J. Laboratory evaluation of the mechanical properties of asphalt mixtures with rubber incorporated by the wet, dry, and semi-wet process. *Constr. Build. Mater.* **2019**, *205*, 164–174. [CrossRef]
64. Wang, J. Application of Direct Investment SBS Modification Technology in Solid Engineering. *J. China Foreign Highw.* **2017**, *37*, 253–256. (In Chinese)
65. Fang, C.; Wu, C.; Hu, J.; Yu, R.; Zhang, Z.; Nie, L.; Zhou, S.; Mi, X. Pavement Properties of Asphalt Modified with Packaging-Waste Polyethylene. *J. Vinyl Addit. Technol.* **2014**, *20*, 31–35. [CrossRef]
66. Lu, J.; Zhang, Z. Study and Practice on Road Performance of Waste Polyethylene Plastic Modified Asphalt. *Highw. Eng.* **1996**, *76*, 62–64. (In Chinese)
67. Dalhat, M.A.; Al-Abdul Wahhab, H.I.; Al-Adham, K. Recycled Plastic Waste Asphalt Concrete via Mineral Aggregate Substitution and Binder Modification. *J. Mater. Civ. Eng.* **2019**, *31*, 04019134. [CrossRef]
68. Khan, I.M.; Kabir, S.; Alhussain, M.A.; Almansoor, F.F. Asphalt Design Using Recycled Plastic and Crumb-rubber Waste for Sustainable Pavement Construction. *Procedia Eng.* **2016**, *145*, 1557–1564. [CrossRef]
69. Ho, S.; Church, R.; Klassen, K.; Law, B.; MacLeod, D.; Zanzotto, L. Study of recycled polyethylene materials as asphalt modifiers. *Can. J. Civ. Eng.* **2006**, *33*, 968–981. [CrossRef]
70. Punith, V.S.; Veeraragavan, A.; Amirkhanian, S.N. Evaluation of Reclaimed Polyethylene Modified Asphalt Concrete Mixtures. *Int. J. Pavement Res.* **2011**, *4*, 1–10.
71. Otuoze, H.S.; Ejeh, S.P.; Amartey, Y.D.; Joel, M. Rheology and Simple Performance Test (SPT) Evaluation of High-Density Polypropylene (HDPP) Waste-Modified Bituminous Mix. *Jordan J. Civ. Eng.* **2018**, *12*, 35–44.
72. Al-Hadidy, A.I.; Tan, Y. Mechanistic approach for polypropylene-modified flexible pavements. *Mater. Des.* **2009**, *30*, 1133–1140. [CrossRef]
73. Cheng, Y.; Fu, Q.; Fang, C.; Zhang, Q.; Lu, C. Preparation, Structure, and Properties of Modified Asphalt with Waste Packaging Polypropylene and Organic Rectorite. *Adv. Mater. Sci. Eng.* **2019**, *2019*, 5362795. [CrossRef]
74. Ziari, H.; Nasiri, E.; Amini, A.; Ferdosian, O. The effect of EAF dust and waste PVC on moisture sensitivity, rutting resistance, and fatigue performance of asphalt binders and mixtures. *Constr. Build. Mater.* **2019**, *203*, 188–200. [CrossRef]
75. Ghabchi, R.; Dharmarathna, C.P.; Mihandoust, M. Feasibility of using micronized recycled Polyethylene Terephthalate (PET) as an asphalt binder additive: A laboratory study. *Constr. Build. Mater.* **2021**, *292*, 123377. [CrossRef]
76. Esfandabad, A.S.; Motevalizadeh, S.M.; Sedghi, R.; Ayar, P.; Asgharzadeh, S.M. Fracture and mechanical properties of asphalt mixtures containing granular polyethylene terephthalate (PET). *Constr. Build. Mater.* **2020**, *259*, 120410. [CrossRef]
77. Fang, C.; Jiao, L.; Hu, J.; Yu, Q.; Guo, D.; Zhou, X.; Yu, R. Viscoelasticity of Asphalt Modified With Packaging Waste Expended Polystyrene. *J. Mater. Sci. Technol.* **2014**, *30*, 939–943. [CrossRef]
78. González, O.; Muñoz, M.E.; Santamaría, A.; García-Morales, M.; Navarro, F.J.; Partal, P. Rheology and stability of bitumen/EVA blends. *Eur. Polym. J.* **2004**, *40*, 2365–2372. [CrossRef]
79. Ameri, M.; Mansourian, A.; Sheikhmotevali, A.H. Investigating effects of ethylene vinyl acetate and gilsonite modifiers upon performance of base bitumen using Superpave tests methodology. *Constr. Build. Mater.* **2012**, *36*, 1001–1007. [CrossRef]

80. De la Colina Martínez, A.L.; Martínez Barrera, G.; Barrera Díaz, C.E.; Ávila Córdoba, L.I.; Ureña Núñez, F.; Delgado Hernández, D.J. Recycled polycarbonate from electronic waste and its use in concrete: Effect of irradiation. *Constr. Build. Mater.* **2019**, *201*, 778–785. [CrossRef]
81. Colbert, B.W.; You, Z. Properties of Modified Asphalt Binders Blended with Electronic Waste Powders. *J. Mater. Civ. Eng.* **2012**, *24*, 1261–1267. [CrossRef]
82. Ameri, M.; Yeganeh, S.; Erfani Valipour, P. Experimental evaluation of fatigue resistance of asphalt mixtures containing waste elastomeric polymers. *Constr. Build. Mater.* **2019**, *198*, 638–649. [CrossRef]
83. AASHTO. *Mechanistic-Empirical Pavement Design Guide*; American Association of State Highway and Transportation Officials: Washington, DC, USA, 2008.
84. Colbert, B.W.; Diab, A.; You, Z. Using M-E PDG to study the effectiveness of electronic waste materials modifications on asphalt pavements design thickness. *Int. J. Pavement Res. Technol.* **2013**, *6*, 319–326.
85. Bazmara, B.; Tahersima, M.; Behravan, A. Influence of thermoplastic polyurethane and synthesized polyurethane additive in performance of asphalt pavements. *Constr. Build. Mater.* **2018**, *166*, 1–11. [CrossRef]
86. Cong, L.; Yang, F.; Guo, G.; Ren, M.; Shi, J.; Tan, L. The use of polyurethane for asphalt pavement engineering applications: A state-of-the-art review. *Constr. Build. Mater.* **2019**, *225*, 1012–1025. [CrossRef]
87. Sun, M.; Zheng, M.; Qu, G.; Yuan, K.; Bi, Y.; Wang, J. Performance of polyurethane modified asphalt and its mixtures. *Constr. Build. Mater.* **2018**, *191*, 386–397. [CrossRef]
88. Salas, M.Á.; Pérez-Acebo, H.; Calderón, V.; Gonzalo-Orden, H. Bitumen modified with recycled polyurethane foam for employment in hot mix asphalt. *Ing. Investig.* **2018**, *38*, 60–66. [CrossRef]
89. Salas, M.Á.; Pérez-Acebo, H. Introduction of recycled polyurethane foam in mastic asphalt. *Gradevinar* **2018**, *70*, 403–412.
90. Naskar, M.; Chaki, T.K.; Reddy, K.S. Effect of waste plastic as modifier on thermal stability and degradation kinetics of bitumen/waste plastics blend. *Thermochim. Acta* **2010**, *509*, 128–134. [CrossRef]
91. Ameri, M.; Nasr, D. Properties of asphalt modified with devulcanized polyethylene terephthalate. *Pet. Sci. Technol.* **2016**, *34*, 1424–1430. [CrossRef]
92. Hu, C.; Lin, W.; Partl, M.; Wang, D.; Yu, H.; Zhang, Z. Waste packaging tape as a novel bitumen modifier for hot-mix asphalt. *Constr. Build. Mater.* **2018**, *193*, 23–31. [CrossRef]
93. Padhan, R.K.; Sreeram, A.; Gupta, A. Evaluation of trans-polyoctenamer and cross-linking agents on the performance of waste polystyrene modified asphalt. *Road Mater. Pavement Des.* **2018**, *21*, 1170–1182. [CrossRef]
94. Dalhat, M.A.; Al-Abdul Wahhab, H.I. Performance of recycled plastic waste modified asphalt binder in Saudi Arabia. *Int. J. Pavement Eng.* **2017**, *18*, 349–357. [CrossRef]
95. Vargas, C.; El Hanandeh, A. Systematic literature review, meta-analysis and artificial neural network modelling of plastic waste addition to bitumen. *J. Clean. Prod.* **2021**, *280*, 124369. [CrossRef]
96. Brovelli, C.; Crispino, M.; Pais, J.; Pereira, P. Using polymers to improve the rutting resistance of asphalt concrete. *Constr. Build. Mater.* **2015**, *77*, 117–123. [CrossRef]
97. García-Morales, M.; Partal, P.; Navarro, F.J.; Gallegos, C. Effect of waste polymer addition on the rheology of modified bitumen. *Fuel* **2006**, *85*, 936–943. [CrossRef]
98. Lai, Z.; Liu, K.; Yang, X.; Yang, D. Research on Performance of Asphalt Modified with Waste Plastics. *J. Shandong Univ. Sci. Technol.* **2010**, *29*, 73–77. (In Chinese)
99. Nasr, D.; Pakshir, A.H. Rheology and storage stability of modified binders with waste polymers composites. *Road Mater. Pavement Des.* **2017**, *20*, 773–792. [CrossRef]
100. Formela, K.; Sulkowski, M.; Saeb, M.R.; Colom, X.; Haponiuk, J.T. Assessment of microstructure, physical and thermal properties of bitumen modified with LDPE/GTR/elastomer ternary blends. *Constr. Build. Mater.* **2016**, *106*, 160–167. [CrossRef]
101. Yi, X.; Wu, G.; Lu, F.; Tang, A. Hydrolysis of aromatic heterocyclic polymers in high temperature water. *J. Appl. Polym. Sci.* **2001**, *82*, 907–915. [CrossRef]
102. Wen, G.; Zhang, Y.; Zhang, Y. Improved Properties of SBS-Modified Asphalt with Dynamic Vulcanization. *Polym. Eng. Sci.* **2002**, *42*, 1070–1081.
103. Wang, S.; Ma, P.; Ouyang, C.; Zhang, Y. Study on stabilization of polyethylene/carbon black modified asphalt. *Pet. Asph.* **2006**, *20*, 22–26. (In Chinese)
104. Nuñez, J.Y.M.; Domingos, M.D.I.; Faxina, A.L. Susceptibility of low-density polyethylene and polyphosphoric acid-modified asphalt binders to rutting and fatigue cracking. *Constr. Build. Mater.* **2014**, *73*, 509–514. [CrossRef]
105. Yadollahi, G.; Sabbagh Mollahosseini, H. Improving the performance of Crumb Rubber bitumen by means of Poly Phosphoric Acid (PPA) and Vestenamer additives. *Constr. Build. Mater.* **2011**, *25*, 3108–3116. [CrossRef]
106. Polacco, G.; Berlincioni, S.; Biondi, D.; Stastna, J.; Zanzotto, L. Asphalt modification with different polyethylene-based polymers. *Eur. Polym. J.* **2005**, *41*, 2831–2844. [CrossRef]
107. Guru, M.; Cubuk, M.K.; Arslan, D.; Farzanian, S.A.; Bilici, I. An approach to the usage of polyethylene terephthalate (PET) waste as roadway pavement material. *J. Hazard Mater.* **2014**, *279*, 30210. [CrossRef]
108. Köfteci, S.; Ahmedzade, P.; Kultayev, B. Performance evaluation of bitumen modified by various types of waste plastics. *Constr. Build. Mater.* **2014**, *73*, 592–602. [CrossRef]

109. Roja, K.L.; Masad, E. Influence of Chemical Constituents of Asphalt Binders on Their Rheological Properties. *Transp. Res. Rec. J. Transp. Res. Board* **2019**, *2673*, 458–466. [CrossRef]
110. Huang, S.C.; Benedetto, H.D. *Advances in Asphalt Materials*, 1st ed.; Woodhead Publishing: Abington, UK, 2015; pp. 31–57.
111. Behnood, A.; Modiri Gharehveran, M. Morphology, rheology, and physical properties of polymer-modified asphalt binders. *Eur. Polym. J.* **2019**, *112*, 766–791. [CrossRef]
112. Masad, E.; Roja, K.L.; Rehman, A.; Abdala, A.A. A Review of Asphalt Modification Using Plastics: A Focus on Polyethylene. Available online: https://www.researchgate.net/publication/341152175_A_Review_of_Asphalt_Modification_Using_Plastics_A_Focus_on_Polyethylene (accessed on 30 August 2021).
113. Lesueur, D. The colloidal structure of bitumen: Consequences on the rheology and on the mechanisms of bitumen modification. *Adv. Colloid Interface Sci.* **2009**, *145*, 42–82. [CrossRef]
114. Giavarini, C.; Filippis, P.D.; Santarelli, M.L.; Scarsella, M. Production of stable polypropylene-modified bitumens. *Fuel* **1996**, *75*, 681–686. [CrossRef]
115. Rivera, C.; Caro, S.; Arámbula-Mercado, E.; Sánchez, D.B.; Karki, P. Comparative evaluation of ageing effects on the properties of regular and highly polymer modified asphalt binders. *Constr. Build. Mater.* **2021**, *302*, 124163. [CrossRef]
116. Fuentes-Audén, C.; Sandoval, J.A.; Jerez, A.; Navarro, F.J.; Martínez-Boza, F.J.; Partal, P.; Gallegos, C. Evaluation of thermal and mechanical properties of recycled polyethylene modified bitumen. *Polym. Test.* **2008**, *27*, 1005–1012. [CrossRef]
117. Fernandes, S.; Costa, L.; Silva, H.; Oliveira, J. Effect of incorporating different waste materials in bitumen. *Ciência Tecnol. Mater.* **2017**, *29*, e204–e209. [CrossRef]
118. Karmakar, S.; Roy, T.K. Effect of Waste Plastic and Waste Tires Ash on Mechanical Behavior of Bitumen. *J. Mater. Civ. Eng.* **2016**, *28*, 04016006. [CrossRef]
119. Lu, X.; Lsacsson, U. Modification of road bitumens with thermoplastic polymers. *Polym. Test.* **2000**, *20*, 77–86. [CrossRef]
120. Alberto, G.G.E.; Rafael, H.N. Effect of Stirring Speed in Hot Mixing Process of Modified Asphalt with SBS Copolymer on Polymeric Distribution and its Rheological Properties. In Proceedings of the MRS Materials Research Society, Florence, Italy, 28 March–1 April 2016; 1813.
121. Babalghaith, A.M.; Koting, S.; Ramli Sulong, N.H.; Karim, M.R. Optimization of mixing time for polymer modified asphalt. *IOP Conf. Ser. Mater. Sci. Eng.* **2019**, *512*, 012030. [CrossRef]
122. Liang, M.; Xin, X.; Fan, W.; Wang, H.; Jiang, H.; Zhang, J.; Yao, Z. Phase behavior and hot storage characteristics of asphalt modified with various polyethylene: Experimental and numerical characterizations. *Constr. Build. Mater.* **2019**, *203*, 608–620. [CrossRef]
123. Rojo, E.; Fernández, M.; Peña, J.J.; Peña, B.; Muñoz, M.E.; Santamaría, A. Rheological aspects of blends of metallocene-catalyzed atactic polypropylenes with bitumen. *Polym. Eng. Sci.* **2004**, *44*, 1792–1799. [CrossRef]
124. Padhan, R.K.; Mohanta, C.; Sreeram, A.; Gupta, A. Rheological evaluation of bitumen modified using antistripping additives synthesised from waste polyethylene terephthalate (PET). *Int. J. Pavement Eng.* **2018**, *21*, 1083–1091. [CrossRef]
125. Li, J.; Zhang, Y.; Zhang, Y. The research of GMA-g-LDPE modified Qinhuangdao bitumen. *Constr. Build. Mater.* **2008**, *22*, 1067–1073.
126. Yeh, P.-H.; Nien, Y.-H.; Chen, J.-H.; Chen, W.-C.; Chen, J.-S. Thermal and rheological properties of maleated polypropylene modified asphalt. *Polym. Eng. Sci.* **2005**, *45*, 1152–1158. [CrossRef]
127. Ahmedzade, P.; Fainleib, A.; Günay, T.; Grygoryeva, O. Modification of bitumen by electron beam irradiated recycled low density polyethylene. *Constr. Build. Mater.* **2014**, *69*, 1–9. [CrossRef]
128. Ahmedzade, P.; Fainleib, A.; Günay, T.; Starostenko, O.; Kovalinska, T. Effect of Gamma-Irradiated Recycled Low-Density Polyethylene on the High- and Low-Temperature Properties of Bitumen. *Int. J. Polym. Sci.* **2013**, *2013*, 8062–8070. [CrossRef]
129. Fang, C.; Hu, J.; Zhou, S.; Wang, H.; Zhang, M.; Zhang, Y. Comparative Study of Asphalts Modified by Packaging Waste EPS and Waste PE. *Polym.-Plast. Technol. Eng.* **2011**, *50*, 220–224. [CrossRef]
130. Lastra-González, P.; Calzada-Pérez, M.A.; Castro-Fresno, D.; Vega-Zamanillo, Á.; Indacoechea-Vega, I. Comparative analysis of the performance of asphalt concretes modified by dry way with polymeric waste. *Constr. Build. Mater.* **2016**, *112*, 1133–1140. [CrossRef]
131. Al-Hadidy, A.I. Effect of laboratory aging on moisture susceptibility and resilient modulus of asphalt concrete mixes containing PE and PP polymers. *Karbala Int. J. Mod. Sci.* **2018**, *4*, 377–381. [CrossRef]
132. Arabani, M.; Yousefpour Taleghani, M. Rutting behavior of hot mix asphalt modified by polyvinyl chloride powder. *Pet. Sci. Technol.* **2017**, *35*, 1621–1626. [CrossRef]
133. Špaček, P.; Hegr, Z.; Beneš, J. Practical experiences with new types of highly modified asphalt binders. *IOP Conf. Ser. Mater. Sci. Eng.* **2017**, *236*, 012020. [CrossRef]
134. Yu, B.; Jiao, L.; Ni, F.; Yang, J. Evaluation of plastic–rubber asphalt: Engineering property and environmental concern. *Constr. Build. Mater.* **2014**, *71*, 416–424. [CrossRef]
135. Beena, K.S.; Bindu, C.S. Waste plastic as a stabilizing additive in Stone Mastic Asphalt. *Int. J. Eng. Technol.* **2010**, *2*, 379–387.
136. Ahmadinia, E.; Zargar, M.; Karim, M.R.; Abdelaziz, M.; Shafigh, P. Using waste plastic bottles as additive for stone mastic asphalt. *Mater. Des.* **2011**, *32*, 4844–4849. [CrossRef]
137. Rongali, U.; Singh, G.; Chourasiya, A.; Jain, P.K. Laboratory performance of stone matrix asphalt containing composite. *J. Sci. Ind. Res.* **2013**, *72*, 186–192.

138. Baghaee Moghaddam, T.; Karim, M.R.; Syammaun, T. Dynamic properties of stone mastic asphalt mixtures containing waste plastic bottles. *Constr. Build. Mater.* **2012**, *34*, 236–242. [CrossRef]
139. Ahmadinia, E.; Zargar, M.; Karim, M.R.; Abdelaziz, M.; Ahmadinia, E. Performance evaluation of utilization of waste Polyethylene Terephthalate (PET) in stone mastic asphalt. *Constr. Build. Mater.* **2012**, *36*, 984–989. [CrossRef]
140. Angelone, S.; Cauhapé Casaux, M.; Borghi, M.; Martinez, F.O. Green pavements: Reuse of plastic waste in asphalt mixtures. *Mater. Struct.* **2015**, *49*, 1655–1665. [CrossRef]
141. Giri, J.P.; Panda, M.; Sahoo, U.C. Use of waste polyethylene for modification of bituminous paving mixes containing recycled concrete aggregates. *Road Mater. Pavement Des.* **2018**, *21*, 289–309. [CrossRef]
142. Almeida, A.; Crucho, J.; Abreu, C.; Picado-Santos, L. An Assessment of Moisture Susceptibility and Ageing Effect on Nanoclay-Modified AC Mixtures Containing Flakes of Plastic Film Collected as Urban Waste. *Appl. Sci.* **2019**, *9*, 3738. [CrossRef]
143. Kamal Karim, H.; Ebrahim, S.M. Effects of Additive Materials on Indirect Tensile Strength and Moisture Sensitivity of Recycled Asphalt Pavement (RAP). *Kurd. J. Appl. Res.* **2019**, *4*, 69–79.
144. The Durability Study of the Use of Modified Waste Plastic-Bitumen in Asphalt Concrete Mix Containing Reclaimed Asphalt Pavement. Available online: http://lib.itenas.ac.id/kti/wp-content/uploads/2013/04/15.-MAkalah-Lombok1.pdf (accessed on 18 November 2021).
145. Leng, Z.; Sreeram, A.; Padhan, R.K.; Tan, Z. Value-added application of waste PET based additives in bituminous mixtures containing high percentage of reclaimed asphalt pavement (RAP). *J. Clean. Prod.* **2018**, *196*, 615–625. [CrossRef]
146. Pouranian, M.R.; Shishehbor, M. Sustainability Assessment of Green Asphalt Mixtures: A Review. *Environments* **2019**, *6*, 73. [CrossRef]
147. Poulikakos, L.D.; Papadaskalopoulou, C.; Hofko, B.; Gschösser, F.; Cannone Falchetto, A.; Bueno, M.; Arraigada, M.; Sousa, J.; Ruiz, R.; Petit, C.; et al. Harvesting the unexplored potential of European waste materials for road construction. *Resour. Conserv. Recycl.* **2017**, *116*, 32–44. [CrossRef]
148. Araújo, J.P.C.; Oliveira, J.R.M.; Silva, H.M.R.D. The importance of the use phase on the LCA of environmentally friendly solutions for asphalt road pavements. *Transp. Res. Part D Transp. Environ.* **2014**, *32*, 97–110. [CrossRef]
149. Santos, J.; Pham, A.; Stasinopoulos, P.; Giustozzi, F. Recycling waste plastics in roads: A life-cycle assessment study using primary data. *Sci. Total Environ.* **2021**, *751*, 141842. [CrossRef]
150. Lastra-González, P.; Lizasoain-Arteaga, E.; Castro-Fresno, D.; Flintsch, G. Analysis of replacing virgin bitumen by plastic waste in asphalt concrete mixtures. *Int. J. Pavement Eng.* **2021**, 1–10. [CrossRef]
151. Duarte, G.M.; Faxina, A.L. Asphalt concrete mixtures modified with polymeric waste by the wet and dry processes: A literature review. *Constr. Build. Mater.* **2021**, *312*, 125408. [CrossRef]
152. Yan, K.; Xu, H.; You, L. Rheological properties of asphalts modified by waste tire rubber and reclaimed low density polyethylene. *Constr. Build. Mater.* **2015**, *83*, 143–149. [CrossRef]
153. Ma, Y.; Wang, S.; Zhou, H.; Hu, W.; Huang, B. Compatibility and rheological characterization of asphalt modified with recycled rubber-plastic blends. *Constr. Build. Mater.* **2020**, *270*, 121416. [CrossRef]
154. Zhang, F.; Yu, J.; Han, J. Effects of thermal oxidative ageing on dynamic viscosity, TG/DTG, DTA and FTIR of SBS- and SBS/sulfur-modified asphalts. *Constr. Build. Mater.* **2011**, *25*, 129–137. [CrossRef]
155. Yu, R.; Fang, C.; Liu, P.; Liu, X.; Li, Y. Storage stability and rheological properties of asphalt modified with waste packaging polyethylene and organic montmorillonite. *Appl. Clay Sci.* **2015**, *104*, 1–7. [CrossRef]
156. Wang, H.; Liu, X.; Erkens, S.; Skarpas, A. Experimental characterization of storage stability of crumb rubber modified bitumen with warm-mix additives. *Constr. Build. Mater.* **2020**, *249*, 118840. [CrossRef]
157. Tang, Y.; Ledieu, E.; Cervellere, M.R.; Millett, P.C.; Ford, D.M.; Qian, X. Formation of polyethersulfone membranes via nonsolvent induced phase separation process from dissipative particle dynamics simulations. *J. Membr. Sci.* **2020**, *599*, 117826. [CrossRef]
158. Chen, M.; Geng, J.; Xia, C.; He, L.; Liu, Z. A review of phase structure of SBS modified asphalt: Affecting factors, analytical methods, phase models and improvements. *Constr. Build. Mater.* **2021**, *294*, 123610. [CrossRef]
159. Qian, C.; Fan, W.; Ren, F.; Lv, X.; Xing, B. Influence of polyphosphoric acid (PPA) on properties of crumb rubber (CR) modified asphalt. *Constr. Build. Mater.* **2019**, *227*, 117094. [CrossRef]
160. Vamegh, M.; Ameri, M.; Chavoshian Naeni, S.F. Performance evaluation of fatigue resistance of asphalt mixtures modified by SBR/PP polymer blends and SBS. *Constr. Build. Mater.* **2019**, *209*, 202–214. [CrossRef]
161. Galooyak, S.S.; Dabir, B.; Nazarbeygi, A.E.; Moeini, A. Rheological properties and storage stability of bitumen/SBS/montmorillonite composites. *Constr. Build. Mater.* **2010**, *24*, 300–307. [CrossRef]
162. Gao, G.; Zhang, Y. Effect of Carbon Black on Storage Stability of Polymer Modified Asphalt. *China Rubber Ind.* **2008**, *55*, 226–230. (In Chinese)
163. Put, J.; van Duin, M.; Reedijk, J.; Haasnoot, J.G.; Versloot, P. Sulfur Vulcanization of Simple Model Olefins, Part IV: Vulcanizations of 2,3-Dimethyl-2-Butene with TMTD and Activated Zinc Dithiocarbamate/Xanthate Accelerators at Different Temperatures. *Rubber Chem. Technol.* **1995**, *68*, 563–572.
164. Wang, S.; Yuan, C.; Deng, J. Crumb tire rubber and polyethylene mutually stabilized in asphalt by screw extrusion. *J. Appl. Polym. Sci.* **2014**, *131*, 81–86. [CrossRef]
165. Karahrodi, M.H.; Jazani, O.M.; Paran, S.M.R.; Formela, K.; Saeb, M.R. Modification of thermal and rheological characteristics of bitumen by waste PET/GTR blends. *Constr. Build. Mater.* **2017**, *134*, 157–166. [CrossRef]

166. Dessouky, S.; Contreras, D.; Sanchez, J.; Papagiannakis, A.T.; Abbas, A. Influence of hindered phenol additives on the rheology of aged polymer-modified bitumen. *Constr. Build. Mater.* **2013**, *38*, 214–223. [CrossRef]
167. Lu, Y.; Yu, L.; Fang, J.; Yun, Q.; Lu, J. Research development of polymer modified asphalt. *New Chem. Mater.* **2020**, *48*, 222–230. (In Chinese) [CrossRef]
168. Yu, H.; Leng, Z.; Zhang, Z.; Li, D.; Zhang, J. Selective absorption of swelling rubber in hot and warm asphalt binder fractions. *Constr. Build. Mater.* **2020**, *238*, 117727. [CrossRef]
169. Fang, C.; Zhou, S.; Zhang, M.; Zhao, S. Modification of waterproofing asphalt by PVC packaging waste. *J. Vinyl Addit. Technol.* **2009**, *15*, 229–233. [CrossRef]
170. Tan, Y.; Li, G.; Shan, L.; Lyu, H.; Meng, A. Research progress of bitumen microstructures and components. *J. Traffic Transp. Eng.* **2020**, *20*, 1–17. (In Chinese)
171. Guo, F.; Zhang, J.; Pei, J.; Zhou, B.; Falchetto, A.C.; Hu, Z. Investigating the interaction behavior between asphalt binder and rubber in rubber asphalt by molecular dynamics simulation. *Constr. Build. Mater.* **2020**, *252*, 118956. [CrossRef]
172. Zheng, W.; Wang, H.; Chen, Y.; Ji, J.; Zhang, Y. A review on compatibility between crumb rubber and asphalt binder. *Constr. Build. Mater.* **2021**, *297*, 123820. [CrossRef]
173. Pérez-Lepe, A.; Martínez-Boza, F.; Gallegos, C. Influence of Polymer Concentration on the Microstructure and Rheological Properties of High-Density Polyethylene (HDPE)-Modified Bitumen. *Energy Fuels* **2005**, *19*, 1148–1152. [CrossRef]
174. Ghuzlan, K.; Al-Khateeb, G.G.; Qasem, Y. *Rheological Properties of Polyethylene-Modified Asphalt Binder, Proceedings of the 3rd Annual International Conference on Civil Engineering (ATINER), Athens, Greece, 20–23 June 2013*; Athens Institute for Education and Research: Athens, Greece, 2013.

materials

MDPI

Article

The Performance Evaluation of Asphalt Mortar and Asphalt Mixture Containing Municipal Solid Waste Incineration Fly Ash

Xiaowen Zhao [1], Dongdong Ge [2,*], Jiaqing Wang [3], Dianwen Wu [1] and Jun Liu [4,*]

1 Hunan Provincial Communications Planning, Survey & Design Institute Co., Ltd., Changsha 410200, China; zhaoxiaowen2013@163.com (X.Z.); dianwenwu@163.com (D.W.)
2 National Engineering Laboratory of Highway Maintenance Technology, School of Traffic & Transportation Engineering, Changsha University of Science & Technology, Changsha 410114, China
3 College of Civil Engineering, Nanjing Forestry University, Nanjing 210037, China; jiaqingw@njfu.edu.cn
4 Louisiana Transportation Research Center, Louisiana State University, 4101 Gourrier Ave, Baton Rouge, LA 70808, USA
* Correspondence: dge1@csust.edu.cn (D.G.); junliu@lsu.edu (J.L.)

Abstract: The aim of the research is to quantify the property of asphalt mortar and asphalt mixture containing municipal solid waste incineration (MSWI) fly ash. The potential of partially replacing mineral fillers with MSWI fly ash in asphalt mixture production was investigated. Five different MSWI fly ash replacement ratios, which include 0%, 25%, 50%, 75%, and 100%, were adopted to assess the influence of fly ash dosage, and the optimum fly ash replacement ratio was proposed. The rheological characteristics of asphalt mortar with MSWI fly ash were assessed with the dynamic shear rheometer (DSR) and bending beam rheometer (BBR). The high temperature properties of the mixture with MSWI fly ash were assessed with the Marshall stability test and the rutting test. The low temperature cracking property was determined with the indirect tensile strength test at low temperatures. The moisture stability property was identified with the immersed Marshall test and the freeze-thaw cycles conditioned indirect tensile strength test. Based on the test results, the addition of fly ash and mineral filler remarkably increased the $|G^*|$ of the asphalt mortar. The δ of asphalt decreased as the dosage of fly ash and mineral filler increased. The addition of fly ash and mineral filler degraded the low temperature characteristics of the mortar. Fly ash improved the high temperature characteristics of the asphalt mixture. The asphalt mixture with MSWI fly ash was more susceptible to thermal cracking than the control sample. The addition of fly ash weakened the moisture stability of the asphalt mixture. In order to guarantee the low temperature characteristics and the moisture susceptibility of the asphalt mixture, the fly ash replacement ratio was recommended to be set around 25%. With proper mixture design and fly ash dosage, the asphalt mixture would have adequate performance, as well as reduced environmental impact.

Keywords: asphalt mortar; asphalt mixture; fly ash; high temperature performance; low temperature property; moisture susceptibility

Citation: Zhao, X.; Ge, D.; Wang, J.; Wu, D.; Liu, J. The Performance Evaluation of Asphalt Mortar and Asphalt Mixture Containing Municipal Solid Waste Incineration Fly Ash. *Materials* **2022**, *15*, 1387. https://doi.org/10.3390/ma15041387

Academic Editor: Francesco Canestrari

Received: 31 December 2021
Accepted: 8 February 2022
Published: 14 February 2022

Publisher's Note: MDPI stays neutral with regard to jurisdictional claims in published maps and institutional affiliations.

Copyright: © 2022 by the authors. Licensee MDPI, Basel, Switzerland. This article is an open access article distributed under the terms and conditions of the Creative Commons Attribution (CC BY) license (https://creativecommons.org/licenses/by/4.0/).

1. Introduction

With city expansion and increasing populations in municipal cities, the amount of new solid waste generated bring a significant issue to cities [1]. Using incineration to process solid waste was the most effective method. Handling solid waste with landfill method wasted precious land and threatened groundwater. After the incineration process, the solid waste included fly ash and bottom ash, based on the location where the ash was collected [2]. Bottom ash is the residue in the bottom of the incinerator, which consisted of slag, glass, metal, and incompletely burned organic components [3]. The fly ash is the fine particles received from the gas purification system of the incinerator, which accounts for 10–20% of the solid waste incineration residue [4]. The characteristics of the fly ash from different incinerators vary because the components of solid waste are different [5]. However, the untreated fly ash normally had low moisture content and had grey or white appearance. During the incineration procedure, the heavy metal components were concentrated in the fly ash [6]. If treated improperly, the fly ash will contaminate the environment and influence the health of human beings.

Fly ash has long been adopted as road construction material. By adopting fly ash in the asphalt mixture and concrete construction, the demanding using natural aggregate and mineral fillers could be reduced [7]. With the consumption of natural materials, the reduction of the yield of original materials increased the cost of construction. Using fly ash in pavements will reduce the cost and relieve the pressure of the fly ash on the environment. Fly ash can be applied as supplementary materials in the concrete production [8]. The main components of fly ash were similar to cement materials, which proved the commercial potential to partially replace cement in the concrete production [9]. The fly ash contributed to the strength of concrete because of the hydraulic and pozzolanic reaction [10,11]. During concrete production, fly ash can replace part of the fine aggregate, but the heavy metal stabilization efficiency was not sufficient. The heavy metal in fly ash was still easily leached out from the concrete. Due to the good bonding property of asphalt binder, asphalt binder was proved to be more effective to stabilize the heavy metal in fly ash [12].

Fly ash has been treated as filler in the asphalt mixture production [13,14]. When mixing fly ash with asphalt mixture, the fly ash interacted with the asphalt binder, and the stiffness of the asphalt mastic was improved [15,16]. The fly ash absorbed the light components and finally reduced the flowability of the asphalt binder [17]. The interface condition between asphalt and aggregate was affected by the addition of the fly ash [18]. The asphalt film thickness of the aggregate was increased and thus increased the modulus of the asphalt mixture [19]. The production property and the serviceability of the asphalt mixture were improved with the adoption of fly ash during the design and the placement of the asphalt mixture [20,21]. The stiffness was enhanced with the adoption of fly ash, and the stability of the asphalt mixture was influenced by the dosage of fly ash in the asphalt mixture [22,23]. The physical property of asphalt mixture was enhanced by using the fly ash in the asphalt mixture, and the improved efficiency was affected by the asphalt binder and fly ash resource [24,25]. The moisture damage resistance of the asphalt mixture was influenced by the mixture with fly ash design. Some research proved that fly ash could be utilized as effective anti-stripping agents to improve the moisture damage resistance of the asphalt mixture [26]. The moisture resistance of the asphalt mixture was influenced by the dosage of the fly ash and the mixture design [27].

The objective of the research is to quantify the properties of asphalt mortar and asphalt mixture containing municipal solid waste incineration (MSWI) fly ash. The rheological characteristics of asphalt mortar with MSWI fly ash were assessed with the dynamic shear rheometer (DSR) and bending beam rheometer (BBR). The high temperature property of the asphalt mixture was assessed with the Marshall stability test and the wheel tracking test. The low temperature performance of the mixture was assessed with the indirect tensile strength test.

2. Materials and Methods

2.1. Materials

The asphalt binder used in the paper was penetration grade #70. Table 1 lists the binder performance.

Table 1. The performance of base asphalt.

Performance	Results
Softening point (°C)	47.1
Penetration (25 °C, 100 g, 5 s)/0.1 mm	71.7
Ductility (15 °C, 5 cm/min)/cm	>150
Flashing point (°C)	262
Viscosity (135 °C)/Pa·s	0.435

The coarse aggregates used were limestone from Hunan in China, and the basic performances of the coarse aggregate are presented in Table 2.

Table 2. The basic performance of coarse aggregate.

Performance	Results
Crushing value (%)	20.7
Soundness (%)	4.0
Moisture content (%)	0.4
Water absorption (%)	0.5
Apparent specific gravity	2.736
LA abrasion loss (%)	16.7

The properties of limestone mineral fillers and MSWI fly ash used in this study are shown in Table 3. The MSWI fly ash has lower apparent specific gravity than the limestone mineral filler. When mixed with asphalt binder, the MSWI fly ash was more likely distributed uniformly and had less tendency to separate from the asphalt binder than limestone mineral filler. The surface area of the MSWI fly ash was 4.66 times higher than that of limestone mineral filler. The larger surface area guaranteed strong adhesion between the MSWI fly ash and the asphalt binder.

Table 3. The properties of limestone mineral filler.

Properties		Test Results	
		Limestone Mineral Filler	MSWI Fly Ash
Apparent specific gravity		2.719	2.284
Moisture content (%)		0.4	0
Water absorption (%)		0.79	0.65
Specific surface area (m^2/g)		1.39	6.48
Percentage passing	0.6 mm (#30)	100	100
	0.15 mm (#100)	97.8	95.7
	0.075 mm (#200)	78.3	77.5

The element component of the MSWI fly ash was measured with the AA-6800 Atomic Absorption Spectrophotometer, and the results are presented in Table 4. The MSWI fly ash had many heavy metals, which included lead (Pb), copper (Cu), cadmium (Cd), nickel (Ni), chromium (Cr), and zinc (Zn), which are harmful to the biological system and the environment. By mixing MSWI fly ash with asphalt binder, the toxic heavy metals can be effectively stabilized and solidified, which mitigated the burden of heavy metals as hazardous pollutants to the environment [28].

Table 4. The element composition of MSWI fly ash.

Component	O	Na	Mg	Al	Si	P	S	Cl	K	Ca	Ti	Cr
Percentage (%)	26.254	2.029	0.569	0.56	1.752	0.163	2.569	22.265	3.397	37.888	0.168	0.022
Component	Mn	Fe	Ni	Cu	Zn	Br	Sr	Cd	Sn	Ba	Pb	
Percentage (%)	0.022	0.725	0.021	0.076	1.052	0.092	0.063	0.026	0.036	0.04	0.212	

2.2. Test Methods

2.2.1. Asphalt Mortar Production

In order to assess the impact of MSWI fly ash and mineral filler on the properties of asphalt binder, asphalt mortar was prepared by mixing asphalt binder with fly ash and mineral filler with three different ratios of fly ash and mineral filler to asphalt (F/A, 0.6, 0.8, and 1.0 by weight). Different fly ash and mineral filler ratios were adopted to explore the optimum fly ash dosage, Table 5 shows the material matrix.

Table 5. The dosage of fly ash and mineral filler in asphalt mortar.

Symbols	F/A Ratios	Fly Ash	Mineral Filler
R0	0.6, 0.8, and 1.0	0	100%
R25		25%	75%
R50		50%	50%
R75		75%	25%
R100		100%	0%

The base asphalt was heated to 135 °C, and the fly ash and mineral filler were heated to 140 °C. Fly ash and mineral filler were added into asphalt binder based on the F/A ratios and the dosage of fly ash and mineral filler in Table 5. The speed of the high shear mixer was 1000 RPM during the materials adding process. The mortar was then mixed at 3000 RPM for 10 min. Finally, the mortar was sheared at 1000 RPM for 5 min to remove air bubbles imported during the high shear mixing process. Each test condition had three replicate samples.

2.2.2. Asphalt Mortar Properties

The rheological properties of asphalt binder under intermedia and high temperatures were assessed with the dynamic shear rheometer (DSR) [29]. The temperature sweeps were conducted on the asphalt mortar at the frequency of 10 rad/s from 30 to 76 °C. The frequency sweeps were conduct on the asphalt mortar under 11 frequencies (1 rad/s, 1.58 rad/s, 2.51 rad/s, 3.98 rad/s, 6.31 rad/s, 10 rad/s, 15.8 rad/s, 25.1 rad/s, 39.8 rad/s, 63.1 rad/s, and 100 rad/s) at 60 °C.

The low temperature flexural creep stiffness of asphalt mortar was evaluated with the bending beam rheometer (BBR). Three different test temperatures (−18 °C, −12 °C, and −6 °C) were adopted.

2.2.3. Asphalt Mixture Production

The asphalt mixture was prepared with different fly ash contents (0%, 25%, 50%, 75%, and 100%) to replace the mineral filler. Based on previous research results that the optimum binder to aggregate ratio was not significantly influenced by the fly ash and mineral filler percentage, the binder to aggregate ratio was set as 4.5% for different types of asphalt mixture. The gradation of the aggregates is presented in Figure 1.

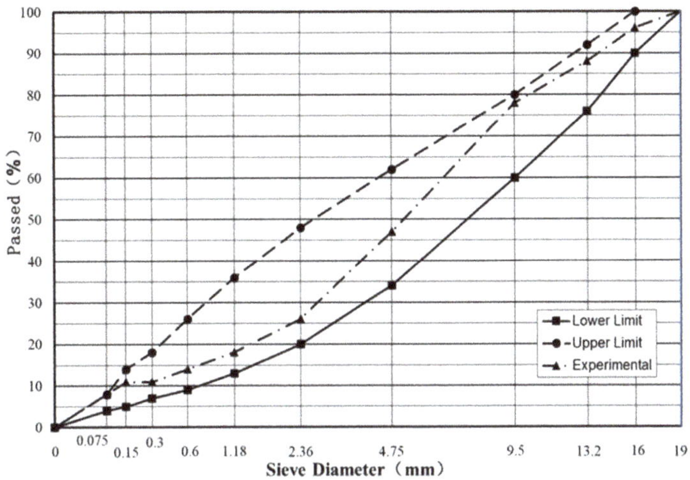

Figure 1. The gradation of the asphalt mixture.

2.2.4. Asphalt Mixture Properties

The high temperature property of the asphalt mixture was assessed with the Marshall stability test and the wheel tracking test. The Marshall stability measured the deformation resistance of the asphalt mixture under the cylindrical axis load with the Marshall apparatus. The test was conducted by following the ASTM D6927 standard [30]. The wheel tracking test generated the relationship between the rutting depth and the wheel tracking cycles. The samples were compacted by the rubber tire at 60 °C for 1 h at the speed of 42 passes/min [31]. Three replicate samples were conducted for each test condition. The dynamic stability was calculated with Equation (1).

$$DS = \frac{42 \times 15}{D_{60} - D_{45}} \tag{1}$$

where, DS is the dynamic stability (cycle/mm), D_{60} is the tracking depth at 60 min (mm), and D_{45} is the tracking depth at 45 min (mm).

The low temperature performance of the mixture was assessed with the indirect tensile strength test. The indirect tensile strength test was run at −10 °C, and the loading rate was 50 mm/min [32]. The test was operated according to the ASTM D6931 standard. Three replicate samples were conducted on each test condition.

The moisture stability characteristics of the asphalt mixture was assessed with the immersed Marshall test and the indirect tensile strength test after freeze-thaw cycles [33]. The Marshall stability of unconditioned sample ($MS_{Control}$) and water bath conditioned sample ($MS_{Conditioned}$) were evaluated, and the retained strength ratio (RSR) was calculated with Equation (2).

$$RSR = \frac{MS_{Conditioned}}{MS_{Control}} \times 100 \tag{2}$$

The indirect tensile strength test was operated by following the AASHTO T283. The conditioned samples were 70–80% saturated under the vacuum. The samples were then placed in a freezer at −18 °C for 16 h, after which the samples were thawed at 60 °C for 24 h [34]. The indirect tensile strength test of the samples under two conditions was compared to quantify the moisture sensitivity of the mixture.

3. Results and Discussions

3.1. Asphalt Mortar

3.1.1. The Intermediate and High Temperatures Performance

The temperature sweeps result of asphalt mortar under different test temperatures with the F/A ratios of 0.6, 0.8, and 1.0 are presented in Figure 2, Figure 3, and Figure 4, respectively. The complex shear modulus (|G*|) decreased as the test temperature increased. The asphalt mortar stiffness decreased with the increase of the temperature since the asphalt mortar changed from elastic state to the viscous state, thus decreased the |G*|. The phase angle (δ) increased with the increase of the test temperature, and the increasing speed was decreased at the high temperatures. The elastic components decreased, and the viscous components increased as the test temperature increased, thus increased the δ. The asphalt mortar was softened, the flowability increased at the high temperatures, and the temperature sensitivity of the δ was decreased.

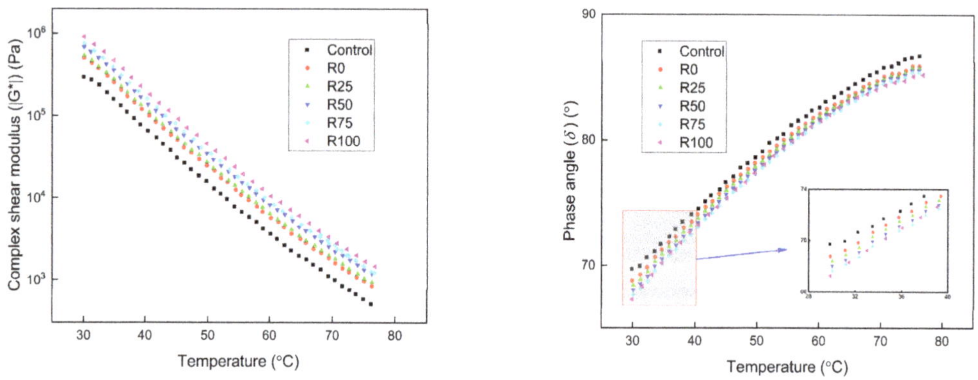

(a) (b)

Figure 2. The temperature sweep results of asphalt mortar under the F/A of 0.6. (**a**) The complex shear modulus |G*| under different conditions; (**b**) the phase angle (δ) under different conditions.

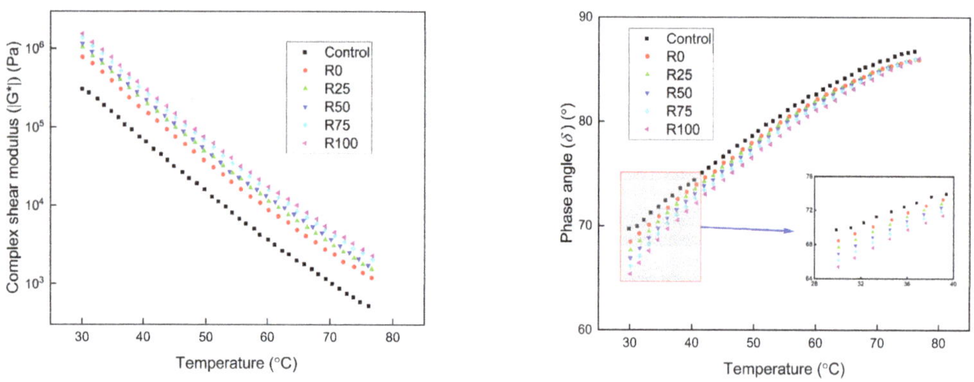

(a) (b)

Figure 3. The temperature sweep results of asphalt mortar under the F/A of 0.8. (**a**) The complex shear modulus |G*| under different conditions; (**b**) the phase angle (δ) under different conditions.

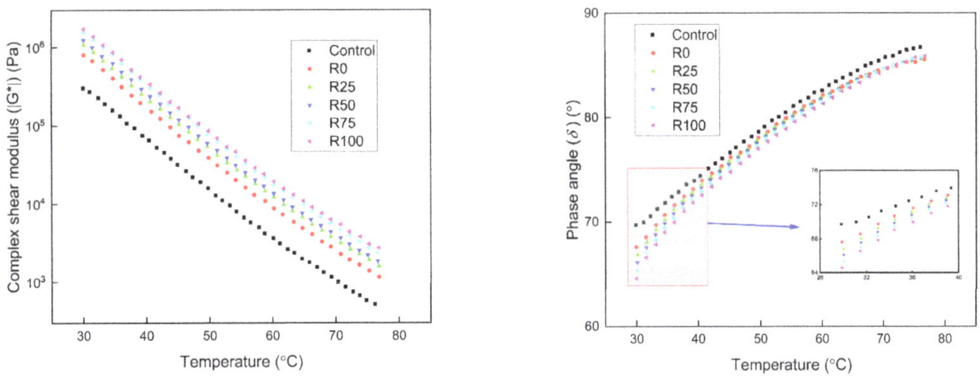

(a) (b)

Figure 4. The temperature sweep results of asphalt mortar under the F/A of 1.0. (**a**) The complex shear modulus |G*| under different conditions; (**b**) the phase angle (δ) under different conditions.

At the same F/A ratio, the |G*| of the asphalt mortar was significantly influenced by the addition of mineral filler and fly ash. The improvement was more significant for asphalt mortar with increased fly ash content. At the F/A ratio of 0.6, the |G*| of asphalt mortar with fly ash was 80% higher than that with mineral filler. The improvement increased to 90–100% higher at the F/A value of 0.8. The value was 110–130% higher at the F/A value of 1.0. The addition of fly ash had more obvious improvement than mineral filler in the |G*| of asphalt mortar. The high temperature performance of asphalt mortar with fly ash was better than that with mineral filler. Based on the fly ash and mineral filler properties in Table 3, the fly ash had higher surface area and lower water absorption rate. As a result of the higher surface area, more asphalt was absorbed by the fly ash, which enhanced the adhesion of asphalt. The interaction between the components of fly ash and asphalt binder also improved the adhesion and shear resistance of asphalt mortar. The |G*| of asphalt mortar increased significantly as the F/A ratio changed from 0.6 to 0.8. The improvement was not apparent as the F/A ratio changed from 0.8 to 1.0. The high F/A ratio decreased the elasticity of asphalt mortar. The low F/A ratio weakened the shear resistance characteristics of asphalt mortar. The F/A ratio of 0.8 was the optimum value, which guaranteed the asphalt mortar had sufficient shear resistance and elasticity.

The δ of asphalt decreased as the fly ash and mineral filler was added. The impact of F/A ratio on the δ of asphalt mortar was reduced with the increase of F/A ratio. The impact of mineral filler on the δ of asphalt mortar was more significant than fly ash. The improved adhesion of asphalt mortar with fly ash increased the elastic components of asphalt mortar. The inclusion of fly ash and mineral filler did not remarkably influence the δ of asphalt mortar.

During the service period of the pavement, the pavement presented a different response to different types of traffic load frequencies. The pavement materials possessed different viscoelastic properties under different loading frequencies. The performance of asphalt mortar under different shear frequencies could represent the response of the asphalt pavement during different loading frequencies. The frequency sweep results of asphalt mortar under different F/A ratios are shown in Figure 5, Figure 6, and Figure 7, respectively. The |G*| of asphalt mortar increased when the loading frequency increased, and the |G*| had a linear relation with the frequency under the log scale. At the low frequency, the loading duration was longer, and the asphalt mortar was more prone to deform under the high temperature. The asphalt mortar had high sensitivity to the loading frequency. Under the same F/A ratio and loading frequency, the |G*| of asphalt mortar with 100% fly ash (R100) was twice the value of asphalt mortar with 100% mineral filler

(R0). The asphalt mortar with fly ash had higher shear resistance at high temperatures than asphalt mortar with mineral filler. The |G*| of asphalt mortar was significantly increased as the F/A ratio changed from 0.6 to 0.8. The F/A ratio had limited improvement as the F/A ratio changed from 0.8 to 1.0. The δ of asphalt decreased as the frequency increased, and the δ of asphalt mortar with mineral filler was higher than that with fly ash. The δ of asphalt mortar was reduced when the F/A ratio increased, but the influence was minimal.

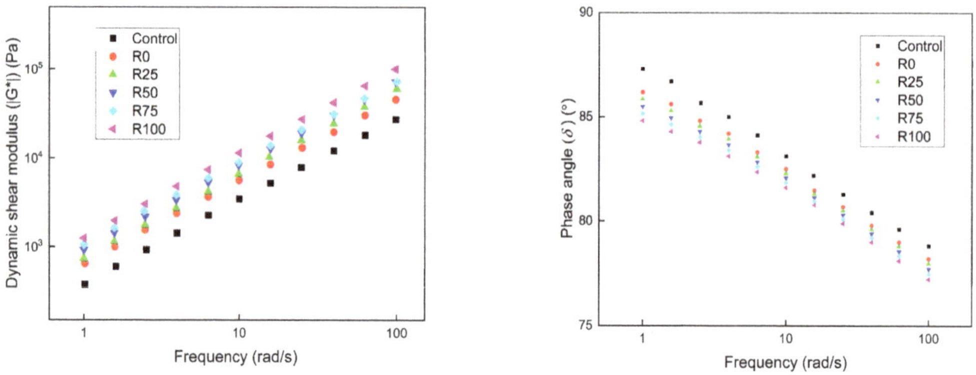

(a) (b)

Figure 5. The frequency sweep results of asphalt mortar under the F/A of 0.6. (**a**) The complex shear modulus |G*| under different conditions; (**b**) the phase angle (δ) under different conditions.

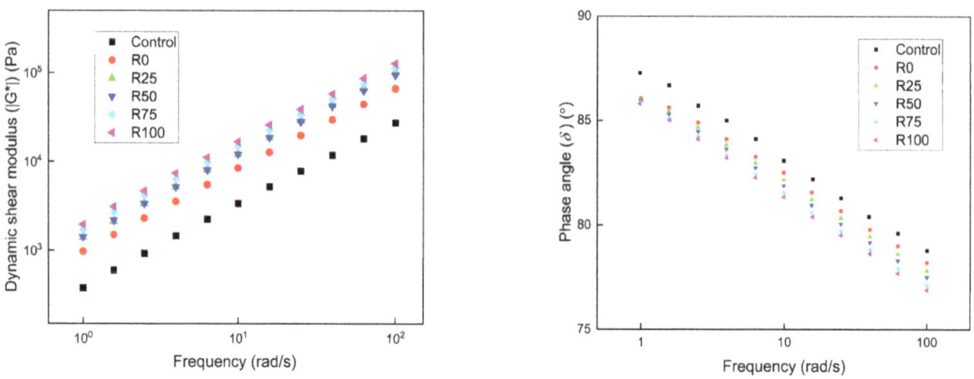

(a) (b)

Figure 6. The frequency sweep results of asphalt mortar under the F/A of 0.8. (**a**) The complex shear modulus |G*| under different conditions; (**b**) the phase angle (δ) under different conditions.

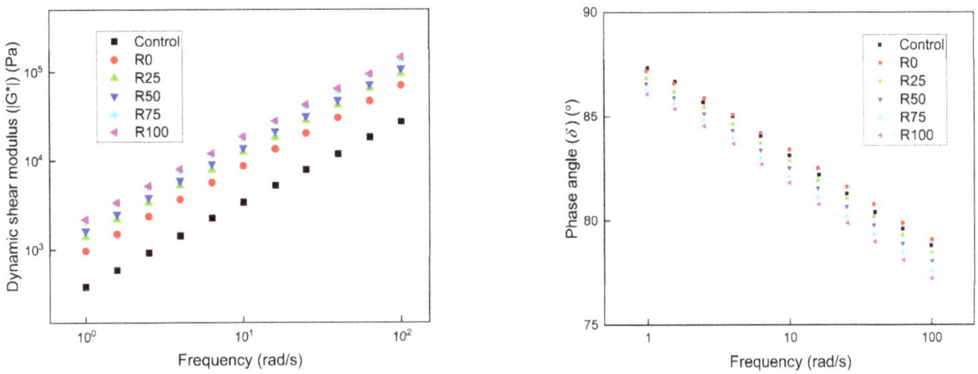

(a) (b)

Figure 7. The frequency sweep results of asphalt mortar under the F/A of 1.0. (**a**) The complex shear modulus |G*| under different conditions; (**b**) the phase angle (δ) under different conditions.

3.1.2. The Low Temperatures Performance

At low temperatures, the asphalt mortar and asphalt mixture became hardened and brittle. The BBR test was adopted to identify the properties of asphalt mortar at low temperatures. The flexural creep stiffness (S) reflected the deformation resistance of the asphalt mortar. The m-value represented the relaxation of asphalt mortar. The cracking resistance of asphalt mortar could be characterized. The S and m-value of asphalt mortar with the F/A ratios of 0.6, 0.8, and 1.0 are shown in Figure 8, Figure 9, and Figure 10, respectively.

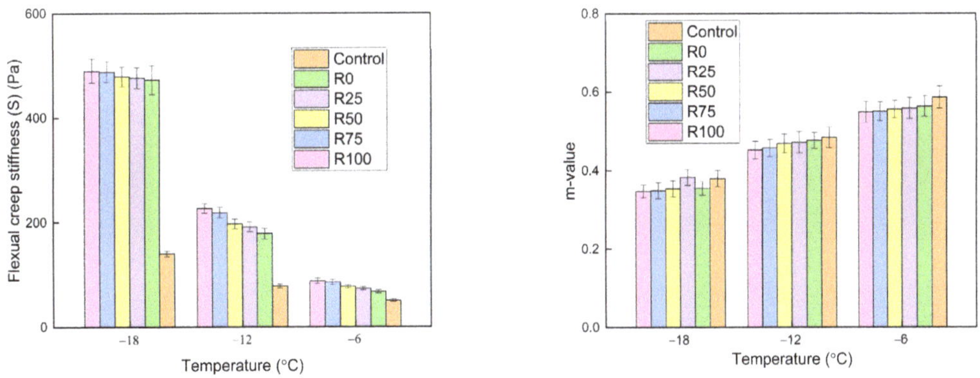

(a) (b)

Figure 8. The BBR test results of asphalt mortar under the F/A of 0.6. (**a**) The stiffness under different conditions; (**b**) the m-value under different conditions.

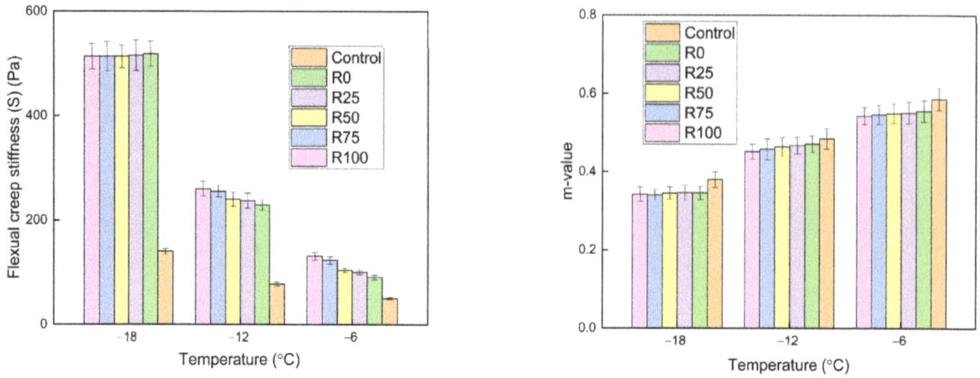

(**a**) (**b**)

Figure 9. The BBR test results of asphalt mortar under the F/A of 0.8. (**a**) The stiffness under different conditions; (**b**) the m-value under different conditions.

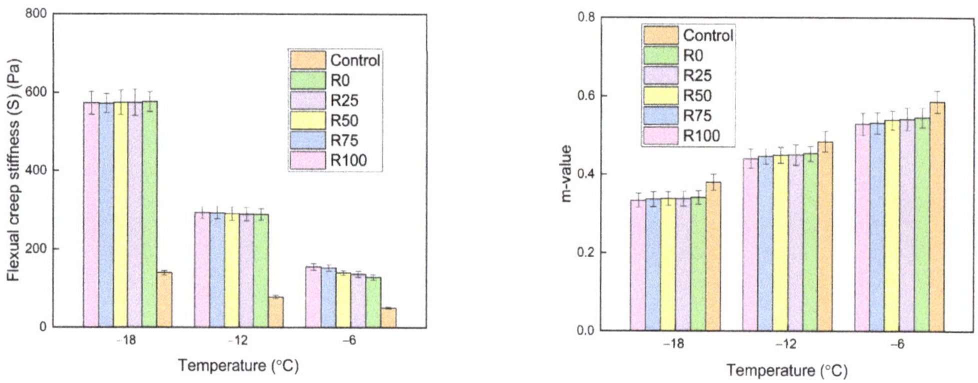

(**a**) (**b**)

Figure 10. The BBR test results of asphalt mortar under the F/A of 1.0. (**a**) The stiffness under different conditions; (**b**) the m-value under different conditions.

The asphalt without fly ash and mineral filler had the lowest stiffness and the highest m-value. After adding fly ash and mineral filler, the deformation resistance was decreased. The stiffness is enhanced and the m-value reduces as the temperature decreased. The temperature had a significant impact on the characteristics of asphalt mortar at low temperatures. At the same temperatures and F/A ratio, the stiffness and m-value linearly changed as the fly ash content increased. At a certain temperature, the stiffness was increased and the m-value decreased, when the F/A ratio increased. The stiffness decreased dramatically when the temperature decreased from −12 to −18 °C, which proved that the temperature sensitivity of stiffness was improved with the decrease of temperature. The differences between the stiffness of fly ash asphalt mortar and mineral filler asphalt mortar were reduced as the temperature decreased. At low temperatures, the filler type had minimal impact on the low temperature characteristics of asphalt mortar.

3.2. Asphalt Mixture

3.2.1. The High Temperature Stability

The high temperature stability of the asphalt mixture reflected the permanent deformation resistance of the asphalt mixture under the impact of high temperature and traffic load [35]. The high temperature permanent deformation resistance was contributed to by the aggregate structure and the cohesive force of the asphalt binder. The physical effect and chemical reaction of the fillers influenced the shear strength, which provided lower contribution than the previous effects. The fly ash and mineral filler enhanced the high temperature property of the asphalt binder, thus influencing the permanent deformation resistance of the asphalt mixture. The high temperature property of the mixture was assessed with the Marshall stability and the wheel tracking test.

The Marshall stability reflected the peak load during the test, and the Marshall flow defined the change of the mixture from the Marshall stability test. The Marshall stability and flow number of different asphalt mixtures are shown in Figure 11. The Marshall stability increased, and the flow number decreased, when the fly ash content increased, which enhanced the high temperature property of the asphalt mixture. The Marshall stability rate increase was higher when the ratio increased from 0 to 50%. Considering the burden to the environment of the asphalt mixture with too much fly ash during the production procedure, the fly ash replacement ratio was suggested to be less than 50%. The higher surface area of fly ash prompted a thicker asphalt film thickness, reduced the light component of asphalt in the mixture, thus increasing the high temperature performance of the mixture.

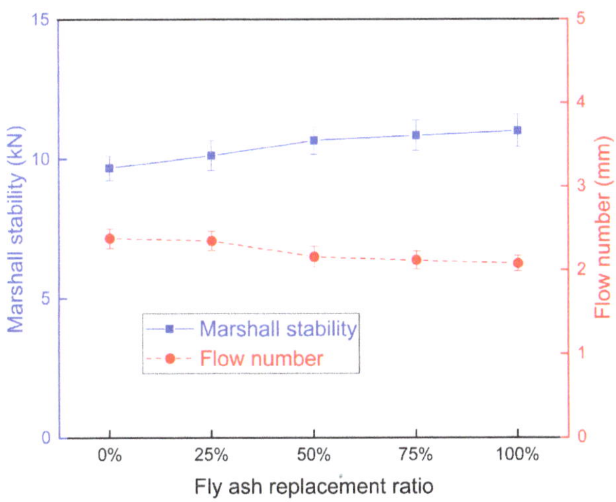

Figure 11. The Marshall stability and flow number of different mixtures.

The wheel tracking test simulated the rutting resistance of the mixture under high temperature and load [36]. The wheel tracking test was conducted on the asphalt mixture with different fly ash replacement ratios. The dynamic stability and total deformation were obtained from the wheel tracking test, as presented in Figure 12. The dynamic stability was increased, and the deformation of the asphalt mixture was reduced, when the fly ash percentage increased. The adoption of fly ash improved the high temperature property of the asphalt mixture, which is consistent with the Marshall stability results.

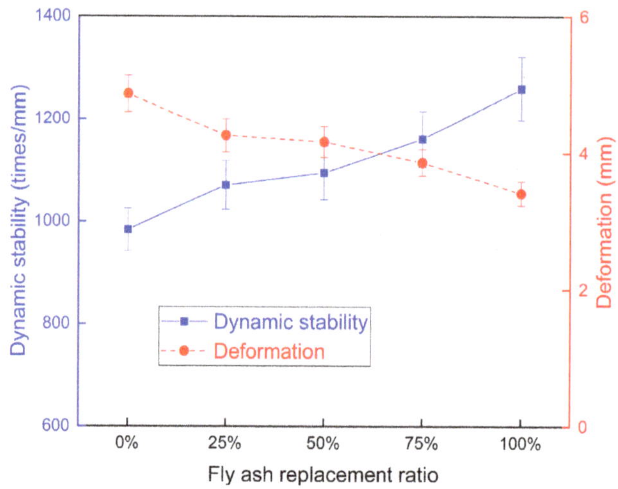

Figure 12. The dynamic stability and deformation of different mixtures.

3.2.2. The Low Temperature Cracking Property

The asphalt mixture became hardened and brittle as the temperature decreased, the bonding failure between the asphalt binder and aggregate was the main cause for the asphalt mixture cracking at low temperatures. The cracking performance of the asphalt mixture was highly related to the performance of asphalt mortar. Based on previous property evaluation of asphalt mortar at low temperatures, the addition of fly ash slightly decreased the low temperature property. The low temperature indirect tensile strength test was adopted to quantify asphalt mixtures with different fly ash replacement ratios, as presented in Figure 13. The tensile strength of the asphalt mixture was linearly increased as the fly ash replacement ratio increased (Figure 13a). The addition of fly ash made the asphalt mixture stiffer and more hardened. The higher surface area of fly ash reduced the light components of the asphalt binder. The tensile strain of the asphalt mixture was decreased when the fly ash replacement ratio increased (Figure 13b). The elasticity of the asphalt mixture decreased as the hardness of the asphalt mixture was increased. The stiffness modulus of the asphalt mixture was increased as the fly ash replacement ratio increased (Figure 13b). One drawback of the hardened asphalt mixture was that the asphalt mixture became more brittle. Thus, the asphalt mixture was more susceptible to cracking at low temperatures. In order for the asphalt mixture to have sufficient cracking resistance at low temperatures, the fly ash replacement ratio should not be too high.

3.2.3. The Moisture Susceptibility of the Asphalt Mixture

The moisture deterioration happened when the cohesion between the binder and aggregate was insufficient. The immersed Marshall test results are presented in Figure 14. The Marshall stability of unconditioned asphalt mixture improved as the fly ash replacement ratio increased. The Marshall stability of conditioned asphalt mixture reduced when the fly ash replacement ratio increased. Fly ash weakened the bonding of the asphalt mixture. The chemical components of fly ash may interact with water, and the intermolecular force was weakened. The bonding between the asphalt and aggregate was damaged, and stripping on the interface between asphalt and aggregate occurred. The *RSR* of the asphalt mixture reduced when the fly ash replacement ratio increased. Based on the standard, the *RSR* should be higher than 80%. The *RSR* decreased to lower than 80% when the fly ash replacement ratio was higher than 25%. In order to guarantee the moisture stability of the asphalt mixture, the fly ash replacement ratio should be set around 25%. If higher fly ash replacement ratio is demanded, an anti-stripping agent could be adopted, or the fly ash could be pretreated by washing with water to eliminate the influence of chemical components of fly ash on the asphalt mixture.

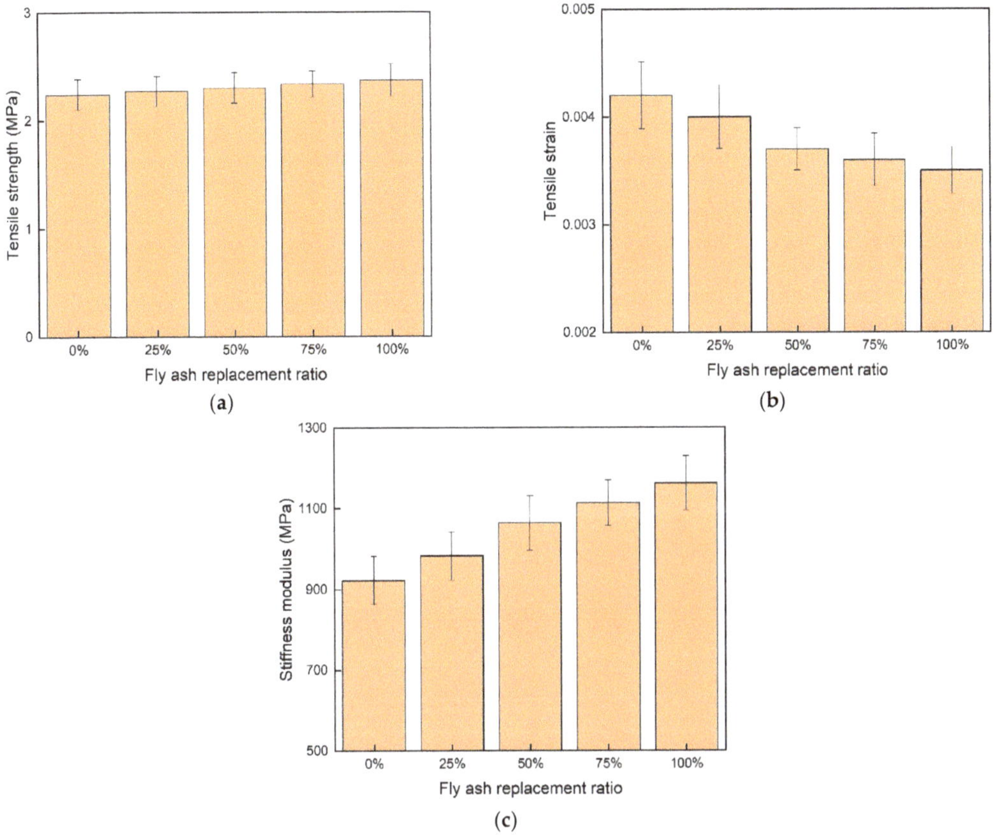

Figure 13. The low temperature cracking test results of asphalt mixture. (**a**) The tensile strength of different mixtures; (**b**) the tensile strain of different mixtures; (**c**) the stiffness modulus of different mixtures.

The indirect tensile strength test condition was more severe than the immersed Marshall stability test. The samples experienced high percentage saturation, low temperature freeze, and high temperature thaw. The indirect tensile strength of different mixtures is presented in Figure 15. The indirect tensile strength of unconditioned samples increased slightly with the increase of the fly ash replacement ratio. After experiencing freeze-thaw cycles, the indirect tensile strength decreased significantly when the fly ash replacement ratio increased. The ratio of the conditioned sample to the unconditioned sample was reduced when fly ash replacement ratio increased. The decreasing trend was similar to that of the immersed Marshall stability test. All the samples with fly ash failed to meet the standard restriction. The fly ash replacement ratio of 25% was slightly lower than the standard value, due to the severe moisture condition. The indirect tensile strength test results were almost consistent with the immersed Marshall stability test results.

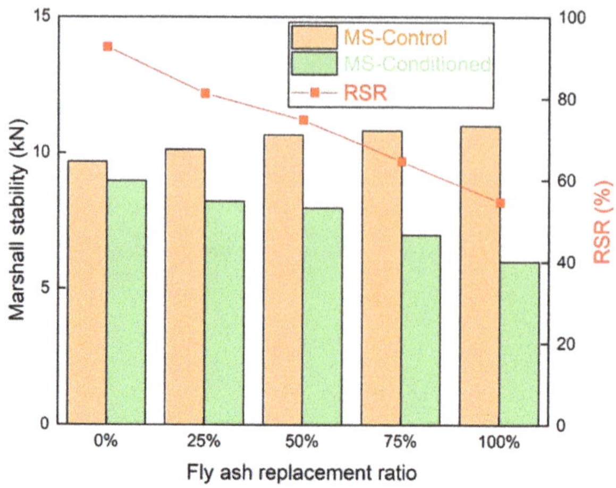

Figure 14. The immersed Marshall stability test results.

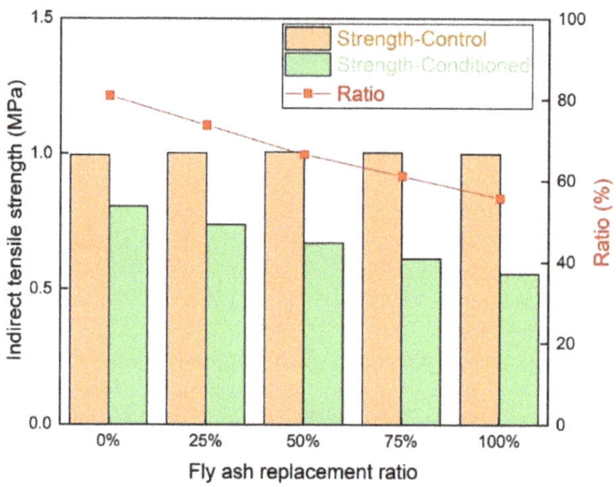

Figure 15. The indirect tensile strength of different mixtures.

4. Conclusions

The impact of fly ash on the properties of asphalt mortar and asphalt mixture were evaluated. Mineral fillers were partially replaced with fly ash to produce asphalt mortar and asphalt mixture. The characteristics of asphalt mortar at different temperatures were assessed. The high temperature, low temperature, and the moisture susceptibility characteristics of the asphalt mixtures were assessed. The following conclusions can be obtained:

- The addition of fly ash enhanced the $|G^*|$ and decreased the δ of the asphalt mortar. The high temperature performance of asphalt mortar with fly ash was better than asphalt mortar with mineral filler and base asphalt. The F/A ratio of 0.8 was the optimum value, which guaranteed that the asphalt mortar had sufficient elasticity and shear resistance.
- The asphalt without fly ash and mineral filler had the lowest stiffness and the highest m-value. After the addition of fly ash and mineral filler, the cracking resistance of asphalt at low temperatures was decreased. The filler type had insignificant impact on the characteristics of asphalt mortar at low temperatures.

- The Marshall stability and the wheel tracking tests' results reached similar findings. The high temperature property of the mixture was enhanced with the increase of fly ash replacement ratio.
- The tensile strength and the stiffness modulus of the asphalt mixture were enhanced as the fly ash replacement ratio increased. The asphalt mixture with higher fly ash replacement ratio was more prone to low temperature cracking. The fly ash replacement ratio should not be too high, to ensure an asphalt mixture with sufficiently low temperature cracking resistance.
- The addition of fly ash weakened the moisture stability of the asphalt mixture, as predicted by the immersed Marshall test and the indirect tensile strength test results. In terms of sufficient moisture stability of the asphalt mixture, the fly ash replacement ratio should be set around 25%. If higher fly ash replacement ratio was demanded, the anti-stripping agent could be adopted, or the fly ash could be pretreated by washing with water to reduce the influence of chemical components of fly ash to the asphalt mixture. Even though the conclusions of the paper were based on limited mixture designs, the results could supply some experience for the mixture design with higher fly ash content and improved low temperature and moisture stability.

Author Contributions: Conceptualization, D.G. and J.L.; methodology, X.Z.; validation, D.G., J.W. and J.L.; formal analysis, X.Z.; investigation, D.W.; data curation, D.G.; writing—original draft preparation, X.Z.; writing—review and editing, J.W., D.W. and J.L.; visualization, D.G.; supervision, D.G.; project administration, D.G. All authors have read and agreed to the published version of the manuscript.

Funding: This research was financed by the Natural Science Foundation of Jiangsu Province (Grant No. BK20210617) and the Science and Technology Development Fund Project of Department of Transportation of Hunan Province (201706).

Institutional Review Board Statement: Not applicable.

Data Availability Statement: The data that support the findings of this study are available on request from the corresponding authors.

Acknowledgments: This authors are sincerely thankful for the funding support.

Conflicts of Interest: The authors declare no conflict of interest.

References

1. Wang, J.; Dai, Q.; Si, R. Experimental and Numerical Investigation of Fracture Behaviors of Steel Fiber–Reinforced Rubber Self-Compacting Concrete. *J. Mater. Civ. Eng.* **2022**, *34*, 04021379. [CrossRef]
2. Ali, N.; Chan, J.S.; Simms, S.; Bushman, R.; Bergan, A.T. Mechanistic evaluation of fly ash asphalt concrete mixtures. *J. Mater. Civ. Eng.* **1996**, *8*, 19–25. [CrossRef]
3. Colonna, P.; Berloco, N.; Ranieri, V.; Shuler, S.T. Application of bottom ash for pavement binder course. *Procedia-Soc. Behav. Sci.* **2012**, *53*, 961–971. [CrossRef]
4. Yan, K.; Gao, F.; Sun, H.; Ge, D.; Yang, S. Effects of municipal solid waste incineration fly ash on the characterization of cement-stabilized macadam. *Constr. Build. Mater.* **2019**, *207*, 181–189. [CrossRef]
5. Likitlersuang, S.; Chompoorat, T. Laboratory investigation of the performances of cement and fly ash modified asphalt concrete mixtures. *Int. J. Pavement Res. Technol.* **2016**, *9*, 337–344. [CrossRef]
6. Yan, K.; Li, L.; Ge, D. Research on properties of bitumen mortar containing municipal solid waste incineration fly ash. *Constr. Build. Mater.* **2019**, *218*, 657–666. [CrossRef]
7. Saride, S.; Avirneni, D.; Javvadi, S.C.P.; Puppala, A.J.; Hoyos, L.R. Evaluation of fly ash treated reclaimed asphalt pavement for base/subbase applications. *Indian Geotech. J.* **2015**, *45*, 401–411. [CrossRef]
8. Yilmaz, A.; Degirmenci, N. Possibility of using waste tire rubber and fly ash with Portland cement as construction materials. *Waste Manag.* **2009**, *29*, 1541–1546. [CrossRef]
9. Milad, A.; Ali, A.S.B.; Babalghaith, A.M.; Memon, Z.A.; Mashaan, N.S.; Arafa, S. Utilisation of Waste-Based Geopolymer in Asphalt Pavement Modification and Construction—A Review. *Sustainability* **2021**, *13*, 3330. [CrossRef]
10. Chaunsali, P.; Ardeshirilajimi, A.; Mondal, P. On the interaction of Class C fly ash with Portland cement-calcium sulfoaluminate cement binder. *Mater. Struct.* **2018**, *51*, 131. [CrossRef]

11. Wang, J.; Dai, Q.; Si, R.; Ma, Y.; Guo, S. Fresh and mechanical performance and freeze-thaw durability of steel fiber-reinforced rubber self-compacting concrete (SRSCC). *J. Clean. Prod.* **2020**, *277*, 123180. [CrossRef]
12. Mohammadinia, A.; Arulrajah, A.; Horpibulsuk, S.; Chinkulkijniwat, A. Effect of fly ash on properties of crushed brick and reclaimed asphalt in pavement base/subbase applications. *J. Hazard. Mater.* **2017**, *321*, 547–556. [CrossRef] [PubMed]
13. Antunes, V.; Freire, A.C.; Quaresma, L.; Micaelo, R. Evaluation of waste materials as alternative sources of filler in asphalt mixtures. *Mater. Struct.* **2017**, *50*, 254. [CrossRef]
14. Jin, D.; Ge, D.; Chen, S.; Che, T.; Liu, H.; Malburg, L.; You, Z. Cold In-Place Recycling Asphalt Mixtures: Laboratory Performance and Preliminary M-E Design Analysis. *Materials* **2021**, *14*, 2036. [CrossRef]
15. Faheem, A.F.; Bahia, H.U. Modelling of asphalt mastic in terms of filler-bitumen interaction. *Road Mater. Pavement Des.* **2010**, *11*, 281–303. [CrossRef]
16. Ge, D.; You, Z.; Chen, S.; Liu, C.; Gao, J.; Lv, S. The performance of asphalt binder with trichloroethylene: Improving the efficiency of using reclaimed asphalt pavement. *J. Clean. Prod.* **2019**, *232*, 205–212. [CrossRef]
17. Sobolev, K.; Vivian, I.F.; Saha, R.; Wasiuddin, N.M.; Saltibus, N.E. The effect of fly ash on the rheological properties of bituminous materials. *Fuel* **2014**, *116*, 471–477. [CrossRef]
18. Cho, B.H.; Nam, B.H.; An, J.; Youn, H. Municipal Solid Waste Incineration (MSWI) Ashes as Construction Materials—A Review. *Materials* **2020**, *13*, 3143. [CrossRef]
19. Çelik, Ö. Influence of fly ash on the properties of asphalt. *Pet. Sci. Technol.* **2008**, *26*, 1571–1582. [CrossRef]
20. Tapkın, S. Mechanical evaluation of asphalt–aggregate mixtures prepared with fly ash as a filler replacement. *Can. J. Civ. Eng.* **2008**, *35*, 27–40. [CrossRef]
21. Ge, D.; Jin, D.; Liu, C.; Gao, J.; Yu, M.; Malburg, L.; You, Z. Laboratory Performance and Field Case Study of Asphalt Mixture with Sasobit Treated Aramid Fiber as Modifier. *Transp. Res. Rec.* **2021**, 03611981211047833. [CrossRef]
22. Al-Hdabi, A.; Al Nageim, H.; Ruddock, F.; Seton, L. Development of sustainable cold rolled surface course asphalt mixtures using waste fly ash and silica fume. *J. Mater. Civ. Eng.* **2013**, *26*, 536–543. [CrossRef]
23. Al-Abdul Wahhab, H.I.; Hussein, I.A.; Parvez, M.A.; Shawabkeh, R.A. Use of modified oil fly ash to enhance asphalt concrete performance. *Mater. Struct.* **2015**, *48*, 3231–3240. [CrossRef]
24. Hoy, M.; Rachan, R.; Horpibulsuk, S.; Arulrajah, A.; Mirzababaei, M. Effect of wetting–drying cycles on compressive strength and microstructure of recycled asphalt pavement–Fly ash geopolymer. *Constr. Build. Mater.* **2017**, *144*, 624–634. [CrossRef]
25. Sharma, V.; Chandra, S.; Choudhary, R. Characterization of fly ash bituminous concrete mixes. *J. Mater. Civ. Eng.* **2010**, *22*, 1209–1216. [CrossRef]
26. Xie, J.; Wu, S.; Pang, L.; Lin, J.; Zhu, Z. Influence of surface treated fly ash with coupling agent on asphalt mixture moisture damage. *Constr. Build. Mater.* **2012**, *30*, 340–346. [CrossRef]
27. Huang, B.; Shu, X.; Dong, Q.; Shen, J. Laboratory evaluation of moisture susceptibility of hot-mix asphalt containing cementitious fillers. *J. Mater. Civ. Eng.* **2010**, *22*, 667–673. [CrossRef]
28. Xue, Y.; Hou, H.; Zhu, S.; Zha, J. Utilization of municipal solid waste incineration ash in stone mastic asphalt mixture: Pavement performance and environmental impact. *Constr. Build. Mater.* **2009**, *23*, 989–996. [CrossRef]
29. ASTM International. *Standard Test Method for Determining the Rheological Properties of Asphalt Binder Using a Dynamic Shear Rheometer*; ASTM International: West Conshohocken, PA, USA, 2015.
30. ASTM International. *ASTM D6927-15 Standard Test Method for Marshall Stability and Flow of Asphalt Mixtures*; ASTM International: West Conshohocken, PA, USA, 2015.
31. Hao, P.; Hachiya, Y. Evaluation indicator of asphalt mixture rutting susceptibility. *J. Test. Eval.* **2004**, *32*, 194–201. [CrossRef]
32. Ge, D.; Chen, S.; You, Z.; Yang, X.; Yao, H.; Ye, M.; Yap, Y.K. Correlation of DSR results and FTIR's carbonyl and sulfoxide indexes: Effect of aging temperature on asphalt rheology. *J. Mater. Civ. Eng.* **2019**, *31*, 04019115. [CrossRef]
33. Airey, G.D.; Collop, A.C.; Zoorob, S.E.; Elliott, R.C. The influence of aggregate, filler and bitumen on asphalt mixture moisture damage. *Constr. Build. Mater.* **2008**, *22*, 2015–2024. [CrossRef]
34. Wang, S.; Yan, K.; Ge, D.; Hong, Z. Laboratory research on the performance of stress-absorption interlayer (SAI) of waste tire rubber and amorphous ploy alpha olefin modified asphalt. *Constr. Build. Mater.* **2019**, *223*, 830–840. [CrossRef]
35. Liu, W.; Yan, K.; Ge, D.; Chen, M. Effect of APAO on the aging properties of waste tire rubber modified asphalt binder. *Constr. Build. Mater.* **2018**, *175*, 333–341. [CrossRef]
36. Rodriguez-Alloza, A.M.; Gallego, J. Mechanical performance of asphalt rubber mixtures with warm mix asphalt additives. *Mater. Struct.* **2017**, *50*, 147. [CrossRef]

materials

MDPI

Article

Preparation of Artificial Pavement Coarse Aggregate Using 3D Printing Technology

Weixiong Li [1,2], Duanyi Wang [1], Bo Chen [1,2,*], Kaihui Hua [3], Zhiyong Huang [1], Chunlong Xiong [1] and Huayang Yu [1]

1 School of Civil Engineering and Transportation, South China University of Technology, Guangzhou 510006, China; 201810101689@mail.scut.edu.cn (W.L.); tcdywang@scut.edu.cn (D.W.); 201910101471@mail.scut.edu.cn (Z.H.); cthgclx@mail.scut.edu.cn (C.X.); huayangyu@scut.edu.cn (H.Y.)
2 Guangzhou Xiaoning Roadway Engineering Technology Research Office Co., Ltd., Guangzhou 510641, China
3 School of Environment and Civil Engineering, Dongguan University of Technology, Dongguan 523808, China; huakh@dgut.edu.cn
* Correspondence: chenb@scut.edu.cn; Tel.: +86-132-8868-2471

Abstract: Coarse aggregate is the main component of asphalt mixtures, and differences in its morphology directly impact road performance. The utilization of standard aggregates can benefit the standard design and performance improvement. In this study, 3D printing technology was adopted to prepare artificial aggregates with specific shapes for the purpose of making the properties of artificial aggregates to be similar to the properties of natural aggregates. Through a series of material experiments, the optimal cement-based material ratio for the preparation of high-strength artificial aggregates and corresponding manufacturing procedures have been determined. The performance of the artificial aggregates has been verified by comparing the physical and mechanical properties with those of natural aggregates. Results indicate that using 3D printing technology to generate the standard coarse aggregate is feasible, but its high cost in implementation cannot be ignored. The 3D shape of the artificial aggregate prepared by the grouting molding process has a good consistency with the natural aggregate, and the relative deviation of the overall macro-scale volume index of the artificial aggregate is within 4%. The average Los Angeles abrasion loss of artificial cement-based aggregate is 15.2%, which is higher than that of diabase aggregate, but significantly lower than that of granite aggregate and limestone aggregate. In a nutshell, 3D printed aggregates prepared using the optimized cement-based material ratio and corresponding manufacturing procedures have superior physical and mechanical performance, which provides technical support for the test standardization and engineering application of asphalt pavements.

Keywords: coarse aggregate; 3D printing; natural aggregate; cement-based aggregate; sustainable construction

Citation: Li, W.; Wang, D.; Chen, B.; Hua, K.; Huang, Z.; Xiong, C.; Yu, H. Preparation of Artificial Pavement Coarse Aggregate Using 3D Printing Technology. *Materials* **2022**, *15*, 1575. https://doi.org/10.3390/ma15041575

Academic Editor: Francesco Canestrari

Received: 16 January 2022
Accepted: 18 February 2022
Published: 20 February 2022

Publisher's Note: MDPI stays neutral with regard to jurisdictional claims in published maps and institutional affiliations.

Copyright: © 2022 by the authors. Licensee MDPI, Basel, Switzerland. This article is an open access article distributed under the terms and conditions of the Creative Commons Attribution (CC BY) license (https://creativecommons.org/licenses/by/4.0/).

1. Introduction

Natural aggregate, which typically accounts for over 90% of the mass of asphalt pavement, is the most widely used material in pavement construction, indicating its considerable impact on road performance. With the rapid development of the economy, up until 2021, the total mileage of roadways and highways in China reached 5,198,100 km and 168,000 km, respectively. During the construction period, more than 2 billion tons of coarse aggregates have been used for road construction per year. However, due to the large number of aggregate processing manufacturers in China, the backward technical management, and the poor quality of the processed stone, the variability of the particle composition of aggregates is often significantly different [1,2]. It is easy to find that inhomogeneity seriously affects the stability of an asphalt mixture. It has to be acknowledged that aggregate is the most unstable and difficult part of the construction, significantly affecting the quality of asphalt pavement.

In the past decades, regarding pavement performance, a stereotype rooted in academia and industry has been the notion that "asphalt binder, which only accounts for 5% to 10% of asphalt mixtures, determines the quality of pavement construction". However, with the rapid development of asphalt technology and higher demand for binder quality, modified asphalt binders with superior properties are commonly used in current pavement construction, implying marginal differences of pavement performance considering asphalt binders [3,4]. Beyond that, the pavement design in China follows the Marshall design principle which focuses on the relationship between asphalt content and pavement performance and gives limited consideration to the quality of aggregates which represent a large portion of the whole asphalt mixture [5–9]. To revisit the traditional design drawbacks and improve the asphalt mixture performance from the aggregate side, it is necessary to comprehensively investigate the influence of aggregates' properties on pavement performance.

It needs to be mentioned that the uniformity of the coarse aggregate quality directly affects the mechanical properties, water resistance, and durability of the asphalt mixture. Specifically, two aspects including the homogeneity (e.g., density, hardness, etc.) of the material and the shape characteristics of the aggregate (e.g., contour shape, angularity, surface texture, etc.) control the uniformity of the coarse aggregate quality [10–12]. Based on the study of Wang et al. [13], the crystal size of the aggregate parent rock is inversely related to the abrasion value. In other words, as the crystal grain size decreases, the coarse aggregate becomes denser, and the abrasion value increases. In further, the morphological characteristics of coarse aggregate include three factors which are shape, angularity and surface texture, respectively. All of them have significant effects on the high temperature stability, low temperature crack resistance, water stability, and skid resistance of asphalt pavements [14,15]. The needle flakiness index, described by the ratio of the length of the long axis to the short axis of the aggregate, is a common engineering index for evaluating the form of coarse aggregates. It is easy to imagine that as the proportion of needle-shaped aggregates increases, the voids in mineral aggregate (VMA) of the mixture would increase and the needle-shaped aggregates are easily crushed during the molding process negatively affecting the gradation stability of the mixture [16].

In the manufacturing of coarse aggregates, the edges and corners of each coarse aggregate particle produced by mechanical crushing vary a lot. Different roughness of the aggregate determines the internal friction angle of the mixture. Varied mechanical cohesion forces result in different mixing energy and compaction pressures required by different sample mixtures [17,18]. The interlocking effect between the grading aggregates of mineral aggregates becomes more obvious as the shape of the coarse aggregates gets closer to a cube. A good angularity of aggregates play an important role in the improvement of the high temperature stability of the hot-mix asphalt mixture [19]. Coarse aggregate edges and corners possess macro-scale (0.5 mm and above) morphological parameters. In addition, the micro-texture morphology of the crushed surface of the aggregate (ranging from 0.001 to 0.5 mm) should also be considered. Behiry et al. [20] selected coarse aggregates with different textures to perform asphalt mixture modulus tests and found that the rebound modulus of the mixture and the texture roughness of the aggregate crushing surface showed a good positive correlation. Based on the report of the United States Highway Strategic Research (SHRP) center [21], among the above three properties (i.e., shape, angularity and surface texture) of aggregates, the angularity of coarse aggregates is considered the most important in the engineering construction.

There are evitable differences between indoor tests and in-field construction because the shape of the coarse aggregate is affected by the characteristics of the parent rock and the manufacturing procedures used. Notably, even within the same quarry, coarse aggregates with different shapes and angularities could be generated without quality control. Even worse, in addition to the operation errors, greater variability in test results would be produced. The traditional test methods are unlikely to obtain the desired mechanical test results with the same raw material under different loading conditions and thus cannot accurately and intuitively perform the grading design with optimized procedures. Following the uniformity evaluation idea of material standardization [22–24], the standardized coarse aggregate materials with different shapes and angularities can eliminate errors caused by varied materials and address the issues of poor repeatability of asphalt mixture tests.

Leveraging the advancement of the fourth industrial revolution with the rapid development of high-performance material design, building module printing, and additive manufacturing technology etc., this study proposes to use the emerging 3D printing technology to generate standard artificial coarse aggregates for stable material properties and sustainable pavement development. Specifically, 3D printing technology [25–28] is an approach of fabricating building materials or construction components into designated entities via layer-by-layer bonding and in accordance with a standard three-dimensional digital model which is completely a new shape. For example, Oak Ridge National Laboratory (ORNL) has developed a novel cable-driven construction process which can be used to 3D print entire buildings as shown in Figure 1 [29] as well as laser-wire-based directed energy deposition (DED) process [30]. It is a digital manufacturing technology without more complex tools and owns the characteristics of personalized customization and rapid prototyping. At the national level, it becomes an important way for countries to realize the reflow of manufacturing and increase the industrial competitiveness. It is receiving exponential attention nowadays and showcases are available in construction areas such as the 3D printing building and concretes. Impressively, in 2012, Sustainable Oceans International (SOI) used 3D printing technology to create a sandstone reef with a height of 1 m and a weight of 500 kg for the first time [31].

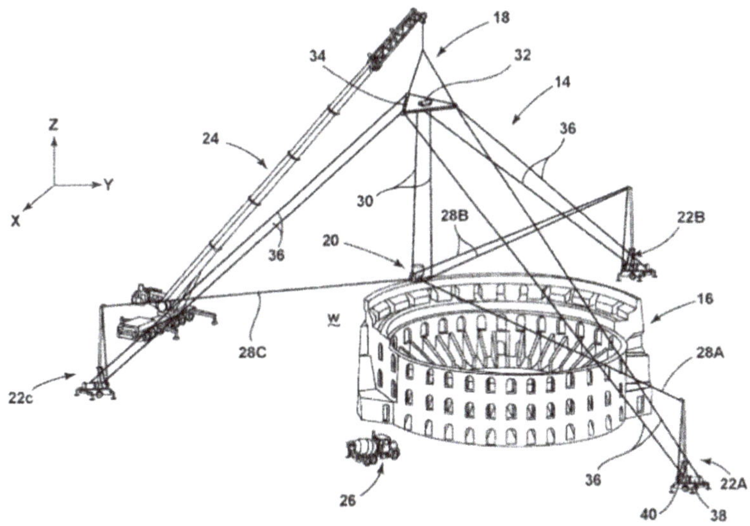

Figure 1. The schematic of the ORNL 'SkyBAAM' 3D printing system.

The same year, a research group [31] at the University of Washington used materials composed of silicon oxide, aluminum, calcium, iron, and magnesium to print a three-dimensional entity that simulates the surface rock of the moon. Their experiments run successfully, and the test results show that 3D printing technology can mimic material properties and generate rock-like materials which are closer to the original entities in shape. In 2017, the research team of Hua from Nanjing University collaborated with other researchers and used raw materials such as polylactic acid, gypsum and resin to generate rocks using a 3D printing method. They found that the printed model has lower compressive strength and stronger ductility than actual rock samples [32]. Fereshtenejad, Vogler, Zhou and Zhu have contributed to the following findings with 3D printing materials [33–35]:

- Adjusting the printing parameters and adopting some post-processing techniques can increase the strength of the sample, reduce the ductility of the sample, and increase materials' brittleness.
- 3DP (powder layer and inkjet head printing technology) printed specimens are very close to sandstone in roughness and tensile strength. Meanwhile, the material failure process is similar between printed specimen and the prototype.
- Resin materials printed by stereo light curing technology (SLA) has good brittleness and is most suitable for simulating physical tests of rocks.

In 2017, Ju and Xie et al. [36] scanned the concrete structure model industrial CT, adopted transparent resin material as the main body, and non-transparent material as the aggregates for 3D printing reconstruction, realizing the visualization of the internal fissure structure. Significantly, the aforementioned technologies provide innovative technical support for the production and testing of standard coarse aggregate models with different 3D forms.

In summary, this study proposes to use 3D printing technology to generate standard aggregate models, discusses the feasibility of 3D printing technology to directly prepare standard aggregates from the perspective of technical indicators and economics, and clarifies the development direction of artificial aggregates. Through a series of indoor material tests, a cement-based material suitable for preparing high-strength man-made aggregates has been developed, and a set of preparation processes have been determined. In the end, the physical and mechanical properties of natural aggregates have been compared to verify the technical performance of man-made aggregates and their engineering feasibility.

2. Methodology and Experiments

2.1. 3D Printing Technology

3D printing has experienced a phenomenal explosion in recent years. Not only it can build physical objects from a geometrical shape, but also it has the potential to revolutionize industries and change the production line [37–40]. For instance, it will reduce the use of manufacturing labors and can greatly affect the economic development. 3D printing technology, computer simulation, and 3D digital modeling have all progressed with the rapid development of science and technology, and they can bring a more flexible and responsive manufacturing process. The general process of 3D printing includes the following steps:

(1) Acquisition of digital point clouds.
(2) Preprocessing of point cloud.
(3) Entity reconstruction of 3D model.
(4) Model optimization and local processing.
(5) Import to the printing computer system.
(6) Track optimization and support addition.
(7) Printer printing.
(8) Post-printing processing, etc.

Figure 2 shows the simplified three stages of 3D printing. According to the different printing materials and printing forging methods, it can be divided into several categories which include Fused Deposition Modeling (FDM), Stereo Lithography Apparatus (SLA), Selecting Laser Sintering (SLS), Digital Light Processing (DLP), Three-Dimensional Printing and Gluing (3DPG), 3D inkjet technology (Ploy Jet) technology, etc. Combined with the current mainstream 3D printing technology development and equipment application, the technical parameters and characteristics are summarized in Table 1.

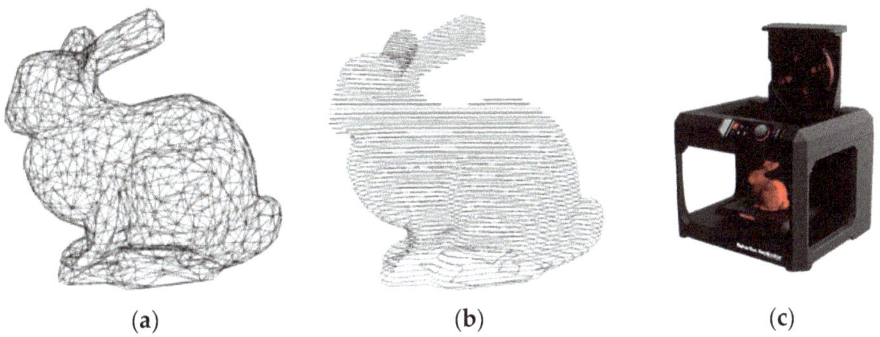

Figure 2. Simplified 3D printing procedure (**a**) Digital model; (**b**) Layered processing; (**c**) Printing process.

Table 1. Summary of major 3D printing techniques.

Type	Accuracy (mm)	Material	Features
FDM	0.025~0.762	Thermoplastic materials: PC, ABS, PLA, etc.	Simple operation, low cost, high material utilization, Simple to support
SLA	0.025~0.1	Transparent photosensitive resin, milky white photosensitive resin, oligomer, reactive diluent, photoinitiator	High degree of automation, high precision, excellent appearance quality, and can make complex structure models
SLS	0.1~0.2	Nylon, ABS, resin coated sand, polycarbonate, metal, ceramic powder, etc.	Using a variety of materials, simple production process, no supporting structure, high material utilization rate
DLP	0.04	Liquid photopolymer	Good stability, support offline printing; high printing accuracy; adjustable printing layer thickness, the system can automatically generate support software
3DPG	0.013~0.1	Powder materials (gypsum powder, etc.)	Low cost, wide range of materials, fast forming speed, can produce complex shape parts
3D inkjet	0.1	Photosensitive resin polymer material	High-quality printing of 3D digital models; short design cycle; wide selection of materials; easy support removal

2.2. Preparation of 3D Printed Coarse Aggregate with Resign-Based Material

2.2.1. Material

Taking equipment accuracy, printing cost, and efficiency into account, this study adopts SLA and DLP which are mature and adoptable for our experimental design and implementation to generate the standard coarse aggregates models. Specifically, DLP mainly uses acrylonitrile butadiene styrene (ABS) resin as the raw material. It is a thermoplastic polymer material with a density of about 1.2 g/cm^3, high strength and easy to mold. With respect to the bonding power, it has a good interlayer bonding effect with other different materials, and is more suitable for surface printing, coating and plating treatments. In addition, ABS has superior impact resistance and can be used under extremely harsh low-temperature conditions. The technical indicators of the selected ABS resin material with a black appearance are listed in Table 2.

Table 2. Properties of ABS resin.

Test Program	Test Value
Tensile strength/MPa	42–62
Notched impact strength/(J·m^{-1})	60–80
Flexural strength/MPa	68–80
Glass transition temperature/°C	100

The SLA technique mainly utilizes photosensitive resin as the raw material for 3D printing to generate the product. Photosensitive resin can be rapidly formed under the action of light and is composed of oligomer, photo initiator, diluent, etc. It is generally used in a liquid form, also named liquid photosensitive resin. Due to the excellent fluidity of photosensitive liquid resin, photocuring-based 3D printing is convenient for complicated construction tasks. In addition, the photosensitive resin is a colloidal substance composed of a chain-linked fence-like fragment polymer. Under the action of ultraviolet light or other specific light, these dispersed polymers will bond with long and long cross-linked polymer polymers, and finally convert from colloid to hard solid. Light-curing resin has the characteristics of low viscosity, small curing shrinkage, fast curing rate, low swelling, high light sensitivity, high curing degree, and high wet strength t. The photosensitive resin used in this study is ZR710. It has a white appearance, and the technical indicators shown in Table 3.

Table 3. Properties of photosensitive resin.

Test Program	Test Value
Flexural modulus/MPa	2813~3520
Notched impact strength/(J·m^{-1})	42~50
Flexural strength/MPa	83~90
Mohr's hardness	87~92

To fairly compare the printed aggregate and the natural aggregates, we choose diabase natural gravel as the prototype which was produced by the Shiniuling Stone Material Processing Plant (Guigang, Guangxi, China). Eight representative gravels (see Figure 3) with a particle size in the range of 5–10 mm were selected. Specifically, the lithology is almond glass-based basalt and presents a porphyritic structure. Besides, the matrix is a glass crystal intertwined structure with volcanic glass (50%), labradorite (10%) and pyroxene (5%). The phenocrysts contain labradorite (16%) and pyroxene (10%).

Figure 3. Original coarse aggregate.

2.2.2. 3D Printing Procedure

The printing instruments used in the experiment are DLP-3D printers (SprintRay, MoonRay, Dongguan, China) and SLA light-curing 3D printers (ZRapid, iSLA500, Suzhou, China). The printing process is illustrated in Figure 4.

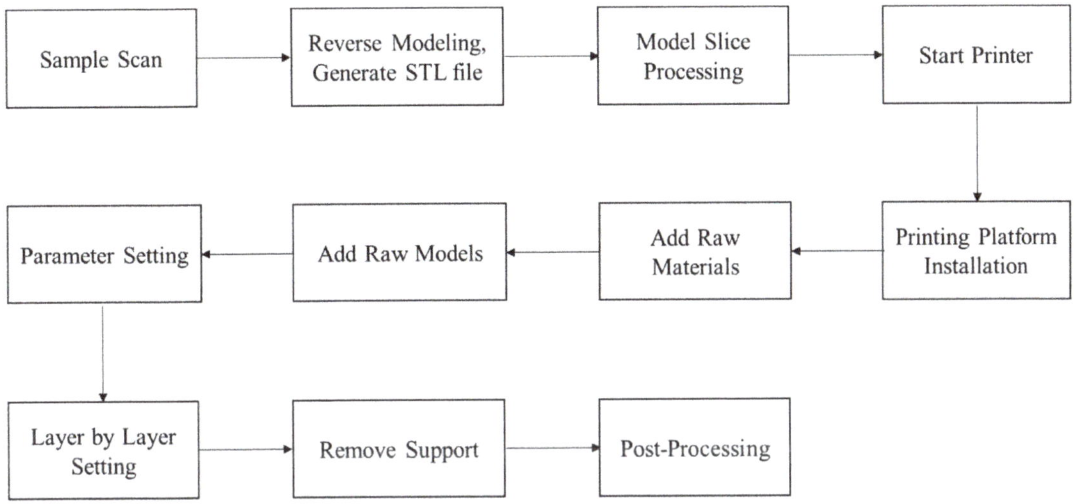

Figure 4. Flow-chart of 3D printing procedure.

To get high accuracy and abundant details of the original coarse aggregate. The three-dimensional scanner (SHINING 3D, AutoScan DS-EX, Dongguan, China) was used to perform the three-dimensional shape tests. Specifically, the camera resolution is 1.3 million pixels. The scanning accuracy is less than 0.015 mm, and the scanning time is around 30 s. After scanning from different perspectives, the device can perform splicing processing, and finally obtain high-precision three-dimensional morphological STL or OBJ data of different aggregate standard models. The scanning operation can be referred to Figure 5. As we mentioned, for standard aggregate preparation, the two methods of DLP and SLA were selected to print the aggregates with the sizes of 2.36–4.75 mm, 4.75–9.5 mm, 9.5–13.2 mm, 13.2–16 mm, 16–19 mm, and 19–26.5 mm, respectively. The iSLA500 equipment was adopted to form a photosensitive resin gravel model. The printing layer thickness parameter was set to 0.1 mm. Its spot accuracy was 0.1~0.5 mm, and the printed gravel model had a white appearance as shown in Figure 6a. The printing process was set to a batch of 36 gravel. A single printing process took about 6 h, and the post-processing was about 2 h. MoonRay digital light processing and its corresponding printing equipment were utilized to form the ABS resin gravel model. The printing layer thickness was set to 0.02 mm, and the printed gravel model had a black appearance as shown in Figure 6b. The printing process was set to a batch of 6 gravel. Meanwhile, a single printing took about 2 h, and the post-processing was about 30 min.

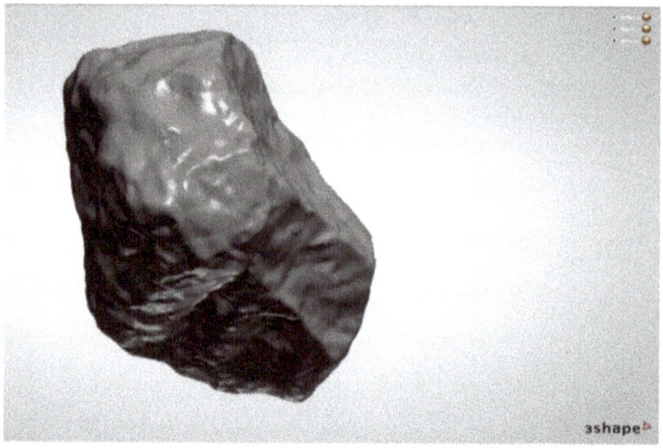

Figure 5. Aggregate scanning.

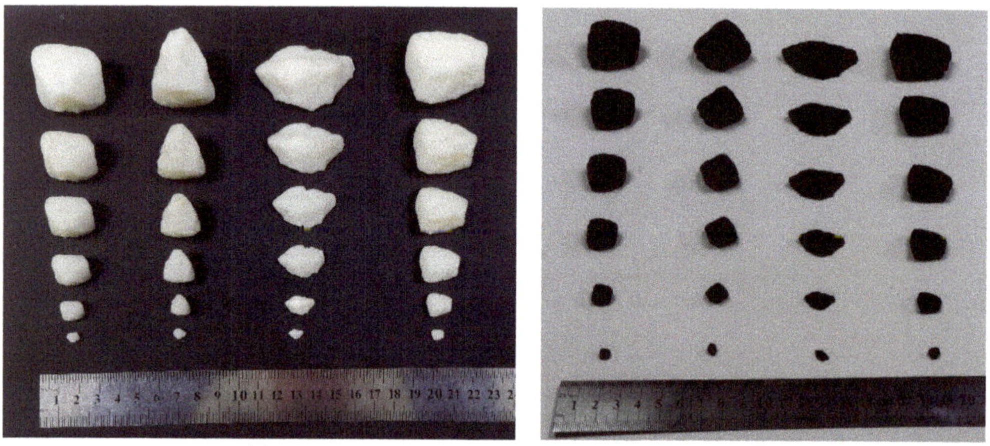

(**a**) (**b**)

Figure 6. 3D printing resign-based models (**a**): photosensitive resin; (**b**): ABS resign model.

2.3. Comparison between 3D Printed Coarse Aggregate Using Resign-Based Material and Natural Aggregate

2.3.1. Physical Performance Analysis and Strength Comparison

The density of the printed resin-based aggregate is around 1.2 g/cm^3, which is slightly higher than that of water and much lower than the 2.89 g/cm^3 of natural aggregate. Considering the strength of the asphalt pavement is largely provided by its structure and the compacted skeleton of the aggregate, the strength, shape, and characteristics of coarse aggregates are therefore particularly critical for pavement performance. It needs to be mentioned that there are lots of required indicators in the production such as crushing value of aggregates, the strength index of abrasion loss, the content of weak rocks, the shape of needle flakes, and the content of dust for construction quality control. In contrast, using 3D printed aggregates can avoid many burdensome requirements of natural aggregates and can simplify the construction procedure. For instance, the printed coarse aggregates contain or include a few tiny particles, micro-cracks, weak joints, weak structural surfaces, etc. which are easily generated in the crushing process, leading to the decline of the strength of prepared asphalt mixtures.

The ABS resin used in our study has a bending strength between 68 MPa and 80 MPa and a Shore hardness between 75 and 80. The photosensitive resin has a bending strength between 83 MPa and 90 MPa and a Shore hardness between 87 and 92. It is easy to find that the photosensitive resign has a better bending strength and Shore hardness compared to ABS resign. Based on the previous studies, the printed model with resin has lower compressive strength and stronger ductility compared to natural rock samples [31]. According to the rock sample evaluation reported by Group 407 of the Hunan Geological Bureau of China, the general Shore hardness of granite and siliceous rock is in the range of 80–90. With the classification of rocks in core drilling reported by Japan, the Shore hardness between 70 and 95 is represented as "hard", and the hardness above 95 is represented as "extremely hard". Thus, the hardness of the ABS resin and photosensitive resin used in this study has reached the "hard" level, which is equivalent to natural rock and indicates better bending resistance compared to natural rock. Even though natural rock possesses excellent hardness and wear resistance, its toughness and flexural and tensile properties are relatively weak. In other words, the rock always presents brittle failure.

2.3.2. Comparison of Morphology and Printing Accuracy Analysis

We compared the scanned models of the printed products of the nature aggregate, the photosensitive resin-based aggregate (SLA printing), and the ABS resign-based aggregate (DLP printing). The appearance of their morphology features can be referred to Figure 7. According to the results (see Table 4), it can be found that the volume deviations of the scan model of the SLA printed aggregate and the scan model of the natural aggregate are all negative deviations, which are ranging from −1.9% to −3.7%, and the average deviation is −2.53%. Meanwhile, the volume deviations of the scan model of the DLP printed aggregate and the scan model of the natural aggregate are also negative deviations, which are ranging from −0.56% to −1.9%, and the average deviation is −1.18%.

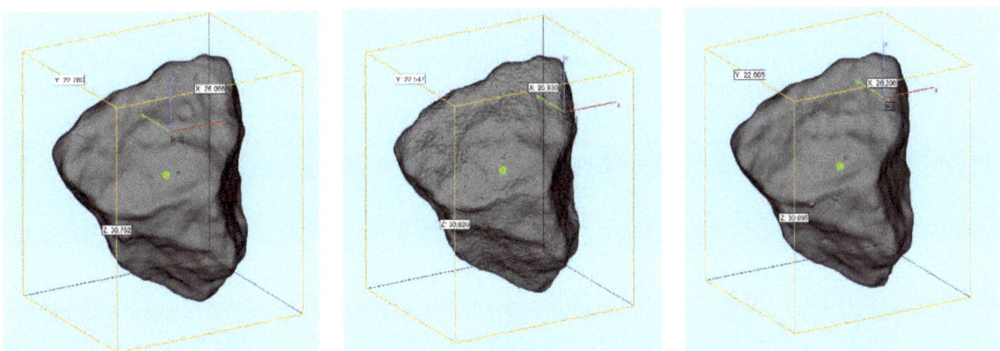

(a) (b) (c)

Figure 7. Three-dimensional appearance of printed aggregates (**a**): natural aggregate; (**b**): photosensitive resin-based aggregate (SLA printing); (**c**): ABS resign-based aggregate (DLP printing).

From the observation, it can be easily found that the macroscopic outline of the 3D printed aggregate is almost the same as the natural aggregate, but from the microscopic texture perspective, the surface texture of the printed resin aggregate is smoother. It can be explained from two perspectives. On the one hand, it is related to the accuracy of the 3D scanning and the accuracy of the 3D printing equipment. On the other hand, it is related to the characteristics of the resin material. Overall, the morphological features of aggregates printed by the two methods are similar, and the volume index deviation of natural aggregates is small. The manufacturing accuracy can reach within 5%. Moreover, the accuracy and stability of the DLP printing technique are slightly better than that of the SLA printing technique.

Table 4. The deviations between resin-based aggregates and natural aggregates.

Samples	Type	Volume Deviation	Surface Area Deviation
1	SLA DLP	−2.80% −1.00%	−0.80% −1.70%
2	SLA DLP	−3.70% 2.70%	−2.00% −0.56%
3	SLA DLP	−2.30% 1.20%	−1.60% −0.72%
4	SLA DLP	−2.10% −1.30%	−0.72% −1.90%
5	SLA DLP	−2.70% 2.30%	−1.20% −1.30%
6	SLA DLP	−1.90% 2.20%	−1.80% −0.85%
7	SLA DLP	−2.10% −1.90%	−1.12% −1.10%
8	SLA DLP	−2.70% 2.30%	−2.20% −1.60%

Based on our initial test results and other reports, the method of directly printing the aggregate model has issues of high cost and low efficiency. According to the market prices in China, the pricing per unit ranged from 2 to 24 CNY per gram. It is easy to find that such a high price significantly restricts field production. To make 3D printing feasible, reduce the production cost, and get the model fulfilling the pavement construction needs, current major techniques such as 3DP, FDM, and SLA can be adopted in the field production and the cement-based material which is cost-effective can be utilized to print the aggregate with 3D printing techniques. In the next section, a cement-based material is developed to improve the artificial aggregates' mechanical properties and reduce the production cost. Meanwhile the high-efficiency manufacturing process is designed to realize large-scale industrial production through direct additive preparation or indirect casting.

2.4. Preparation of 3D Printed Coarse Aggregate Using Cement-Based Material

2.4.1. Cement Material Design

Raw Material

To produce the cement-based printing material, we choose Portland cement (manufactured in Zhucheng, China), pearl river fine sand (from Guangdong, China), water-reducing agent (polycarboxylic acid water-reducing agent, BASF, Ludwigshafen, Germany; the water-reducing rate of the mortar is $\geq$14%), Calcium sulfoaluminate cement expansion agent (obtained from Yantai, China), grade I fly ash (manufactured in Shenzhen, China), silica fume (manufacturing site Luoyang, China), and distilled water as the basic components for material synthesis. The technical index values of each component as mentioned above are listed in Tables 5–9.

Table 5. Properties of Portland cement.

Specific Surface Area (m^2/kg)	Stability	Initial Setting Time (min)	Final Setting Time (min)	Flexural Strength (MPa)		Compressive Strength (MPa)	
				3d	28d	3d	28d
410	Pass	173	214	6.2	9.1	32	59.3

Table 6. Gradation of pearl river fine sand (the fineness modulus is 1.51).

Aggregate Size (mm)	≥4.75	2.36–4.75	1.18–2.36	0.6–1.18	0.3–0.6	0.15–0.3	≤0.15
Mass Ratio (%)	0	0	0	0	62.7	26.0	11.3

Table 7. Technical indicators of calcium sulfoaluminate cement.

Specific Surface Area (m^2/kg)	1.18 mm Screen Sieve Residue (%)	Initial Setting Time (min)	Final Setting Time (min)	7d Limit Expansion Rate in Water	Compressive Strength (MPa)	
					7d	28d
284	0.22	150	230	0.029	31.8	52.0

Table 8. Technical indicators of fly ash.

45 μm Screening Fineness (%)	Stability (mm)	Water Demand Ratio (%)	Sulfur Trioxide (%)	Moisture Content (%)	Ignition Loss (%)	Free Calcium Oxide (%)
11.0	0.5	94	0.74	0.1	4.41	0.8

Table 9. Technical indicators of silica fume.

Specific Surface Area (m^2/kg)	Ignition Loss (%)	Water Demand Ratio (%)	Chloride Ion (%)	Moisture Content (%)	SiO_2 Content (%)	Activity Index (%)
25,100	2.5	113	0.014	1.1	94.05	112

Experimental Procedures

To discuss the influence of each component, the initial ratio (mass ratio) of cement: fly ash: silica fume: sand: water reducing agent: expansion agent: water is set as 100:12:6:46.8:1.35:3:28.5. Based on this initial ratio, the impact of component changes on mechanical properties is tested and evaluated in the following sections. For the convenience, the initial ratio is shown in Table 10.

Table 10. The initial ratios of each component.

<table>
<tr><th>Raw Material</th><th colspan="6">Material Mass Ratio</th><th>Initial Ratio</th></tr>
<tr><td>Cement</td><td colspan="6">100</td><td>100</td></tr>
<tr><td>Fly ash</td><td>6</td><td>9</td><td colspan="2">12</td><td>15</td><td>18</td><td>12</td></tr>
<tr><td>Sand</td><td>15.6</td><td>31.2</td><td colspan="2">46.8</td><td>62.4</td><td>78</td><td>46.8</td></tr>
<tr><td>Silica fume</td><td>0</td><td>2</td><td>4</td><td>6</td><td>8</td><td>10</td><td>6</td></tr>
<tr><td>Water reducing agent</td><td>0</td><td>0.45</td><td>0.9</td><td>1.35</td><td>1.8</td><td>2.25</td><td>1.35</td></tr>
<tr><td>Swelling agent</td><td>0</td><td>1</td><td>2</td><td>3</td><td>4</td><td>5</td><td>3</td></tr>
<tr><td>Water</td><td colspan="6">28.5</td><td>28.5</td></tr>
</table>

Strength Tests and Results Analysis

According to Highway Engineering Cement and Cement Test Regulations (JTGE 30-2005) [41], a cube specimen of 70.7 mm × 70.7 mm × 70.7 mm was formed and cured at a temperature of 20 °C ± 2 °C under standard curing conditions for 54 days in a water bath. A universal testing machine (MTS810, MTS Systems Corporation, Minneapolis, MN, USA) was used to test the strength changes of different materials under different blending ratios. Figure 8 shows the changes of strength of specimen under different ratios of components.

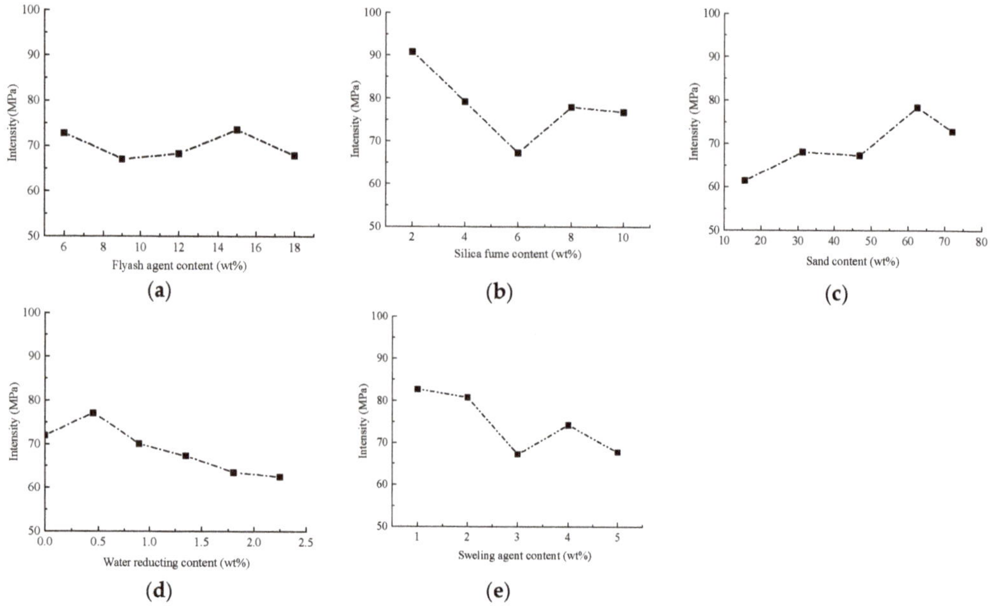

Figure 8. The influence of the content of different components on the compressive strength (**a**): fly ash; (**b**) silica fume; (**c**) sand; (**d**) water reducing agent; (**e**) expansion agent.

From Figure 8, it can be easily found that there are significant and different changes of compressive strength as the contents of fly ash, silica fume, sand, water reducing agent and expansion agent are varied. Fly ash not only can enhance the compressive strength of the material, but also it optimizes the mixing and grading of cement-based materials with its micro-aggregate and active effects. Moreover, with the addition of fly ash, the internal pores of the cement-based material are filled, the fluidity of cement mortar is enhanced, and the grouting efficiency is improved. However, adding excessive fly ash will also cause the SiO_2 powder to agglomerate, which will affect the mixing gradation of cement-based materials. Then the uniform mixing is negatively influenced to introduce strength decrease of cement-based materials [40]. Thus, we prefer to set the cement: fly ash ratio at 100:15.

When the cement: silica fume ratio is 100:2, the cement test samples display the highest strength. This is because fly ash is also present in the formula, providing sufficient pozzolanic activity and micro-aggregate effects for cement-based materials. The positive effect is that it fully filled the internal pores of cement-based materials, but when excessive silica fume is incorporated, it results in agglomeration of SiO_2 powder. Correspondingly, there is uneven mixing, and a significant decrease in the strength of the resulting cement-based materials would happen [42]. The incorporation of sand adjusts the gradation of cement-based materials. Especially, fine sand compensates for the pores between cement-based materials, optimizes the mix ratio, promotes the optimal performance of various materials, and enhances the strength of cement-based materials. However, excessive sand leads to a smaller proportion of cement as a binder, reduces the degree of bonding of cement-based materials, decreases the strength of cement-based materials, and increases the relative density of cement-based materials. We recommend the cement: sand is 100:62.4 for cement-based material preparation.

The addition of polycarboxylic acid superplasticizer reduces the need for a large amount of water in cement mortar. Meanwhile, its water-reducing and viscosity-reducing effect reduces the viscosity of cement mortar and helps improve the strength of cement mortar. Excessive mixing of water reducing agent aggravates the effects of reducing water and viscosity, leading to severe segregation of cement-based materials and reduced strength. When the cement:water reducing agent ratio is 100:0.45, the strength of cement-based materials achieves its optimal value.

When the expansion agent and water are mixed into cement, a large amount of expansive crystalline hydrates are generated. The associated compressive stress can offset the tensile stress when there is cement shrinkage, thereby reducing this shrinkage and the cracking of cement samples. Adding an appropriate amount of expansion agent can prompt prestress through the generated expansion hydrate, protect the cement-based material and improve its early strength. Same with previous results, excessive expansion agent leads to redundant hydrate, and overexpansion of cement-based material. Regarding this, it aversively affects the curing process and strength improvement of cement-based material in the late stage. We recommend a cement:expansion agent ratio of 100:1 to achieve the optimized strength level observed in this study.

Liquidity Tests and Results Analysis

To ensure the molding and workability of cement mortar, the fluidity of cement mortar was measured to achieve a high degree of uniformity between the fluidity and strength of cement mortar. Specifically, a NLD-3 mortar fluidity instrument (Lei Yun Test Instrument Manufacturing Co., LTD, Shanghai, China) was used for testing. The vibration frequency was 1 Hz, and the vibration plane diameter is 300 mm $\pm$ 1 mm. The number of vibrations was 25. The vibration drop distance is 10 mm $\pm$ 0.2 mm. Figure 9 shows the flow test process. Due to the high fluidity of prepared samples, the fluidity exceeds the vibrating platform with a 300 mm diameter. In order to understand the changes in fluidity, its calculation is represented in Equation (1):

$$F = F_1 + (25 - N) \times 10 \quad (1)$$

where F is the fluidity of cement mortar, mm; F_1 is the maximum fluidity value of the equipment 300 mm; N is the number of vibrations corresponding to the state where the cement mortar starts to drip when the vibrating table is covered.

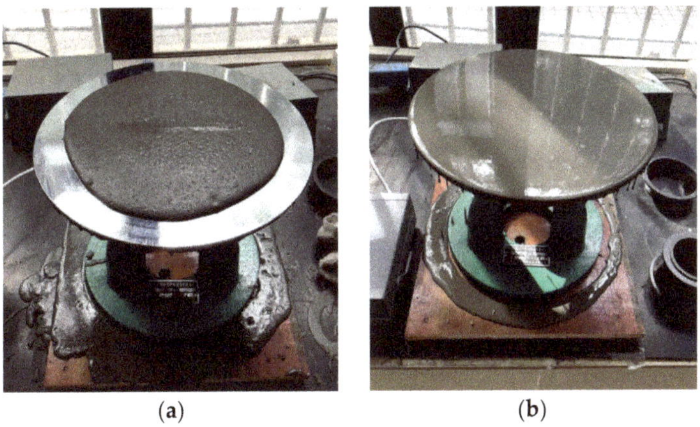

Figure 9. Fluidity test (**a**): low fluidity status; (**b**) high fluidity status.

Test results can be seen in Figure 10, where it is observed that the fluidity of cement mortars with different composition ratios is quite different, as each component contributes variably to the fluidity. With the increase of fly ash content, the fluidity of cement mortar gradually decreases, but the overall fluidity of cement specimens is at a high level. A proper amount of fly ash can increase the fluidity of cement mortar. This can be explained by the fact that fly ash contains many spherical particles, that act as a "ball" between cement particles, reducing the relative slippage between cement particles. As the amount of fly ash increases, it plays a filling role and reduces the water-cement ratio. Therefore, the fluidity of cement mortar shows a decreasing trend as the amount of fly ash increases. With the increase of silica fume content, the fluidity of cement mortar first increases and then decreases. Silica fume is an ultrafine powder collected from the smoke and dust during the smelting of ferrosilicon alloy or metallic silicon in ferrosilicon alloy plants. It is characterized by large fineness and high SiO_2 content. The addition of a small amount of silica fume refines the gradation and improves the fluidity of the slurry. However, due to the early activity of silica fume, it reacts with Ca $(OH)_2$ in a relatively short period of time to form C-S-H condensate with low CaO/SiO_2 ratio. Glue has an adverse effect on the slump. Excessive silica fume can easily lead to a decrease in material fluidity.

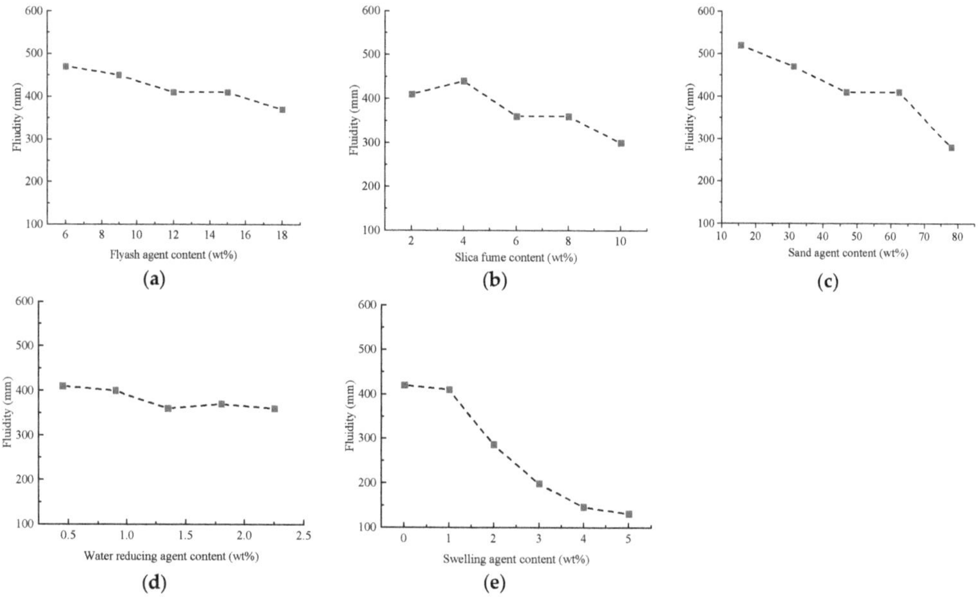

Figure 10. The influence of the content of different components on fluidity (**a**): fly ash; (**b**) silica fume; (**c**) sand; (**d**) water reducing agent; (**e**) expansion agent.

As the sand content increases, the fluidity of the cement mortar also continues to decrease. The addition of a small amount of sand brings a beneficial effect on the fluidity of the cement mortar. As the amount of sand increases, the water absorption of the material approaches saturation. The addition of dry sand in the later stage indirectly reduces the water-cement ratio of the cement material, resulting in a downward trend of fluidity. With the increase of the water-reducing agent content, the fluidity of cement mortar first increases and then tends to be stable. This is because the negatively charged groups such as -COO- and $-SO_3-$ included in the acid-based water-reducing agent can be adsorbed on the surface of cement particles. The electric field with strong repulsive force disperses the cement particles, so the fluidity increases. When the content of the water-reducing agent reaches a certain range, the electric field strength tends to be stable, and the fluidity remains stable as the content of the water-reducing agent increases. As the content of UEA expander increases, the fluidity of cement mortar decreases. This is due to the hydration reaction of the mineral components in UEA expansion agent, producing more high-sulfur hydrated calcium sulfoaluminate (abbreviated as Aft), which leads to increased hydration products of the cement hydration system and increased system consistency. Thus, the fluidity continually becomes smaller.

Based on the experimental results as discussed above, the cement slurry with a fluidity of 290 mm or more can complete the grouting process on a small gravel mold. Therefore, the ratio of cement-based materials requires high material strength and high fluidity. To balance the strength and fluidity, the optimal ratio in this study is determined as follows: cement: fly ash: water reducing agent:silica fume:expanding agent:sand:water = 100:15:0.45:2:1:62.4:28.5. With this ratio, the artificial coarse aggregates using cement-based material are produced.

2.4.2. Mold Preparation and Grouting Process

Mold Preparation

To generate artificial aggregates for large-scale industrial utilization, cost-effective cement-based materials were adopted for sample preparation in this study. On the one hand, the material is easy to acquire, on the other hand, its corresponding mold is easy to prepare with plaster and silica. This section introduces the procedure of mold preparation and compares the mold using plaster and silica, respectively.

In this study, we used plaster ($CaSO_4$) with a fineness of 1500 mesh. Its Mohs hardness is 3. The compressive strength is 7.5 MPa, and an initial setting time ranges from 6 to 8 min. The procedures to generate the plaster mold are as follows:

(1) Weigh the gypsum powder and water in a ratio of 2:1.
(2) Pour water into the plaster powder and manually stir for 30 s until a uniform fluid is formed.
(3) Apply petroleum jelly on the 3D printed aggregate model to prevent the model from sticking to the plaster.
(4) Fix the aggregate model in a container, slowly inject the plaster solution until the liquid level is about 10 mm higher than the top of the model.
(5) Demold after 15 min, and cut the mold into four parts to facilitate the later demolding of cement-based aggregates.

The silica's chemical formula is $mSiO_2 \cdot nH_2O$. Its Shore hardness is 24. The tensile strength is 3.2 MPa. The elongation is 500%, and the tear strength is 24 kh/M. The procedures to generate the silica mold is as follows:

(1) Weigh the silicone and curing agent following the ratio of 100:2.
(2) Mix the silicone and curing agent and stir them manually for 30 s.
(3) Apply petroleum jelly on the 3D printed aggregate model to prevent the model from sticking to the silica.
(4) Fix the aggregate model in the container, slowly inject the silica gel solution until the liquid level is about 10 mm higher than the top of the model.
(5) Demold after 12 h and cut the mold into four parts to facilitate the later demolding of cement-based aggregates.

In Figure 11, we can easily compare the prepared two molds. The plaster mold preparation process is simple, and its molding speed is faster compared to the silica mold. Generally, a batch of molds can be finished in 20–30 min, however, the plaster mold is difficult to demold, especially for the aggregate models which have edges and corners. Moreover, it is easy to drop powders when demold and it is difficult to reuse. In contrast, the silica mold and the aggregate mold are easy to separate, and the demolding process is smooth. In addition, its surface can be stuck to aggregate and can better restore the aggregate. After demolding, the cement will not stick to the mold surface too much, and the mold is easy to clean for reuse. Most importantly, it has better ductility and wouldn't be deformed after each experiment. As stated above, we choose silica mold to generate cement-based samples.

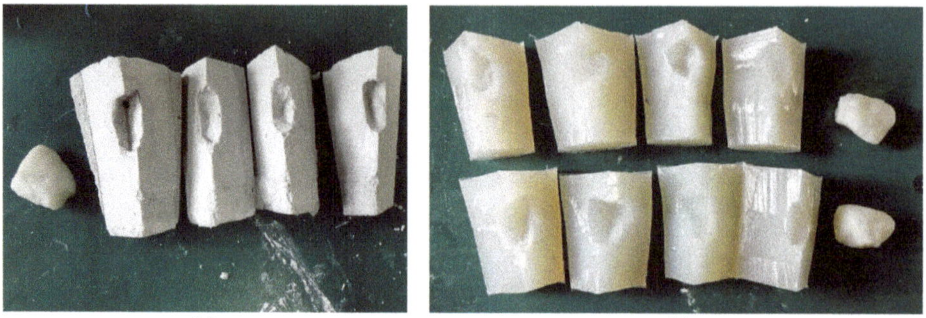

(**a**) (**b**)

Figure 11. The prepared two molds (**a**): plaster mold; (**b**): silica mold.

Grouting Process

The grouting process can be seen in Figure 12. It includes the following major steps:

(1) Prepare cement-based mortar according to the confirmed cement-based ratio.
(2) Use a syringe to inject high-strength cement-based materials into the silicone aggregate mold.
(3) Wait for the cement to dry and form inside and demold after about 48 h.
(4) Use a 55 °C water bath curing for 5 days to shorten the curing time.

(**a**) (**b**) (**c**)

Figure 12. Mold grouting (**a**): preparation of cement mortar; (**b**): mortar injection; (**c**) Demolding.

Note that in the grouting process, there are a lot of bubbles inside, and the internal bubbles need to be eliminated. We have tested the following two approaches to remove bubbles: (1) Vibration defoaming: It extends the vibrator into the mold (internal vibration defoaming), passes the internal vibration stirs the cement mortar to discharge the air bubbles to ensure the morphological characteristics of the artificial aggregate; (2) Stomata defoaming: In this way, the side bottom pores are drilled in the silicone mold, and the needle is used for grouting. The pressure is used to remove the bubbles from the pores during the grouting process. Medium extrusion is used to prepare artificial aggregates with better morphology to ensure the morphological characteristics of artificial aggregates. Based on our experiments, the grouting method adopts the stomata defoaming approach. We drill a small hole on the side, top, and bottom of the mold (a total of three holes), respectively. Grouting is conducted from the small hole at the bottom, which can effectively reduce the air bubbles in the artificial aggregate. The formed artificial aggregates are shown in Figure 13. In appearance, it looks similar to the natural aggregate in both shape and volume.

Figure 13. Prepared artificial aggregates.

2.5. Comparison between 3D Printed Coarse Aggregate Using Cement-Based Material and Natural Aggregate

2.5.1. Morphological Comparison

To compare the morphological differences between artificial cement-based aggregates and natural aggregates, high accuracy 3D scanning has been conducted. The indictors of volume area and surface area have been selected for evaluation. Figure 14 shows the scanning results of natural aggregates (random samples). Figure 15 presents the scanning results of artificial aggregates (random samples). Figure 16 plots the relative differences of these two indicators using natural aggregates and cement-based aggregates.

Based on the 3D scanning results of natural aggregates and cement-based artificial aggregates, it can be found that the 3D shape of cement-based artificial aggregates has good consistency with natural aggregates. According to the calculated volume index, the relative volume deviation of cement-based artificial aggregates and natural aggregates is in the range of −0.54% to 3.11%, while the relative deviation of the surface area of the 3D model is in the range of −3.95% to 2.39%. A larger deviation can be explained by the fact that there are inevitable losses in the molding with the silica mold. However, the overall relative deviation of the macro-scale volume index of cement-based artificial aggregates can be controlled within 5%.

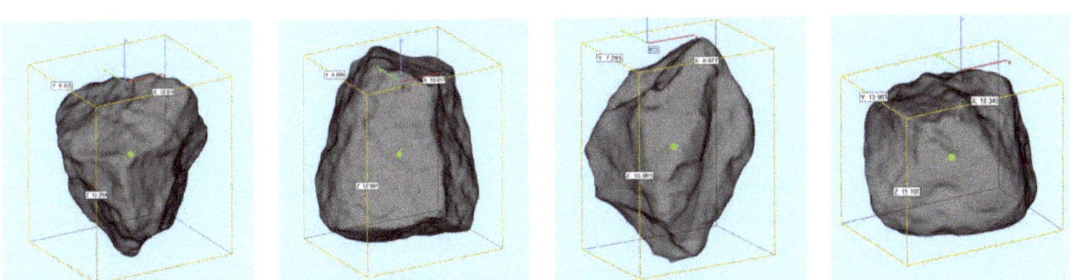

(a) (b) (c) (d)

Figure 14. 3D scanning results of natural aggregates ((**a**–**d**) are random samples).

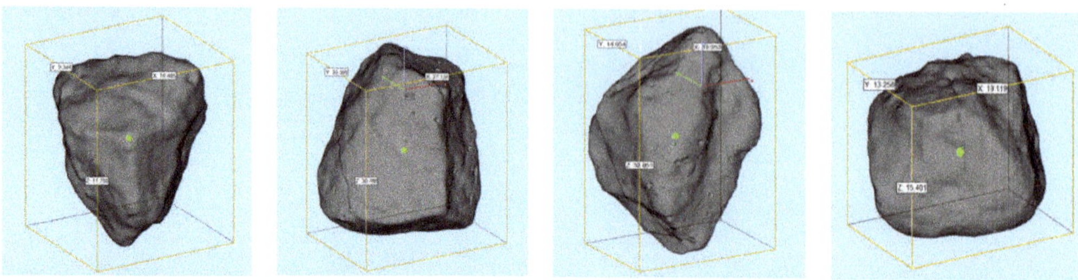

Figure 15. 3D scanning results of cement-based artificial aggregates ((**a**–**d**) are random samples).

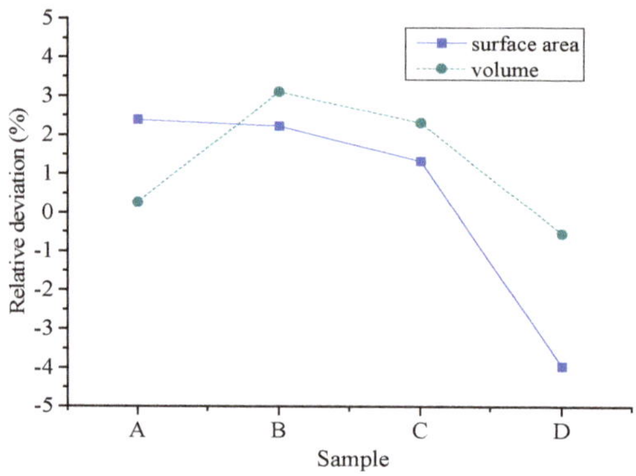

Figure 16. The relative deviation of the volume indicators of natural aggregates and cement-based artificial aggregates.

2.5.2. Physical Indicators Analysis

For a comprehensive evaluation, we compared the density and water absorption of ABS resin-based artificial aggregates, cement-based artificial aggregates, and natural aggregates. Test results are shown in Table 11.

Table 11. Apparent density and water absorption of different materials.

	Apparent Density (g/cm^3)	Moisture Absorption (%)
ABS resin-based artificial aggregates	1.2	0
Cement-based artificial aggregates	2.16	3.6
Natural aggregates (diabase)	2.89	0.5

It can be easily found that ABS resin-based artificial aggregates have the smallest apparent density, followed by cement-based artificial aggregates. Natural aggregates have the largest density (2.89 g/cm^3). Beyond that, the density of cement-based artificial aggregates is around 25% lighter than natural aggregates. This is due to the properties of the raw materials. Based on the test results of moisture absorption, we can find that the surface of ABS resin-based artificial aggregates has no obvious voids, indicating they do not absorb moisture. Since there are holes in the surface of the cement-based artificial aggregate, the moisture absorption rate is around 3.6%. Large moisture absorption can be explained by the residual air bubbles of the cement-based mixture and the material properties after hardening. Since the natural aggregates are denser, the water absorption rate is less than 1%.

2.5.3. Los Angeles Abrasion Test Analysis

Following the ASTM C131, we carried out the Los Angeles abrasion test on cement-based artificial aggregates and natural aggregates. Generally, the Los Angeles abrasion test is used to determine the ability of coarse aggregates to resist friction and impact under standard conditions. It is expressed as abrasion loss (%). The aggregate with small abrasion loss indicates that the asphalt pavement aggregate is hard, wear-resistant, and durable. The calculation formula for abrasion loss can be referred to Equation (2). The test results can be referred to Figure 17:

$$Q = \frac{m_1 - m_2}{m_1} \times 100 \quad (2)$$

where Q is Los Angeles abrasion loss. m_1 is initial sample weight and m_2 is the sample weight after cleaning and drying on a 1.7 mm screen.

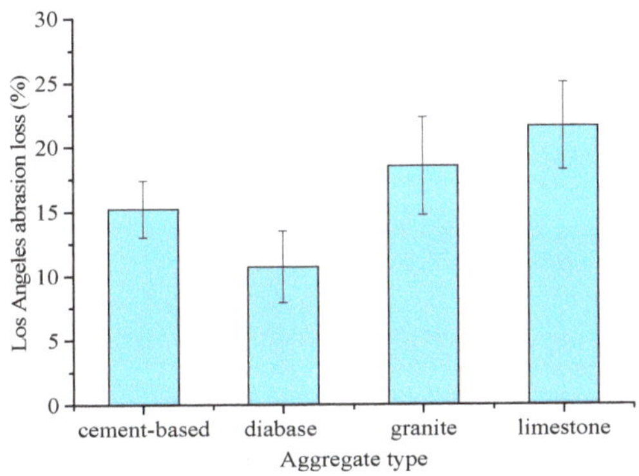

Figure 17. Los Angeles abrasion test results of different aggregates.

Four kinds of materials were used for the tests, which were cement-based artificial aggregates, natural diabase aggregates (Guigang, Guangxi, China), natural limestone aggregates (Yangchun, Guangdong, China), and natural granite aggregates (Maoming, Guangdong, China) with a particle size of 9.5 mm~16 mm. It can be seen that (Figure 17) the Los Angeles abrasion loss of the diabase aggregate is the smallest, indicating that the aggregate is the hardest and most wear-resistant. Generally, it can be used in the top layer of highway asphalt pavement in China. Granite aggregates have a greater loss in Los Angeles abrasion, and limestone aggregates have the largest loss in Los Angeles abrasion test. They can be utilized in the middle and bottom layers of the asphalt pavement structure. The average loss of Los Angeles abrasion test of the cement-based artificial aggregate is 15.2%, which is about 4.5% higher than that of the diabase aggregate, but it is significantly lower than the granite aggregate and limestone aggregate around 3.3% to 6.4%, respectively. It indicates the proposed artificial aggregates have superior strength performance and it is mainly related to the optimized cement-based material ratio in our design. In addition, with the increase of the curing age, the strength of the artificial aggregate will be further improved. Beyond that, the physical shape, mechanical strength, grouting procedures, and the curing method of the artificial aggregate proposed in this study have reached a production level, indicating large-scale industrialization in the near future.

3. Findings and Future Studies

Based on high-precision laser scanning and 3D printing technology, using natural coarse aggregates as the prototype, we designed, prepared, and comprehensively compared the resin-based aggregates and cement-based aggregates. Specifically, the mechanical strength, physical properties, and morphological features of the artificial aggregates and natural aggregates were compared. The findings can be summarized as follows:

- From the perspective of mechanical strength, photosensitive resin-based material and ABS resin-based material have superior bending properties. But their compressive strength and hardness are weaker compared to natural aggregates. The density of resin materials is around 40% of the density of natural aggregates. The deviation of the volume index between natural aggregates and resin-based aggregates can be controlled within 5%.
- The use of 3D printing technology for the preparation of resin-based artificial aggregates is technically feasible, but its high cost seriously restricts the large-scale industrialization in the field implementation. Cement-based material is a low-cost and reliable alternative to prepare artificial aggregates using 3D printing techniques.
- Cement-based artificial aggregates require high standards of fluidity and strength. To balance the strength and fluidity, an optimized cement-based material is designed and prepared. In this study, we recommend that the optimal ratio of the cement-based materials be as follows: cement: fly ash: water reducer: silica fume: expansion agent: sand: water 100:15:0.45:2:1:62.4:28.5.
- The manufacturing procedures of cement-based artificial aggregates are improved and optimized. Silica molds have great superiority in shape stability, easy demolding and cleaning and reusability. We recommend using silica materials to prepare artficial aggregate molds. To reduce the bubbles generated during the molding process, the stomata defoaming method is adopted. In order to shorten the curing time, it is recommended to use high-temperature water bath curing.
- Based on our thorough evaluation, the 3D shape of the prepared cement-based artificial aggregates has good consistency with that of natural aggregates. The relative deviation of the overall macro-scale volume index is smaller than 4%. The apparent density of cement-based artificial aggregates is 2.16 g/cm^3, which is 25% lighter than the density of natural aggregates, but its water absorption rate is slightly larger than that of natural aggregates. The average loss of Los Angeles abrasion loss of cement-based artificial aggregate is 15.2%, which is 4.5% higher than that of diabase aggregate, but it is significantly lower than that of granite aggregate and limestone aggregate. Overall, cement-based artificial aggregates have good mechanical properties.

In sum, this study investigates the feasibility of using 3D printing technology to prepare a standard form of artificial aggregate, and proposes a low-cost, optimized preparation procedure, and a reliable cement-based material design formula. From the perspective of morphological, physical, and mechanical properties, artificial aggregate is a possible solution for future sustainable and green construction. In our next studies, we would like to expand our research in three areas: (1) Improving the production efficiency and shorten the demolding and curing time. (2) Investigating the impacts of artificial aggregates with different morphological features on asphalt mixture performance. (3) Producing artificial aggregates from construction waste. Furthermore, the life cycle cost of the entire process will be considered in the next phase.

Author Contributions: Conceptualization, W.L.; methodology, W.L. and D.W.; formal analysis, B.C. and K.H.; investigation, Z.H. and H.Y.; resources, D.W.; writing—W.L. and D.W.; writing—review and editing, B.C. and C.X.; supervision, H.Y. All authors have read and agreed to the published version of the manuscript.

Funding: This study is funded by the "China Postdoctoral Science Foundation" (2020M672639), the China Natural Science Foundation of Guangdong Province-PhD start (No. 2018A030310684) and the Dongguan Social Science and Technology Development (General) Project (No. 2019507140574). All the help and support are greatly appreciated.

Institutional Review Board Statement: Not applicable.

Informed Consent Statement: Informed consent was obtained from all subjects involved in the study.

Data Availability Statement: The data presented in this study are available on request from the corresponding author.

Acknowledgments: All individuals included in this section have consented to the acknowledgement.

Conflicts of Interest: The authors declare no conflict of interest.

References

1. Ma, H.; Zhang, Z.; Zhao, X.; Wu, S. A Comparative Life Cycle Assessment (LCA) of Warm Mix Asphalt (WMA) and Hot Mix Asphalt (HMA) Pavement: A Case Study in China. *Adv. Civ. Eng.* **2019**, *2019*, 9391857. [CrossRef]
2. Wang, X.D.; Zhang, L.; Zhou, X.Y.; Xiao, Q.; Guan, W.; Shan, L.Y. *Research Progress of RIOHTRACK in China, Accelerated Pavement Testing to Transport Infrastructure Innovation*; Springer: Cham, Switzerland, 2020; pp. 21–31. [CrossRef]
3. Yu, H.; Leng, Z.; Zhou, Z.; Shih, K.; Xiao, F.; Gao, Z. Optimization of preparation procedure of liquid warm mix additive modified asphalt rubber. *J. Clean. Prod.* **2017**, *141*, 336–345. [CrossRef]
4. Yu, H.; Deng, G.; Zhang, Z.; Zhu, M.; Gong, M.; Oeser, M. Workability of rubberized asphalt from a perspective of particle effect. *Transp. Res. Part D Transp. Environ.* **2021**, *91*, 102712. [CrossRef]
5. Li, J.; Shang, M.; Liu, G.; Yang, T.; Pan, Y.; Zhou, J.; Zhao, Y. Two-step improvements of volumetric design method based on multi-point supported skeleton for asphalt mixtures. *Constr. Build. Mater.* **2019**, *217*, 456–472. [CrossRef]
6. Sreedhar, S.; Coleri, E. The effect of long-term aging on fatigue cracking resistance of asphalt mixtures. *Int. J. Pavement Eng.* **2020**, *23*, 308–320. [CrossRef]
7. Sreedhar, S.; Coleri, E.; Obaid, I.A.; Kumar, V. Development of a Balanced Mix Design Method in Oregon to Improve Long-Term Pavement Performance. *Transp. Res. Rec.* **2021**, *2675*, 1121–1137. [CrossRef]
8. Yu, H.; Zhu, Z.; Leng, Z.; Wu, C.; Zhang, Z.; Wang, D.; Oeser, M. Effect of mixing sequence on asphalt mixtures containing waste tire rubber and warm mix surfactants. *J. Clean. Prod.* **2019**, *246*, 119008. [CrossRef]
9. Yu, H.; Leng, Z.; Dong, Z.; Tan, Z.; Guo, F.; Yan, J. Workability and mechanical property characterization of asphalt rubber mixtures modified with various warm mix asphalt additives. *Constr. Build. Mater.* **2018**, *175*, 392–401. [CrossRef]
10. Kamani, M.; Ajalloeian, R. Investigation of the changes in aggregate morphology during different aggregate abrasion/degradation tests using image analysis. *Constr. Build. Mater.* **2021**, *314*, 125614. [CrossRef]
11. Xie, X.; Wang, C.; Wang, D.; Fan, Q.; Oeser, M. Evaluation of Polishing Behavior of Fine Aggregates Using an Accelerated Polishing Machine with Real Tires. *J. Transp. Eng. Part B Pavements* **2019**, *145*, 04019015. [CrossRef]
12. Xu, D.; Zhang, N.; Zhuang, X. Investigation of the Coarse Aggregate Texture Property Using an Improved Laplacian Smoothing Algorithm. *J. Test. Eval.* **2019**, *49*, 1879–1896. [CrossRef]
13. Wang, D.; Chen, X.; Xie, X.; Stanjek, H.; Oeser, M.; Steinauer, B. A study of the laboratory polishing behavior of granite as road surfacing aggregate. *Constr. Build. Mater.* **2015**, *89*, 25–35. [CrossRef]
14. Gao, J.; Wang, H.; Bu, Y.; You, Z.; Zhang, X.; Irfan, M. Influence of Coarse-Aggregate Angularity on Asphalt Mixture Macroperformance: Skid Resistance, High-Temperature, and Compaction Performance. *J. Mater. Civ. Eng.* **2020**, *32*, 04020095. [CrossRef]
15. Kuang, D.; Wang, X.; Jiao, Y.; Zhang, B.; Liu, Y.; Chen, H. Influence of angularity and roughness of coarse aggregates on asphalt mixture performance. *Constr. Build. Mater.* **2019**, *200*, 681–686. [CrossRef]
16. Zhou, C.; Zhang, M.; Li, Y.; Lu, J.; Chen, J. Influence of particle shape on aggregate mixture's performance: DEM results. *Road Mater. Pavement Des.* **2017**, *20*, 399–413. [CrossRef]
17. Liu, P.; Hu, J.; Falla, G.C.; Wang, D.; Leischner, S.; Oeser, M. Primary investigation on the relationship between microstructural characteristics and the mechanical performance of asphalt mixtures with different compaction degrees. *Constr. Build. Mater.* **2019**, *223*, 784–793. [CrossRef]
18. Doyle, J.D.; Sefidmazgi, N.R.; Mejías-Santiago, M.; Rushing, J.F. Relationship of Field versus Laboratory Compaction to Hot- and Warm-Mixed Asphalt Rutting Performance. *J. Transp. Eng. Part B Pavements* **2021**, *147*, 04021014. [CrossRef]
19. Kong, D.; Chen, M.; Xie, J.; Zhao, M.; Yang, C. Geometric Characteristics of BOF Slag Coarse Aggregate and its Influence on Asphalt Concrete. *Materials* **2019**, *12*, 741. [CrossRef]
20. Behiry, A.E.A.E.-M. Optimisation of hot mix asphalt performance based on aggregate selection. *Int. J. Pavement Eng.* **2015**, *17*, 924–940. [CrossRef]
21. Harnaeni, S.R.; Pramesti, F.P.; Budiarto, A.; Setyawan, A. The effect of temperature changes on mechanistic performance of hotmix asphalt as wearing course with different gradation types. *AIP Conf. Proc.* **2018**, *1977*, 030026. [CrossRef]
22. Zhang, J.; Li, X.; Ma, W.; Pei, J. Characterizing Heterogeneity of Asphalt Mixture Based on Aggregate Particles Movements. *Iran. J. Sci. Technol. Trans. Civ. Eng.* **2018**, *43*, 81–91. [CrossRef]
23. Tran, N.T.; Takahashi, O. A comparative analysis of the fracture performance of wearing course mixtures under different geometries, compactions, and aggregate gradations. *Int. J. Pavement Eng.* **2019**, *21*, 1703–1715. [CrossRef]
24. Kakar, M.R.K.; Hamzah, M.O.; Valentin, J. A review on moisture damages of hot and warm mix asphalt and related investigations. *J. Clean. Prod.* **2015**, *99*, 39–58. [CrossRef]
25. Hao, B.; Lin, G. 3D Printing Technology and Its Application in Industrial Manufacturing. *IOP Conf. Ser. Mater. Sci. Eng.* **2020**, *782*, 022065. [CrossRef]

26. Wu, C.; Wu, L.; Shang, G.; Guo, H. Application and Research of 3D Printing Technology in the Field of Architecture. In Proceedings of the 2021 4th International Conference on Electron Device and Mechanical Engineering (ICEDME), Guangzhou, China, 19–21 March 2021; pp. 71–74.
27. Kong, Y. Application of 3D Printing Technology in Jewelry Design in the Era of Artificial Intelligence. In *Cyber Security Intelligence and Analytics*; Springer: Cham, Switzerland, 2021; pp. 162–169. [CrossRef]
28. Quan, H.; Zhang, T.; Xu, H.; Luo, S.; Nie, J.; Zhu, X. Photo-curing 3D printing technique and its challenges. *Bioact. Mater.* **2020**, *5*, 110–115. [CrossRef] [PubMed]
29. Gibson, B.T.; Bandari, Y.K.; Richardson, B.S.; Roschli, A.C.; Post, B.K.; Thornton, M.C.B.A.; Henry, W.C.; Lamsey, M.; Love, L.J. Melt pool monitoring for control and data analytics in large-scale metal additive manufacturing. In *2019 International Solid Freeform Fabrication Symposium*; University of Texas at Austin: Austin, TX, USA, 2019.
30. Fu, Z.; Liu, Q.; Niu, K.M. *Test Method of Cement and Concrete for Highway Engineering (JTG E30-2005)*; Ministry of Communications of the People's Republic of China: Beijing, China, 2005.
31. Jiang, C.; Zhao, G.-F. A Preliminary Study of 3D Printing on Rock Mechanics. *Rock Mech. Rock Eng.* **2014**, *48*, 1041–1050. [CrossRef]
32. Hua, M. Preliminary study on the application of 3D printing technology in the specimen preparation of rock mechanics. Master's Thesis, Nanjing University, Nanjing, China, May 2015. (In Chinese)
33. Fereshtenejad, S.; Song, J.-J. Fundamental Study on Applicability of Powder-Based 3D Printer for Physical Modeling in Rock Mechanics. *Rock Mech. Rock Eng.* **2016**, *49*, 2065–2074. [CrossRef]
34. Vogler, D.; Walsh, S.; Dombrovski, E.; Perras, M. A comparison of tensile failure in 3D-printed and natural sandstone. *Eng. Geol.* **2017**, *226*, 221–235. [CrossRef]
35. Zhou, T.; Zhu, J. An Experimental Investigation of Tensile Fracturing Behavior of Natural and Artificial Rocks in Static and Dynamic Brazilian Disc Tests. *Procedia Eng.* **2017**, *191*, 992–998. [CrossRef]
36. Ju, Y.; Wang, L.; Xie, H.; Ma, G.; Mao, L.; Zheng, Z.; Lu, J. Visualization of the three-dimensional structure and stress field of aggregated concrete materials through 3D printing and frozen-stress techniques. *Constr. Build. Mater.* **2017**, *143*, 121–137. [CrossRef]
37. Li, H.; Fan, W.; Zhu, X. Three-dimensional printing: The potential technology widely used in medical fields. *J. Biomed. Mater. Res. Part A* **2020**, *108*, 2217–2229. [CrossRef] [PubMed]
38. Schweiger, J.; Edelhoff, D.; Güth, J.-F. 3D Printing in Digital Prosthetic Dentistry: An Overview of Recent Developments in Additive Manufacturing. *J. Clin. Med.* **2021**, *10*, 2010. [CrossRef] [PubMed]
39. Chen, W.; Zhang, Y. Development and application of 3D printing technology in various fields. *J. Phys. Conf. Ser.* **2019**, *1303*, 012032. [CrossRef]
40. De Maeijer, P.K.; Craeye, B.; Snellings, R.; Kazemi-Kamyab, H.; Loots, M.; Janssens, K.; Nuyts, G. Effect of ultra-fine fly ash on concrete performance and durability. *Constr. Build. Mater.* **2020**, *263*, 120493. [CrossRef]
41. Bandari, Y.K.; Williams, S.; Ding, J.; Martina, F. Additive manufacture of large structures: Robotic or CNC systems? In *2014 International Solid Freeform Fabrication Symposium*; University of Texas at Austin: Austin, TX, USA, 2015.
42. Fu, C.; Guo, R.; Lin, Z.; Xia, H.; Yang, Y.; Ma, Q. Effect of nanosilica and silica fume on the mechanical properties and microstructure of lightweight engineered cementitious composites. *Constr. Build. Mater.* **2021**, *298*, 123788. [CrossRef]

materials

MDPI

Article

Characterizing and Predicting the Resilient Modulus of Recycled Aggregates from Building Demolition Waste with Breakage-Induced Gradation Variation

Yuanjie Xiao [1,2], Kunfeng Kong [1,3], Umar Faruk Aminu [1], Zhiyong Li [4], Qiang Li [5], Hongwei Zhu [3] and Degou Cai [3,*]

1 School of Civil Engineering, Central South University, Changsha 410075, China; yjxiao@csu.edu.cn (Y.X.); kongkunfeng@csu.edu.cn (K.K.); umaraminu.ck@gmail.com (U.F.A.)
2 Ministry of Education Key Laboratory of Engineering Structures of Heavy Haul Railway, Central South University, Changsha 410075, China
3 State Key Laboratory for High-Speed Railway Track Technology, China Academy of Railway Sciences Corporation Limited, Beijing 100081, China; zhuhongwei@yeah.net
4 Hunan Communications Research Institute Co. Ltd., Changsha 410015, China; zhiyongli_jky@163.com
5 College of Civil Engineering, Nanjing Forestry University, Nanjing 210037, China; liqiang2526@njfu.edu.cn
* Correspondence: cdg@rails.cn

Abstract: Building demolition waste (BDW) has been massively stockpiled due to increasingly rapid urbanization and modernization. The use of recycled BDW as unbound granular base/subbase materials is among the sustainable, cost-effective, and environmentally friendly pavement construction alternatives. The resilient modulus is an important mechanical property of BDW-derived aggregates and mechanistic design input of pavements incorporating BDW. This paper presents the results of a comprehensive laboratory study on the shear strength and resilient modulus characteristics of BDW-derived aggregate materials. A series of monotonic triaxial compression tests and repeated-load triaxial (RLT) tests were conducted with five different gradations representing particle breakage and different stress paths. The apparent cohesion and internal friction angle of recycled BDW aggregates under consolidated drained conditions ranged from 35.3 to 57.5 kPa and from 30.2° to 54.3°, respectively. The apparent cohesion and internal friction angle also increased and decreased non-linearly with the increasing relative content of fine particles, respectively. The resilient modulus of recycled BDW aggregates gradually decreased with increasing relative content of fine particles at the same stress level. Both the deviator stress and confining pressure exhibited significant influences on the resilient modulus, while the effect of confining pressure was more profound. Based on laboratory testing data, a mechanistic-empirical model was developed to predict the resilient modulus of recycled BDW aggregates from gradation and stress-state variables. The findings could be useful for extended engineering applications of BDW in unbound granular pavement base/subbase construction.

Keywords: building demolition waste; unbound granular materials; shear strength; resilient modulus; gradation; prediction model

Citation: Xiao, Y.; Kong, K.; Aminu, U.F.; Li, Z.; Li, Q.; Zhu, H.; Cai, D. Characterizing and Predicting the Resilient Modulus of Recycled Aggregates from Building Demolition Waste with Breakage-Induced Gradation Variation. *Materials* **2022**, *15*, 2670. https://doi.org/10.3390/ma15072670

Academic Editor: Roman Fediuk

Received: 14 March 2022
Accepted: 3 April 2022
Published: 5 April 2022

Publisher's Note: MDPI stays neutral with regard to jurisdictional claims in published maps and institutional affiliations.

Copyright: © 2022 by the authors. Licensee MDPI, Basel, Switzerland. This article is an open access article distributed under the terms and conditions of the Creative Commons Attribution (CC BY) license (https://creativecommons.org/licenses/by/4.0/).

1. Introduction

With the increasingly rapid urbanization and modernization around the globe, large quantities of building demolition waste (BDW) have been stockpiled, from which land use, public safety, and ecological environment concerns arise [1]. In recent years, the increasing awareness of environmental sustainability has demanded the substitution of these waste hazards for non-renewable natural resources in diverse construction industries.

On average, less than 10% of stockpiled BDW is currently recycled for reuse in China and other developing countries, which is considerably lower than the recycling and reusing rate of more than 90% in developed countries [2]. On the other hand, the construction of transportation infrastructure, such as railroad tracks and roadway and airfield pavement, demands the use of stone-based granular materials in massive quantities, thus, causing the increasing depletion of non-renewable natural aggregate resources. Therefore, the efficient recycling and sustainable reusing of non-traditional construction and demolition wastes (CDWs), including BDWs have emerged as resource-conserving, environmentally friendly substituting solutions to alleviate such problems, especially in the new era of carbon peak and carbon neutrality (i.e., zero net-carbon emission) commitments [3,4].

CDWs are usually generated during the operations of construction, maintenance, repair, rehabilitation, and demolition of buildings and other infrastructures [5]. Featuring a heterogeneous nature, these wastes consist of distinct types of materials, such as concrete, bricks, ceramics, wood, glass, plastic, soil, and gypsum-based construction materials. Depending on the different geographical regions, natural environment, economic development, and construction practices, CDWs possess distinct compositions with bulky wastes (e.g., bricks and concrete) accounting for 70–80% of the total weight.

Volodchenko et al. [6] confirmed the potential applicability of nonconventional argillous raw materials in the structurization optimization of building composites from X-ray diffraction and scanning electron microscopy analyses. Azevedo et al. [7] explored the potential use of the Curauá fibers in reinforcing wall-covering mortars, where the consistency, trapped water and air contents, compressive strength, water absorption and wet–dry durability were studied. Lu et al. [8] conducted a comprehensive study to evaluate the mechanical properties, functionalities, and environmental benefits of a new permeable material derived from recycled ceramic aggregates and bio-based polyurethane (PU) binder, and concluded that it could be used as the promising substitution for conventional Porous Asphalt (PA).

Yu et al. [9] investigated the mechanical and engineering properties of asphalt mixtures modified with waste packing tape (WPT), e.g., the mixture workability, moisture susceptibility, modulus and strength, fatigue resistance, and rutting resistance. They found that the aforementioned performance of a WPT-modified mixture was greatly improved as compared to the raw asphalt mixture. Luhar et al. [10] and Mohajerani et al. [11] reviewed and confirmed the promising large-scale use of glass wastes for producing durable, cost-effective, and sustainable geopolymer concretes. Mishra et al. [12] studied the shear strength, California Bearing Ratio (CBR), and indirect tensile strength of subgrade soils stabilized with both recycled polyethylene terephthalate (PET) fibers and fly ash, and reported that the addition of such two waste materials of proper amounts led to improved strength parameters.

Aggregates recycled from bulky wastes (e.g., bricks and concrete) of CDWs have been used mainly in concrete production as constituents, including mortar, concrete, and bricks, as well as in pavement construction as marginal stone-based materials [13]. In the past few decades, extensive studies have been conducted to explore hydro-mechanical performances and assess the engineering suitability of stone-based granular materials derived from BDWs, such as recycled concrete aggregate (RCA), recycled clay masonry brick (RCM), and reclaimed asphalt pavement (RAP), in pavement construction.

The results of triaxial tests conducted by Zhang et al. [14] showed that the recycled construction and demolition waste (CDW) exhibited much higher resilient moduli and lower accumulated permanent deformation as compared to embankment clay soils. This demonstrated that the substitution of recycled wastes for clay soils would improve the structural capacity and reduce rutting damage. Silva et al. [15] presented a review of the most important physical properties of different types of recycled aggregates (RAs) and their comparisons with natural aggregates (NAs) as well as of how these properties affect their hydro-mechanical behavior when compacted.

The results collected from the literature indicate that the engineering performance of most RAs is comparable to that of NAs and can be used in unbound pavement layers or in other applications requiring compaction. Vieira et al. [16] presented a laboratory study performed on two coarse-grained recycled aggregates and discussed their suitability for constructing embankments and capping layers of transport infrastructures, including rural and forest roads. The results showed that the mixed CDW aggregates possessed suitable characteristics for potential use in the construction of embankments, as well as for serving as alternative aggregates for the subbase layers of rural and forest roads.

In addition, Vieira [17] assessed, using direct shear and pullout tests, the possibility of using fine-grained materials recycled from CDW as backfills of geosynthetic-reinforced structures to replace the natural soils otherwise typically used in the construction of these structures. Those studies reported that, even though recycled BDW aggregates contain different combinations of complex constituents, such as asphalt concrete, cement concrete, brick, and glass, their strength and deformation properties are relatively comparable to those of traditional virgin aggregates [18,19].

Saberian et al. [20] conducted a series of large-scale direct shear tests (LDST) to evaluate the shear strength properties of RCA mixed with crumb rubber (derived from used tires) of different sizes and amounts and found that such mixtures of RCA and crumb rubber exhibited an increase in the apparent cohesion due to the addition of fine or coarse crumb rubber and satisfied the shear strength requirements for potential applications in a pavement base/subbase. Arulrajah et al. [21] conducted direct shear tests and consolidated drained triaxial tests to characterize the shear strength properties of recycled BDW materials. They further classified their shear responses into two different categories, i.e., the dilatancy-induced peak strength and dilatancy-associated strain-hardening behaviors.

According to their testing results, RCA and RCM clearly exhibited dilatancy-induced peak strength preceded by the occurrence of the maximum dilatancy ratio, whereas higher dilatancy ratios were associated with greater peak friction angles; on the other hand, RAP showed dilatancy-associated strain-hardening behavior even with a relatively high magnitude of dilatancy. From those studies focusing on the shear strength characteristics of compacted BDW materials, it can be inferred that they have the potential to be used in pavement base/subbase applications as they meet the requirements of the minimum effective friction angles required.

As an important mechanical performance indicator, the resilient modulus (*Mr*) characterizes the ability of compacted unbound granular pavement materials to resist elastic deformation induced by repeated traffic loading. The resilient modulus has a direct impact on pavement performance, such as fatigue cracking and rutting resistance, and thus has become a key input parameter in the current mechanistic pavement analysis and design [22–24]. The resilient behavior of traditional aggregates under different working conditions and material compositions has been a well-studied topic. Diagne et al. [25] studied the performance of RCM and RCA blended in different proportions as a fill material of a granule base, and found that the resilient modulus of this fill material decreased with the increasing moisture content and RCM content.

Silva et al. [6] conducted a literature survey to compare the physical and mechanical properties of recycled CDW aggregates and natural aggregates (NA) and indicated that the material composition and the degree of aggregate particle breakage affected the resilient modulus. Arisha et al. [26] studied the resilient modulus of recycled CDW fill materials with different contents of recycled bricks by following the AASHTO-T307 testing protocol. They found that the resilient modulus exhibited a downward trend with increasing content of recycled bricks, and they proposed a resilient modulus prediction model by incorporating the content of recycled bricks.

Soleimanbeigi et al. [27] evaluated the resilient modulus of RAP samples. According to their testing results, the resilient modulus of RAP samples was greater than that of RCA or NA samples at 20–30 °C; however, when the temperature was raised to 50 °C, the resilient modulus of RAP samples decreased significantly, whereas that of RCA or NA samples remained basically unchanged. Recycled CDW aggregates are prone to particle breakage and degradation, especially under the application of heavy traffic loading, harsh environmental loading (e.g., wet–dry/freeze–thaw cycles), or both. Potentially marked gradation variations could then occur throughout their service life periods and detrimentally affect their hydro-mechanical behavior and in-service performance. However, few previous studies have addressed such gradation variations of recycled CDW aggregates and their possible consequences, which may hinder their reliable and widespread use. It is then of great significance and necessity to perform an in-depth study in this regard.

In light of the above-mentioned research need, this study aims to investigate the feasibility and applicability of recycled BDW materials as unbound granular fill materials for a pavement base/subbase. Five different gradations were designed to represent particle breakage-induced gradation variations. A series of monotonic triaxial compression tests at consolidated drained conditions and repeated-load triaxial (RLT) tests were conducted to investigate the shear strength and resilient modulus characteristics of recycled BDW aggregates with five different gradations. Based on the laboratory testing data obtained, a mechanistic-empirical prediction model of the resilient modulus was then proposed by incorporating gradation variations and stress states. We expect that the findings of this study will be able to provide technical reference and guidance for promoting the use of recycled BDW as unbound granular materials in pavement base/subbase applications.

2. Materials and Methods

2.1. Materials Tested

The materials tested in this study were recycled BDW aggregates that were generated from different building demolition sites in Changsha, China and then transported to a processing center for crushing, screening, and stockpiling. As illustrated in Figure 1, the recycled BDW aggregates mainly consisted of used aggregate, cement mortar, and red brick particles, of which the combined percentage by mass was more than 95%. The recycled BDW aggregate particles were divided into four size groups: 0–5 mm, 5–10 mm, 10–20 mm, and 20–40 mm. The individual percentage by mass of each of the three different aggregate types (i.e., used aggregate, cement mortar, and red brick) within each of the four different size ranges was obtained by random sampling and manual sorting as listed in Table 1.

Figure 1. Aggregate particles recycled from BDW.

Table 1. Percentages by weight of three different types of aggregate particles recycled from BDW (%).

Size	Gravel	Mortar	Brick	Others (Tiles, Wood, etc.)
5–10 mm	62.87	27.82	7.58	1.73
10–20 mm	68.49	25.75	3.88	1.88
20–40 mm	69.87	23.56	4.34	2.23

The morphological characteristics of coarse particles significantly affect the formation of the load-transferring aggregate skeleton structure and thus key hydro-mechanical performance (e.g., the resilient modulus, shear strength, permanent deformation, and permeability) of unbound aggregates recycled from CDW [28]; therefore, it is necessary to characterize and quantify the morphology of individual coarse particles, i.e., form, angularity, and surface texture. A total of 540 particles were randomly selected to represent the three different types of recycled BDW aggregates (i.e., used aggregate, cement mortar, and red brick) within four different size ranges. Those selected particles were then processed by a three-dimensional (3D) laser scanner for shape acquisition, and their 3D models were reconstructed accordingly for calculating the shape indices, including elongation, flatness, and sphericity [29] as shown in Figure 2.

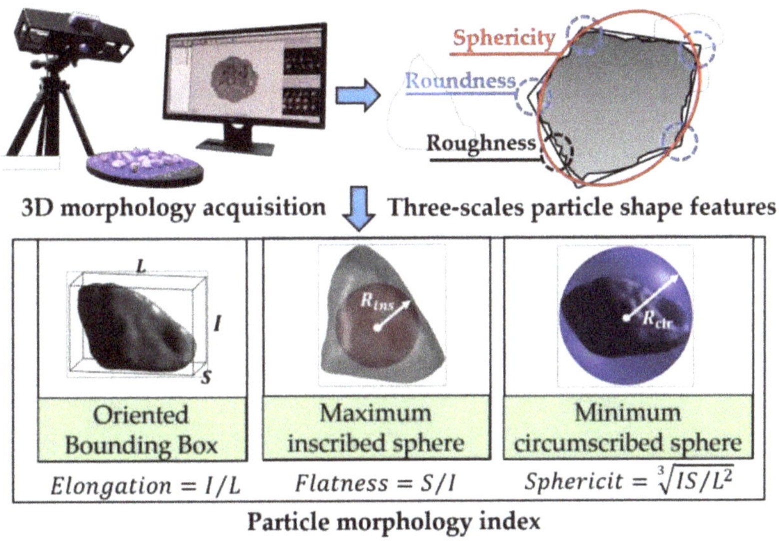

Figure 2. Illustration of the shape quantification of coarse particles recycled from BDW.

Zingg's bivariate method [29] was used to classify the shape of different types of coarse BDW particles, and the classification results are shown in Figure 3. As can be seen from Figure 3, most of the coarse BDW particles resemble blocks or disks, and most of the sphericity values of used aggregate and brick particles range from 0.5 to 1; on the other hand, the shape distribution of cement mortar particles is relatively more concentrated, i.e., there are almost no bar-shaped particles, and their sphericity values range almost from 0.67 to 1.

In addition, there appeared to be no significant difference in the particle morphology among different particle size groups of the same type of recycled BDW aggregates. Therefore, the subsequent laboratory triaxial tests in this study were conducted by controlling the particle shape difference so that the calculated particle shape indices of different specimens remained relatively similar or close to each other; in other words, particle shape was excluded from the list of explanatory variables considered in the testing matrix of this study and will be the subject of future studies.

The basic index properties of unbound aggregate materials derived from recycled BDW aggregates, such as the Atterberg limits, crushing value, water absorption, and specific gravity, were measured by following related testing standards or specifications [30,31], as listed in Table 2. It can be seen from Table 2 that the recycled BDW aggregates possessed a relatively high crushing value and water absorption due to the presence of aged cement mortar fully or partially covering aggregate particles and small cracks on the surface of recycled aggregates. Therefore, recycled BDW aggregates are likely to be more susceptible to particle crushing (or breakage) and moisture-related degradation when compared to virgin aggregates, which may eventually cause differences in the mechanical performance and long-term durability between these two different types of materials.

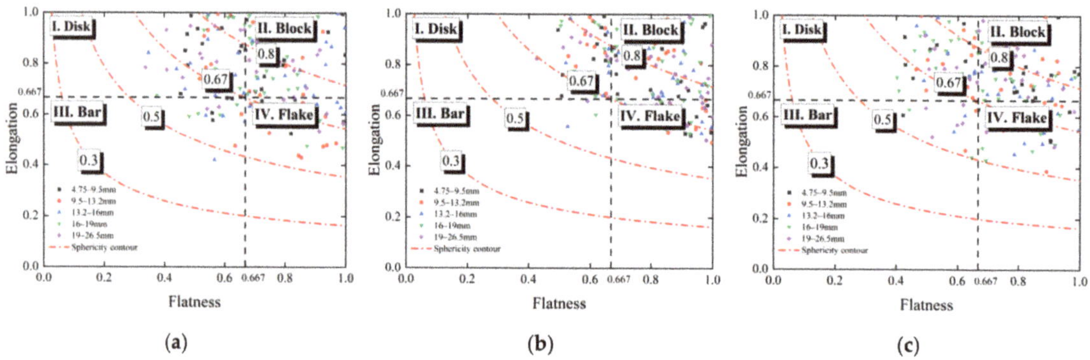

Figure 3. The shape classification results of three different types of aggregate particles recycled from BDW. (**a**) Used aggregate particles. (**b**) Cement mortar particles. (**c**) Brick particles.

Table 2. The basic physical properties of aggregates recycled from BDW.

Liquid Limit ω_L/%	Plastic Limit ω_P/%	Plasticity Index I_P/%	Crushing Value/%	Water Absorption/%	Specific Gravity
23.7~29.6	13.5~17.8	10.2~11.8	18.5~24.5	6.9~9.2	2.53~2.69

2.2. Gradations Designed

The gradation itself has a crucial influence on the hydro-mechanical properties of unbound granular materials, including the resilient modulus, shear strength, permanent deformation, permeability, and durability. In this study, the gravel-to-sand ratio (*G/S*) parameter, which was proposed based on the theory of particle packing, was adopted for the gradation design and control of laboratory triaxial specimens. The *G/S* parameter and its applicability to traditional and reclaimed unbound granular materials have been verified and supported in the literature [32–34]. To be specific, the *G/S* parameter can be calculated from the maximum particle size D_{max} and the shape parameter n of the gradation curve according to Equation (1), wherein the shape parameter n of the gradation curve can be regressed from the mechanical sieve analysis result according to Equation (2).

$$G/S = \frac{1 - p_{4.75}}{p_{4.75} - p_{0.075}} = \frac{{D_{max}}^n - 4.75^n}{4.75^n - 0.075^n} \quad (1)$$

$$P_d = \left(\frac{d_i}{D_{max}}\right)^n \quad (2)$$

where P_d is the percentage of materials passing the sieve with an opening size of d_i; d_i is the sieve opening size (mm); D_{max} is the maximum particle size (mm); and n is the parameter controlling the shape of gradation curves.

The recycled BDW aggregates used in the laboratory testing program can be classified as well-graded gravel (GW). They were oven-dried and sieved apart into different particle size groups. The maximum particle size was set as 26.5 mm due to the specimen size limit of the triaxial testing apparatus, and particles larger than 26.5 mm were picked out and replaced by smaller particles of equal weights. The specimens of the recycled BDW aggregates were prepared according to the five different gradations shown in Figure 4b of which the *G/S* values were 1.0, 1.6, 1.8, 2.0, and 2.5, respectively.

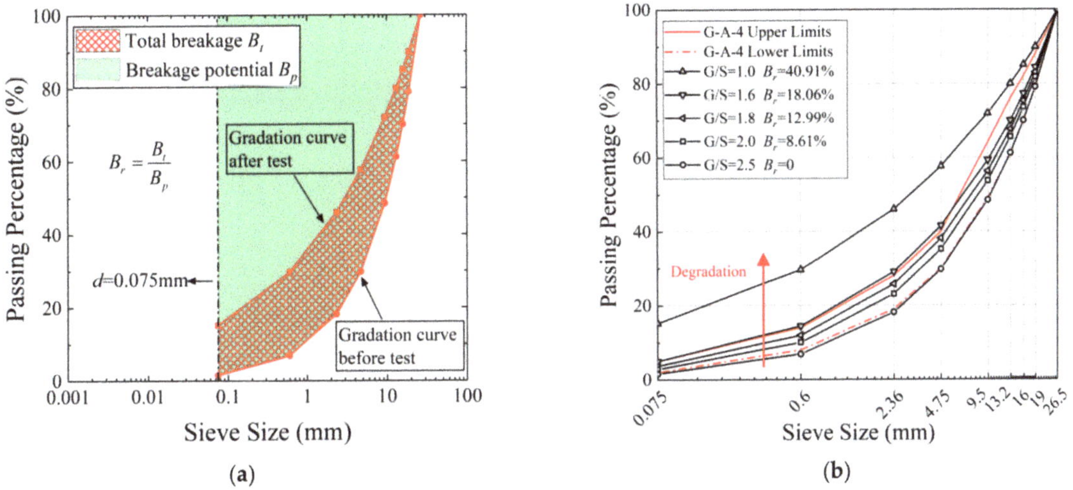

Figure 4. Gradation curves of recycled BDW aggregates representing particle-breakage-induced gradation variations. (**a**) Definition of particle breakage index B_r. (**b**) Gradation curves and their corresponding B_r.

Note that the upper-limit and lower-limit gradation curves denote the G-A-4 gradation band for high-grade highway pavement base/subbase materials as recommended in China's Technical Guidelines for Construction of Highway Roadbases (JTG/TF20-2015). Quantitatively defining the degree of particle breakage is necessary. The effective way is to use the particle size distribution curves measured before and after tests to estimate the amount of particle crushing [35]. A relative breakage index B_r proposed by Hardin [35] and shown in Figure 4a was adopted in the current work, which can be expressed as:

$$B_r = \frac{B_t}{B_p} \tag{3}$$

where B_p is called the crushing potential, i.e., the area between the initial grading (gradation curve before test) and boundary of 0.075 mm; and the parameter B_t is called the total crushing, i.e., the area between the initial grading and current grading (gradation curve after test). According to this definition, the gradation of G/S = 2.5 is considered as the initial state, and then the particle breakage index corresponding to the other four gradations (G/S = 2.0, 1.8, 1.6, and 1.0) can be calculated separately.

These five different gradations were chosen to represent the different stages of particle breakage of recycled BDW aggregates as indicated by the particle breakage index values B_r in Figure 4b. In other words, as the value of the parameter G/S decreases, the relative content of fine particles gradually increases, which indicates a process of degradation. This phenomenon is expressed in the particle size distribution curve as a shift of the gradation curve to the upper left. The correspondence between the gradation parameter G/S and the breakage index values B_r is given in Figure 5 and Equation (4).

$$G/S = 2.46e^{-0.023x} \tag{4}$$

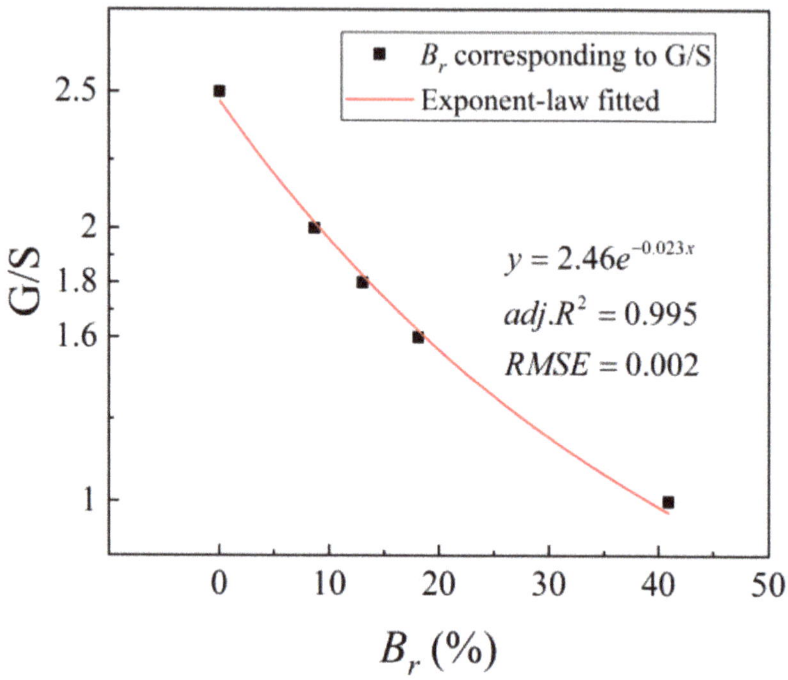

Figure 5. Fitting results of the particle breakage index values B_r against the G/S parameter.

2.3. Description of Laboratory Testing Program

2.3.1. Compaction Tests

Laboratory compaction tests were conducted according to the following Chinese standards (JTG 3430-2020) [30] with the related compaction parameters listed in Table 3, from which the compaction curves of the gravitational moisture content versus achieved dry density were obtained to determine the optimal moisture content (OMC) and maximum dry density (MDD) for each of the five different gradations. Specifically, a cylindrical mold with an inner diameter of 152 mm and an effective height of 120 mm was used for the laboratory compaction tests. The recycled BDW aggregates prepared for compaction tests according to the designated gradations and moisture contents were placed into the mold in three sub-layers, and each sub-layer was then compacted by 98 blows of a 4.5-kg hammer falling freely from a height of 450 mm. This experimental method of compaction is equivalent to the Modified Proctor's compaction (AASHTO T180), and the hitting-work per unit is 2691 KJ/cm^3.

Table 3. The laboratory compaction test parameters as specified in the Chinese standard that was followed.

Test Method	Specimen Height (cm)	Specimen Volume (cm^3)	Sub-Layers	Blows Per Sub-Layer	Maximum Particle Size (mm)
Heavy II-2	12	2177	3	98	40

2.3.2. Monotonic Triaxial Compression Tests

The monotonic triaxial compression tests under consolidated drained (CD) conditions (herein abbreviated as CD triaxial tests) were performed in the laboratory to determine the shear strength properties of recycled BDW aggregates. The use of consolidated drained tests was meant to simulate long-term field conditions. The CD triaxial tests were performed using an automated apparatus shown in Figure 6 and following the testing protocol T0146 specified in the Chinese standard (JTG 3430-2020).

Figure 6. The laboratory triaxial apparatus for monotonic and repeated-loading tests. 1. Vertical loading system, including a hydraulic servo actuator and transducer for axial deformation; 2. triaxial cell (specimens); 3. transducer for axial force; 4. confining pressure servo; 5. oil pressure system, including a high-pressure syringe pump; 6. controller; 7. computer interface system; and 8. specimen.

The specimens were fabricated in a split mold with a 100-mm inner diameter and 200-mm height. According to the optimal moisture content and maximum dry density determined from the compaction tests, the specimens were first prepared at the optimal moisture content, placed into the split mold in five sub-layers, and then compacted to achieve the relative compaction (the ratio of achieved dry density to maximum dry density) of 95%. The specimens were covered by special latex membranes of 0.5 mm thickness to refrain from collapse during the specimen preparation and loading stages.

The isotropic consolidation stage was conducted with three different levels of effective confining pressure, i.e., 50, 100, and 200 kPa. The specimens were then consolidated under drained conditions. The subsequent monotonic compression stage (or shearing stage) was performed at the strain control mode with the pre-selected loading rate of 1 mm per minute. The criterion adopted in the study for terminating the tests was the occurrence of either peak axial (or deviator) stress or the axial strain of 15%, whichever took place first. For the latter case, the axial stress corresponding to the axial strain of 10% was treated as the peak.

2.3.3. Repeated Load Triaxial Tests

The specimen preparation procedure of the repeated-load triaxial (RLT) tests for characterizing the resilient modulus behavior was the same as that for the aforementioned monotonic triaxial compression tests of five different gradations of specimens at optimum moisture and 95% maximum dry density. In this study, the AASHTO-T307 testing protocol was adopted for the RLT tests, i.e., the entire loading process consisted of the initial pre-loading conditioning stage and the subsequent applications of a total of 15 different sequences of repeated loading [36].

The different combinations of confining pressure and deviator stress specified for the conditioning stage and the repeated-loading sequences are summarized in Table 4. The time–history curve of axial stress (σ_1) applied is shown in Figure 7. The loading waveform is a half-sine wave with a frequency of 1 Hz. The triaxial specimens were conditioned for 1000 load applications with the combination of 103.4 kPa confining pressure and 103.4 kPa deviator stress in order to eliminate contact deformation between the loading cap (or base) and the specimens, followed by the 15 different sequences of repeated loading.

Table 4. Loading sequences of the resilient modulus tests of recycled BDW aggregates.

SequenceNo.	Confining Pressure, σ_3/kPa	Contact Stress, σ_c/kPa	Deviator Stress, σ_d/kPa	Axial Stress, σ_{max}/kPa	No. of Load Applications
0	103.4	10.3	93.1	103.4	1000
1	20.7	2.1	18.6	20.7	100
2	20.7	4.1	37.3	41.4	100
3	20.7	6.2	55.9	62.1	100
4	34.5	3.5	31	34.5	100
5	34.5	6.9	62	68.9	100
6	34.5	10.3	93.1	103.4	100
7	68.9	6.9	62	68.9	100
8	68.9	13.8	124.1	137.9	100
9	68.9	20.7	186.1	206.8	100
10	103.4	6.9	62	68.9	100
11	103.4	10.3	93.1	103.4	100
12	103.4	20.7	186.1	206.8	100
13	137.9	10.3	93.1	103.4	100
14	137.9	13.8	124.1	137.9	100
15	137.9	27.6	248.2	275.8	100

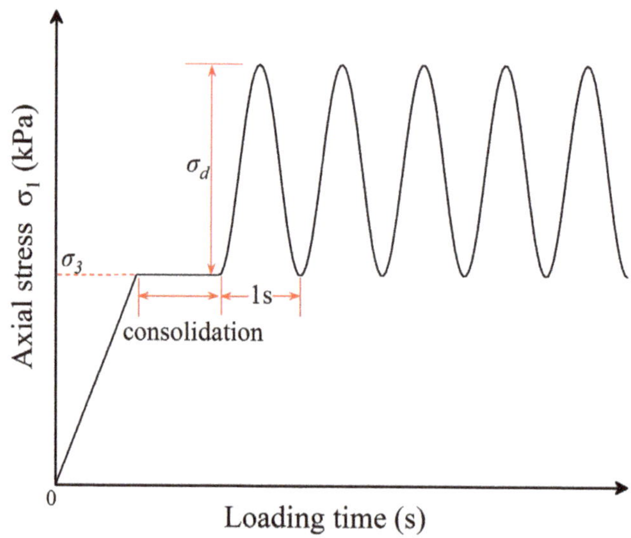

Figure 7. The time–history curve of axial stress applied during the laboratory repeated-load triaxial tests.

Each of the 15 sequences included 100 repeated load applications. If the accumulated plastic axial strain of the specimens exceeded 5% during loading, the tests were then terminated. The resilient modulus corresponding to each of the last five load applications of each repeated-loading sequence was calculated via Equation (5) as illustrated in Figure 8. The average value of the resilient moduli of the last five load applications was taken as the representative value of the resilient moduli of each sequence of repeated loading (or equivalently each combination of stress states).

$$M_r = \sigma_d / \varepsilon_r \tag{5}$$

where M_r is the resilient modulus (MPa); σ_d is the deviatoric stress (MPa); and ε_r is the resilient (or recoverable) axial strain.

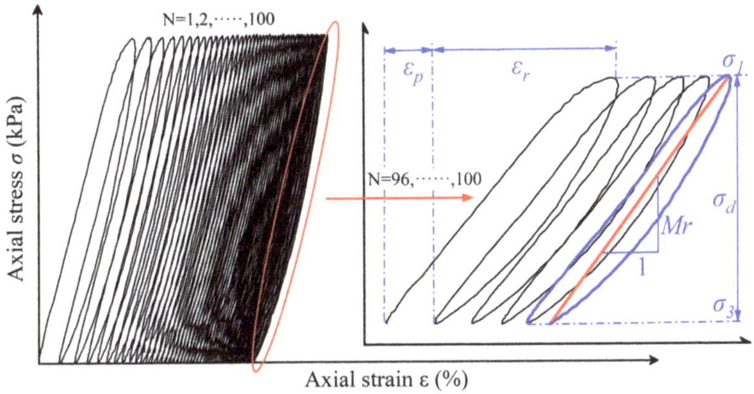

Figure 8. Schematic diagram of the resilient modulus defined from the axial stress–axial strain loops.

3. Testing Results and Analysis

3.1. Laboratory Compaction Results

The laboratory compaction test results (i.e., the OMC and MDD) are shown in Table 5. Those OMC and MDD values corresponding to five different gradations were used subsequently for preparing specimens of monotonic triaxial compression tests for the shear strength properties and repeated-load triaxial tests for the resilient modulus.

Table 5. The laboratory compaction and shear strength properties of recycled BDW aggregates with five different gradations.

Gradation Parameter G/S	Maximum Dry Density (g·cm^{-3})	Optimum Moisture Content (%)	Apparent Cohesion c′ (kPa)	Internal Friction Angle φ′ (°)
2.5	1.929	8.98	35.3	54.3
2.0	1.977	9.02	43.9	49.8
1.8	1.981	9.19	46.8	48.2
1.6	1.978	9.03	51.7	46.4
1.0	1.968	9.21	57.7	30.2

3.2. Shear Strength Properties

The deviator stress versus axial strain curves of the specimens with five gradations were obtained at three different levels of confining pressure, i.e., 50, 100, and 200 kPa, as plotted in Figure 9. It can be seen that, as the G/S value gradually decreases, i.e., the degree of degradation continues to increase, the peak deviatoric stress at the same confining pressure also gradually decreases. From each of the curves in Figure 9, the peak deviatoric stress at failure under each confining pressure was determined.

The Mohr–Coulomb envelopes fitted from the results of monotonic triaxial compression tests are shown in Figure 10. Accordingly, the shear-strength parameters, i.e., the internal friction angle and apparent cohesion, were then calculated for each of the five different gradations as listed in Table 5. It can be seen that, as the relative content of fine particles in the specimens increases, i.e., the gradation parameter *G/S* value decreases, the apparent cohesion gradually increases, and the internal friction angle gradually decreases. This indicates that the degradation or particle breakage weakens the frictional interaction among the particles inside the specimen, resulting in a reduction in overall strength.

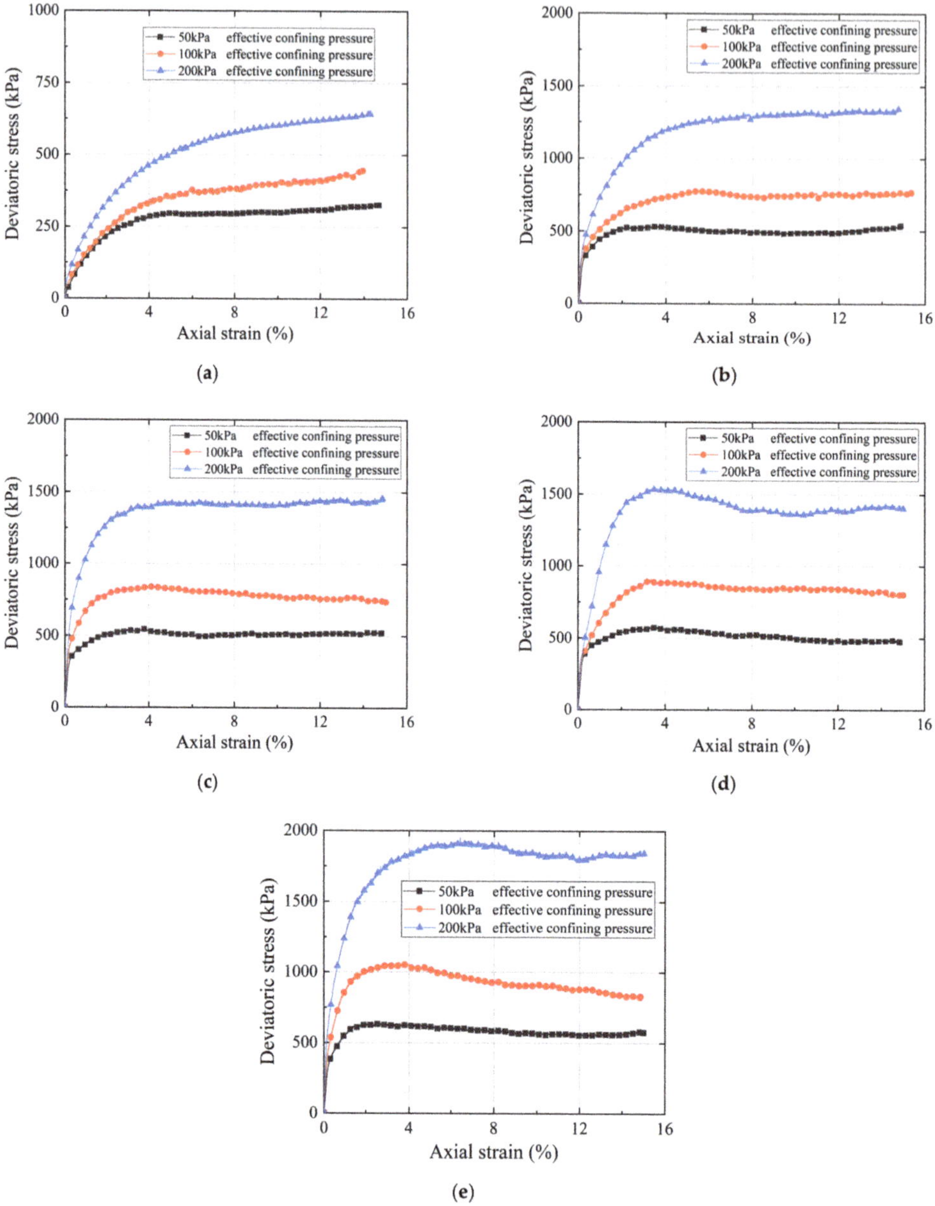

Figure 9. Deviator stress versus axial strain curves determined from the consolidated drained monotonic triaxial compression tests for the recycled BDW aggregates with different gradations: (**a**) G/S = 1.0; (**b**) G/S = 1.6; (**c**) G/S = 1.8; (**d**) G/S = 2.0; and (**e**) G/S = 2.5.

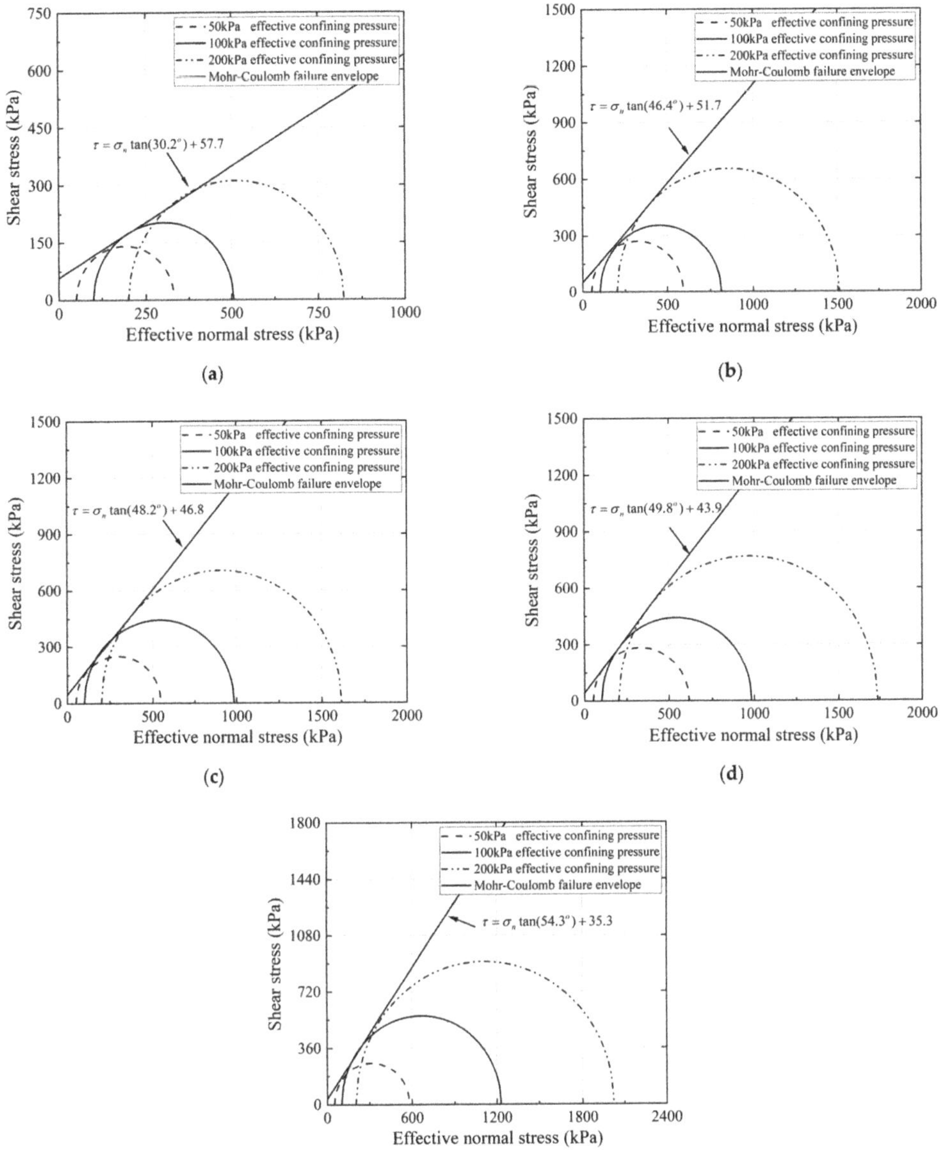

Figure 10. Mohr–Coulomb failure envelopes determined from the consolidated drained monotonic triaxial compression tests for recycled BDW aggregates with different gradations: (**a**) G/S = 1.0; (**b**) G/S = 1.6; (**c**) G/S = 1.8; (**d**) G/S = 2.0; and (**e**) G/S = 2.5.

3.3. Resilient Modulus Properties

3.3.1. Hysteresis Curves of Deviator Stress versus Axial Strain

Figure 11 plots the hysteresis curves of the deviator stress versus axial strain for recycled BDW aggregate specimens with different gradations during the applications of 15 different loading sequences. Each of the 15 groups of dashed lines represents the complete number of 100 repeated-load applications of each loading sequence, while each of the 15 groups of solid lines represents the last five repeated-load applications (i.e., the 96th to 100th load applications) within each loading sequence.

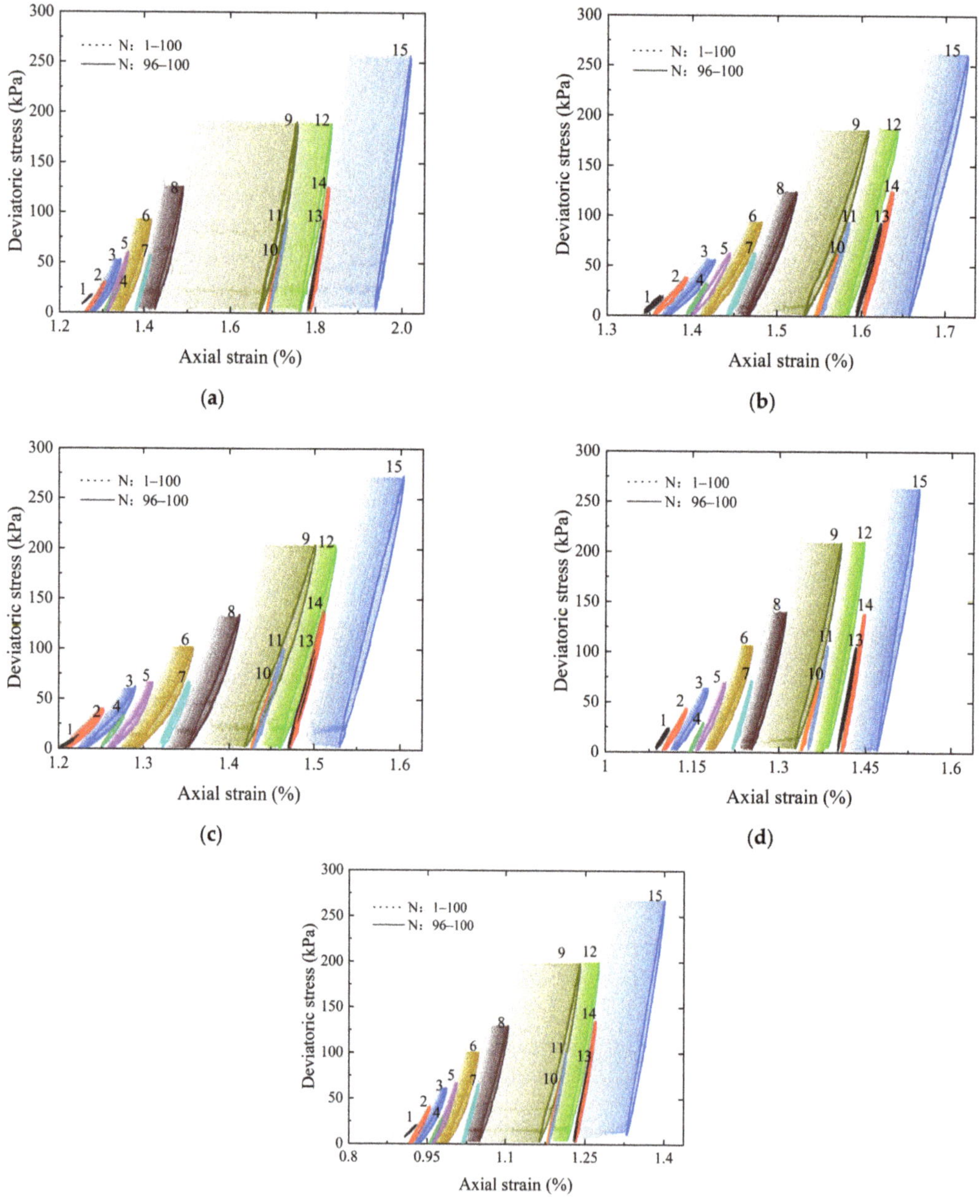

Figure 11. Hysteresis curves of the deviator stress versus axial strain of recycled BDW aggregates with different gradations: (**a**) G/S = 1.0; (**b**) G/S = 1.6; (**c**) G/S = 1.8; (**d**) G/S = 2.0; and (**e**) G/S = 2.5.

It can be observed from Figure 11 that, during each loading sequence of confining pressure and deviator stress, the hysteresis loops of deviator stress versus axial strain tend to overlap as the number of repeated-load applications gradually increases. This indicates that the stability of internal structures of the specimens increases gradually. Further, the axial deformation tends to gradually become completely elastic, i.e., the plastic axial strain gradually becomes stable and ceases to further accumulate.

The final accumulated plastic axial strain values of all the specimens met the specification requirements (i.e., less than 5%). In addition, as the confining pressure increased, the slope of each hysteresis loop of deviator stress versus axial strain tended to "bend" towards the left side; in contrast, the deviator stress level had no obvious effect on the slopes of the hysteresis loops. It can be further seen from Figure 11 that the final accumulated plastic axial strain of the specimens increased with decreasing G/S value (or degradation developed) and that the slope of the hysteresis loop decreased gradually.

3.3.2. Effects of the Stress State on the Resilient Modulus

Figure 12a shows the hysteresis curves of the deviator stress versus axial strain of recycled BDW aggregates with a G/S value of 1.8 under different combinations of confining pressure and deviator stress (or different stress states). It can be seen that the hysteresis curves corresponding to the same level of deviator stress (e.g., 68.9 kPa) clearly exhibit an increasingly downward trend with decreasing confining pressure, thus, indicating that the resilient modulus of the specimen decreases gradually. Figure 12b shows the curves of the resilient modulus (*Mr*) versus deviator stress under different levels of confining pressure for the recycled BDW aggregate specimens with G/S value of 1.8.

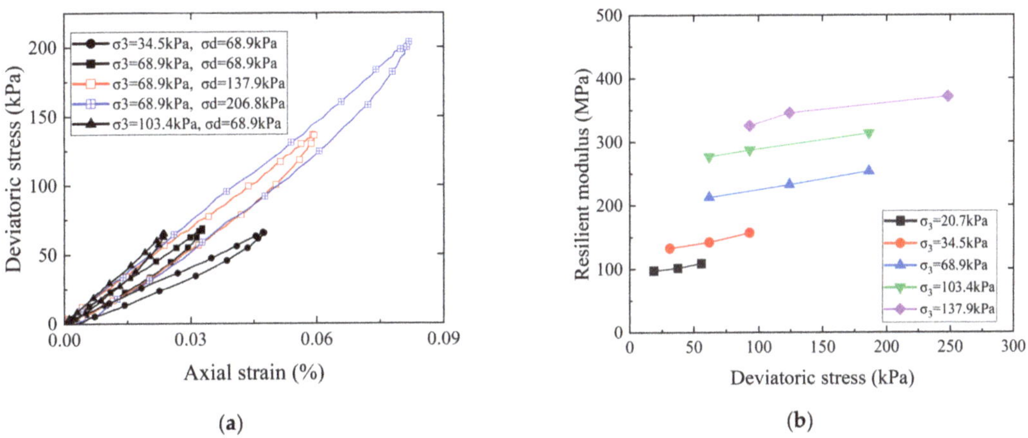

Figure 12. Effects of different stress states on the resilient modulus of recycled BDW aggregates: (**a**) Hysteresis curves of the deviator stress versus axial strain. (**b**) The resilient modulus versus deviator stress.

It can be seen from Figure 12b that the *Mr* increases by 11.6–19.4% with increasing deviator stress under the same level of confining pressure; on the other hand, the *Mr* increases by up to 95.1% with increasing confining pressure under the same level of deviatoric stress (e.g., 68.9 kPa). Therefore, both the deviator stress and confining pressure exert great influences on the *Mr* of recycled BDW aggregates, and the influence of the confining pressure is relatively greater than that of deviator stress. The reason is likely that the friction strength of coarse particles plays a leading role compared to the cohesion strength, and that the strengthening effect of confining pressure on the resilient modulus and stiffness of coarse-grained soils is much greater than the strain-hardening effect of shear displacement caused by increasing deviator stress on the internal structure of coarse-grained soils.

3.3.3. Effect of Gradation on the Resilient Modulus

Figure 13a shows the hysteresis curves of the deviator stress versus axial strain of recycled BDW aggregates with different G/S values under the conditions of optimal moisture content and 95% maximum dry density (i.e., a relative compaction of 95%), where both the confining pressure and deviator stress are 103.4 kPa. It can be seen from Figure 13a that, under the same stress state, the *Mr* of recycled BDW aggregates decreased with the increasing relative content of fine particles (i.e., decreasing G/S value) as indicated by the gradually decreasing slope of hysteresis loops. Figure 13b shows the curves of *Mr* versus gradation parameter G/S under 15 different stress states (i.e., the 15 loading sequences).

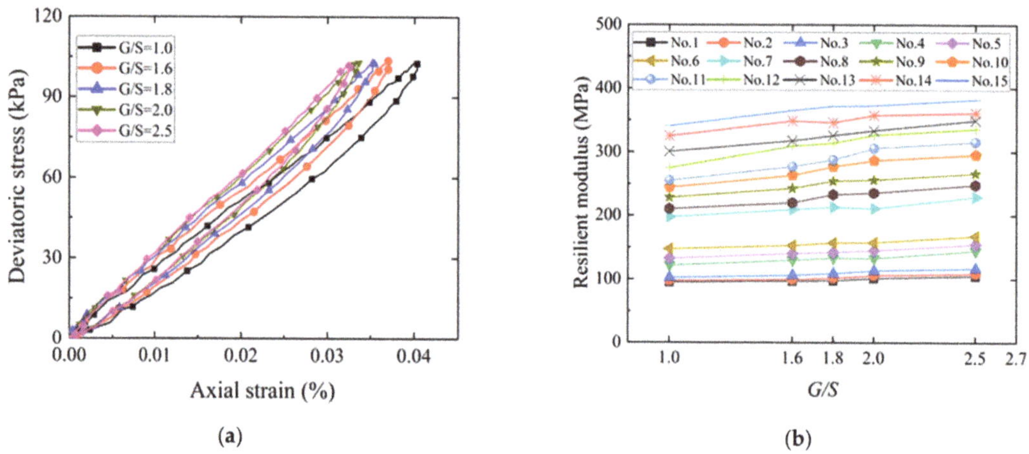

Figure 13. Effect of different gradations (G/S values) on the resilience modulus of unbound granular materials recycled from CDW: (**a**) Hysteresis curves of the deviator stress versus axial strain. (**b**) The resilient modulus versus G/S.

It can be seen that, at lower stress levels (e.g., the first to sixth loading sequences), the effect of gradation on *Mr* is not significant; however, at higher stress levels (e.g., the 7th to 15th loading sequences), the *Mr* increases by 11–23.5% with increasing G/S value (or relative content of coarse particles). It can be inferred from the above results that the relative content of coarse particles in recycled BDW aggregates may need to be increased appropriately when such materials are to be applied in pavement structures sustaining heavier axle loads; in other words, their gradations may need to be optimized. For pavement structures sustaining lower axle loads where recycled BDW aggregates are to be applied, the optimization of other material physical properties rather than gradation may be suitable.

4. Development of the Resilient Modulus Prediction Model

4.1. Evaluation of Existing Prediction Models of the Resilient Modulus

As a key input parameter of mechanistic-empirical pavement design and analysis, the resilient modulus is affected by the stress states and physical properties of materials, such as the moisture and density conditions, gradation, and particle shape. The resilient modulus of traditional unbound granular materials has been commonly predicted from purely empirical or mechanistic-empirical models [37]. A selection of the most commonly used ones is listed in Table 6, where models No. 1 to 3 are univariate models of stress states, and models No. 4 to 6 are bivariate models of stress states.

Table 6. Representative prediction models of the resilient modulus surveyed from the existing literature.

Model No.	Model Equations	Authors
No. 1	$M_R = k_1 {\sigma_3}^{k_2}$	Monismith [38]
No. 2	$M_R = k_1 {\sigma_d}^{k_2}$	Moossazadeh and Witczak [39]
No. 3	$M_R = k_1 \theta^{k_2}$	Seed [40]
No. 4	$M_R = k_1 {\sigma_3}^{k_2} {\sigma_d}^{k_3}$	Pezo [41]
No. 5	$M_R = k_1 \theta^{k_2} {\sigma_d}^{k_3}$	Uzan [42]
No. 6	$M_R = k_1 p_a \left(\frac{\theta}{p_a}\right)^{k_2} \left(\frac{\tau_{oct}}{p_a} + 1\right)^{k_3}$	AASHTO 2004 [36]

p_a is the standard atmospheric pressure and usually taken as 100 kPa; θ is the bulk stress, i.e., $\theta = \sigma_1 + \sigma_2 + \sigma_3$; τ_{oct} is the octahedral shear stress, i.e., $\tau_{oct} = \sqrt{(\sigma_1 - \sigma_2)^2 + (\sigma_1 - \sigma_3)^2 + (\sigma_2 - \sigma_3)^2}/3$; and k_i are the regression coefficients of different models.

In essence, such models were established from nonlinear regression analysis of laboratory testing results and are often not suitable for generalization to other conditions (e.g., material types, stress states, or physical properties) different from the original laboratory tests, i.e., the extrapolated applications of those statistical models could be erroneous and risky. The effect of particle breakage-induced gradation variations on the resilient modulus of recycled unbound aggregates has barely been considered in laboratory testing programs or prediction models. To address this drawback, this section presents the development of a resilient modulus prediction model from the previously mentioned laboratory RLT testing results to take into account gradation variations induced by particle breakage and degradation.

To accomplish this, the six representative resilient modulus prediction models listed in Table 6 were fitted against laboratory RLT data obtained for different gradations and stress levels and described previously in Section 3.3, respectively. The goodness-of-fit indices of such different prediction models, e.g., the adjusted coefficient of determination (adj. R^2) and root mean square error (RMSE), were calculated and comparatively analyzed as shown in Figure 14 for the gradation with G/S = 1.8.

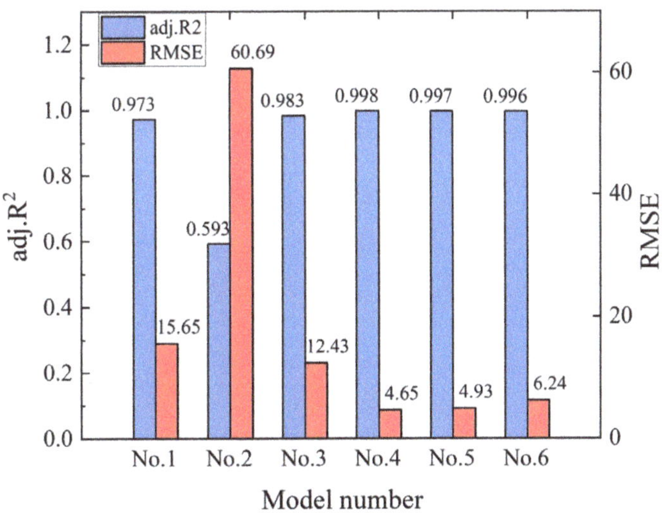

Figure 14. Goodness-of-fit index values of different resilient modulus prediction models for the gradation with G/S = 1.8.

It can be seen that the adj. R^2 value of the univariate prediction model No. 1 considering only the confining pressure was 0.973, which is much larger than that of 0.593 of the univariate prediction model No. 2 considering only deviator stress. This indicates that, compared to the deviatoric stress, the confining pressure is much more statistically significantly correlated with the resilient modulus of recycled BDW aggregates, which is consistent with the aforementioned laboratory testing results. The adj. R^2 value of the univariate prediction model No. 3 considering only bulk stress was insignificantly improved when compared to model No. 1.

The adj. R^2 values of bivariate prediction models No. 4, No. 5, and No. 6, considering stress states, were all greater than 0.973 and showed insignificant differences from each other, which may be explained by the fact that all three prediction models include the effect of confining pressure. By considering the two variables of confining pressure and deviatoric stress separately, the prediction model No. 4 yielded an increased adj. R^2 value of 0.998. Note that the octahedral shear stress is equivalent to deviatoric stress in conventional triaxial compression conditions where the minor principal stress σ_3 and intermediate principal stress σ_2 are equal; therefore, the prediction models No. 5 and No. 6 are essentially equivalent.

Despite the effect of deviatoric stress being repeatedly considered in model No. 6 by the variables of bulk stress and octahedral shear stress, the goodness-of-fit was not improved compared to the model No. 5. In addition, the incorporation of standard atmospheric pressure (Pa) and the constant of 1 in the model No. 6 failed to improve the goodness-of-fit significantly. Therefore, in terms of the adj. R^2 and RMSE values, the model No. 4 can be regarded as the best starting formulation to develop a new improved M_R prediction model for recycled BDW aggregates with breakage-induced gradation variation incorporated as an additional explanatory variable.

4.2. Development of New Improved Prediction Model

For the five different levels of gradation variations induced by particle breakage and degradation (i.e., five different G/S levels) of recycled BDW aggregates, the optimal model No. 4 selected from the six candidates (see Table 6) was then fitted against the aforementioned laboratory testing results of the resilient modulus. Table 7 summarizes the regressed model parameters k_1, k_2, and k_3, as well as the calculated adj. R^2 and RMSE values. It can be seen from Table 7 that the fitted model parameter k_1 exhibited relatively greater differences among different gradations (or G/S values) than the model parameters k_2 and k_3.

Table 7. Regression parameters k_1, k_2, and k_3 of resilient modulus prediction model No. 4 for different gradations.

G/S Value	k_1	k_2	k_3	adj. R^2	RMSE
1.0	11.946	0.544	0.119	0.996	5.078
1.6	11.076	0.559	0.128	0.996	5.559
1.8	12.265	0.550	0.127	0.998	4.654
2.0	12.638	0.564	0.112	0.993	8.349
2.5	14.567	0.550	0.114	0.995	6.943

Specifically, the model parameter k_1 increased with increasing G/S level (or reduced level of breakage and degradation), while the model parameters k_2 and k_3 exhibited much fewer variations among different G/S levels (i.e., their values are distributed in relatively narrow ranges). In order to establish the relationship between the resilient modulus and particle-breakage-induced gradation variation, the regressed model parameter k_1 of the model No. 4 was fitted against gradation parameter G/S by exponent- and power-law functions, respectively, as shown in Figure 15.

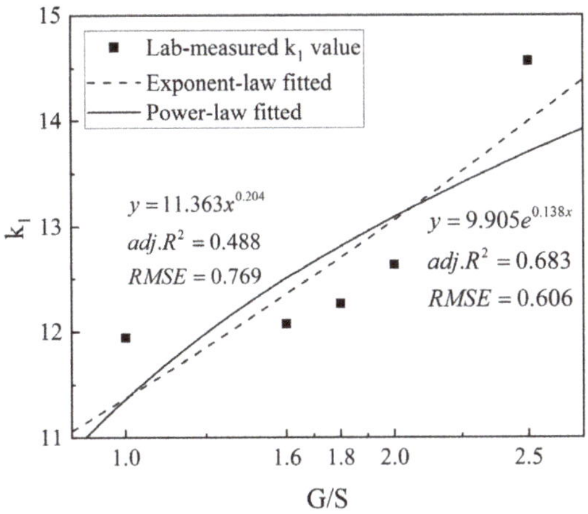

Figure 15. Fitting results of model parameter k_1 against the G/S parameter.

According to the fitting results, the exponent-law model was used to relate the resilient modulus with the G/S parameter (or particle breakage index), and the additional gradation parameter G/S (or particle breakage index) was introduced into prediction model No. 4 to integrate the effects of stress states and breakage-induced gradation variation on the resilient modulus of recycled BDW aggregates. The new improved prediction model of the resilient modulus developed in this study is shown in Equation (6).

The resilient modulus values predicted by the new improved model (i.e., Equation (6)) for different gradations were compared against their laboratory-measured counterparts, as shown in Figure 16. From the adj. R^2 and RMSE values, it can be indicated that the new improved model incorporating the exponential relation between MR and G/S can better predict the resilient modulus of recycled BDW aggregates with particle-breakage-induced gradation variations at varying stress states.

$$\begin{gathered} M_R = k_1 e^{k_2 (G/S)} {\sigma_3}^{k_3} {\sigma_d}^{k_4} \\ M_R = k_1 e^{k_2 (2.46 e^{-0.023Br})} {\sigma_3}^{k_3} {\sigma_d}^{k_4} \end{gathered} \quad (6)$$

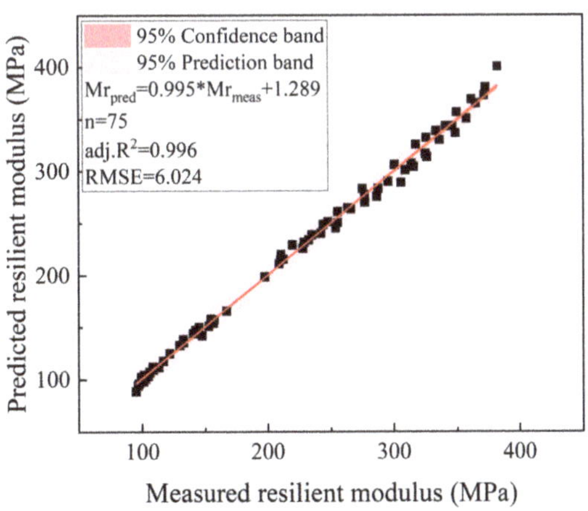

Figure 16. Comparison of the laboratory-measured resilient modulus values against those predicted from the new improved model (Equation (6)).

5. Discussions

The application of recycled BDW aggregates in unbound pavement granular layers has been extensively studied. Recycled BDW particles commonly possess low-strength and highly breakable features, while loading-related particle breakage could lead to significant changes in the initial gradation and thus severely affect the hydromechanical properties, such as strength, deformation, and permeability. However, it can be seen from the above-mentioned literature survey that few previous studies have addressed particle breakage and the resulting gradation variations of recycled CDW aggregates as well as their possible consequences on the strength and resilient modulus behavior.

This knowledge gap may hinder their reliable and widespread use in sustainable and long-lasting pavement construction. Therefore, in this study, five different gradations were designed to represent different levels of particle breakage (i.e., the breakage-induced gradation variations). The effects of particle breakage on the shear strength and resilient modulus properties of recycled BDW aggregates were investigated from laboratory monotonic triaxial compression tests and repeated-load triaxial tests.

According to the laboratory testing results, the apparent cohesion and internal friction angle of recycled BDW aggregates under consolidated drained conditions ranged from 35.3 to 57.5 kPa and from 30.2° to 54.3°, respectively. Such results are in reasonably well agreement with the cohesion value of 41 to 46 kPa and friction angle values of 49° to 51° reported by Arulrajah et al. [21,43] for the mixtures of recycled concrete aggregates, crushed bricks, and waste gravels. Therefore, it can be inferred that recycled BDW aggregates have the potential to substitute natural aggregates for use in unbound pavement granular base/subbase layers as they could meet the requirement of the minimum effective friction angle.

In addition, both the deviator stress and confining pressure exhibited significant influences on the resilient modulus, while the effect of confining pressure was more pronounced. This is consistent with previous studies [44–47]. Accordingly, the stress- and breakage-dependent resilient modulus prediction model was developed based on the experimental data [45,47].

It can be seen from the monotonic triaxial compression test results that the apparent cohesion and internal friction angle also increased and decreased non-linearly with increasing relative content of fine particles, respectively. This indicates that particle breakage and degradation weaken the inter-particle frictional interactions among recycled BDW aggregates, thus, resulting in a reduction in the overall shear strength. In addition, the results from laboratory RLT tests revealed that the resilient modulus of recycled BDW aggregates gradually decreased with the increasing relative content of fine particles (or equivalently increasing particle breakage and degradation) at the same stress level.

Overall, the test results show that the breakage-induced gradation variation, (i.e., the increase in the relative content of fine particles) weakened the shear strength and the ability to resist resilient and permanent deformation of recycled BDW aggregates subjected to the repeated applications of moving wheel loads. The findings could be useful for extended engineering applications of recycled BDW aggregates in unbound granular pavement base/subbase construction.

6. Summary and Conclusions

In view of the low-strength and breakage-susceptible nature of unbound aggregate particles recycled from building demolition waste (BDW), five different gradations were designed with the gravel-to-sand-ratio concept in this study to represent gradation variations induced by particle breakage and degradation. The shear strength and resilient modulus characteristics of the recycled BDW aggregates were investigated by conducting a series of laboratory monotonic triaxial compression tests and repeated-load triaxial (RLT) tests under different gradation and stress state conditions. The following major conclusions were drawn from the findings of this study.

- The results of particle morphology analysis showed that there appeared to be no significant differences in the particle morphology among different particle size groups of the same type of recycled BDW aggregates. Therefore, the effect of particle shape on the test results was excluded in this study by controlling the particle shape index to remain relatively similar or close to each other.
- The testing results of laboratory monotonic triaxial compression tests conducted under consolidated drained conditions on recycled BDW aggregates indicated that, as the relative content of fine particles increased, the apparent cohesion and internal friction angle showed a non-linear increase and decrease, respectively. This indicates that particle breakage and degradation weakened the inter-particle frictional interactions among recycled BDW aggregates, resulting in a reduction in the overall shear strength.
- The results from the laboratory RLT tests revealed that the resilient modulus of recycled BDW aggregates gradually increased with the decreasing relative content of fine particles or equivalently decreasing particle breakage and degradation) at the same stress level.
- Both the deviatoric stress and confining pressure exerted great influences on the *Mr* of recycled BDW aggregates, while a greater influence was exerted by the confining pressure when compared to the deviatoric stress.
- Based on the laboratory testing data, an improved resilient modulus prediction model that takes into account particle-breakage-induced gradation variation and stress states was proposed, and its better prediction accuracy was statistically confirmed.

In conclusion, this paper investigated the effects of breakage-induced gradation variation on the shear strength and resilient properties of recycled aggregates derived from building demolition waste (BDW) through a series of monotonic triaxial compression tests and repeated-load triaxial tests. We proposed a resilient-modulus prediction model incorporating particle breakage. The model provides a useful guide for predicting the resilient properties of recycled BDW aggregates with significant low-strength and a breakage-susceptible nature. In addition, considering the high water-absorbing potential of recycled BDW aggregates, further work is currently underway to investigate their water-sensitive properties for more a comprehensive understanding of the application of recycled BDW aggregates in pavement granular layers.

Author Contributions: Conceptualization, K.K., Y.X., U.F.A., Z.L., Q.L., H.Z., and D.C.; methodology, K.K., Y.X., U.F.A., Z.L., and Q.L.; validation, K.K., Y.X., Z.L., Q.L., H.Z., and D.C.; formal analysis, K.K., Y.X., U.F.A., Z.L., Q.L., and D.C.; writing—original draft preparation, K.K., Y.X., Z.L., and Q.L.; writing—review and editing, Y.X., U.F.A., Z.L., Q.L., H.Z., and D.C.; visualization, K.K., Y.X., and U.F.A.; supervision, Y.X., Z.L., Q.L., H.Z., and D.C.; funding acquisition, Y.X., Z.L., Q.L., H.Z., and D.C. All authors have read and agreed to the published version of the manuscript.

Funding: The work presented herein is supported jointly by the National Natural Science Foundation of China (Grant No. 52178443, 51878673, U1934209 & 52078485), the National Key R&D program of China (Grant No. 2019YFC1904704), the Key R&D Program of Chinese Academy of Railway Sciences (Grant No. 2020YJ022 & 2019YJ026), the key open-fund research program of the State Key Laboratory of Mechanical Behavior and System Safety of Traffic Engineering Structures (Shijiazhuang Tiedao University) (Grant No. KF2020-03), the Hunan Provincial Natural Science Foundation (Grant No. 2020JJ5740), the Key Project of Vocational Education Highland of Hunan Department of Education (Grant No. ZJGD2021144), the Project of Hunan Department of Science and Technology (Grant No. S2021JJSSLH0088), and the Open Foundation of MOE Key Laboratory of Engineering Structures of Heavy Haul Railway (Central South University) (Grant No. 2021JZZ01 & 2021JZZ02).

Institutional Review Board Statement: Not applicable.

Informed Consent Statement: Not applicable.

Data Availability Statement: Not applicable.

Conflicts of Interest: The authors declare no conflict of interest.

References

1. Zheng, L.; Wu, H.; Zhang, H.; Duan, H.; Wang, J.; Jiang, W.; Dong, B.; Liu, G.; Zuo, J.; Song, Q. Characterizing the generation and flows of construction and demolition waste in China. *Constr. Build. Mater.* **2017**, *136*, 405–413. [CrossRef]
2. Huang, B.; Wang, X.; Harnwei, K.; Geng, Y.; Raimund, B.; Ren, J. Construction and demolition waste management in China through the 3R principle-ScienceDirect. *Resour. Conserv. Recycl.* **2018**, *129*, 36–44. [CrossRef]
3. Shi, X.; Mukhopadhyay, A.; Zollinger, D.; Grasley, Z. Economic input-output life cycle assessment of concrete pavement containing recycled concrete aggregate. *J. Clean. Prod.* **2019**, *225*, 414–425. [CrossRef]
4. Maury-Ramírez, A.; Illera-Perozo, D.; Mesa, J.A. Circular Economy in the Construction Sector: A Case Study of Santiago de Cali (Colombia). *Sustainability* **2022**, *14*, 1923. [CrossRef]
5. Vieira, C.S.; Pereira, P.M. Use of recycled construction and demolition materials in geotechnical applications: A review. *Resour. Conserv. Recycl.* **2015**, *103*, 192–204. [CrossRef]
6. Volodchenko, A.A.; Lesovik, V.S.; Cherepanova, I.A.; Volodchenko, A.N.; Zagorodnjuk, L.H.; Elistratkin, M.Y. Peculiarities of non-autoclaved lime wall materials production using clays. *IOP Conf. Ser. Mater. Sci. Eng.* **2018**, *327*, 022021. [CrossRef]
7. Azevedo, A.R.G.d.; Klyuev, S.; Marvila, M.T.; Vatin, N.; Alfimova, N.; Lima, T.E.S.d.; Fediuk, R.; Olisov, A. Investigation of the Potential Use of Curauá Fiber for Reinforcing Mortars. *Fibers* **2020**, *8*, 69. [CrossRef]
8. Lu, G.; Liu, P.; Wang, Y.; Faßbender, S.; Wang, D.; Oeser, M. Development of a sustainable pervious pavement material using recycled ceramic aggregate and bio-based polyurethane binder. *J. Clean. Prod.* **2019**, *220*, 1052–1060. [CrossRef]
9. Yu, H.; Zhu, Z.; Zhang, Z.; Yu, J.; Oeser, M.; Wang, D. Recycling waste packaging tape into bituminous mixtures towards enhanced mechanical properties and environmental benefits. *J. Clean. Prod.* **2019**, *229*, 22–31. [CrossRef]
10. Luhar, S.; Cheng, T.W.; Nicolaides, D.; Luhar, I.; Panias, D.; Sakkas, K. Valorisation of glass wastes for the development of geopolymer composites—Durability, thermal and microstructural properties: A review. *Constr. Build. Mater.* **2019**, *222*, 673–687. [CrossRef]
11. Mohajerani, A.; Vajna, J.; Cheung, T.; Kurmus, H.; Arulrajah, A.; Horpibulsuk, S. Practical recycling applications of crushed waste glass in construction materials: A review. *Constr. Build. Mater.* **2017**, *156*, 443–467. [CrossRef]
12. Mishra, B.; Gupta, M.K. Use of randomly oriented polyethylene terephthalate (pet) fiber in combination with fly ash in subgrade of flexible pavement. *Constr. Build. Mater.* **2018**, *190*, 95–107. [CrossRef]
13. Bernardo, M.; Gomes, M.C.; de Brito, J. Demolition waste generation for development of a regional management chain model. *Waste Manag.* **2016**, *49*, 156–169. [CrossRef]
14. Zhang, J.; Gu, F.; Zhang, Y. Use of building-related construction and demolition wastes in highway embankment: Laboratory and field evaluations. *J. Clean. Prod.* **2019**, *230*, 1051–1060. [CrossRef]
15. Silva, R.V.; Cardoso, R.; Brito, J.; Dhir, R. Use of recycled aggregates from construction and demolition waste in geotechnical applications: A literature review. *Waste Manag.* **2016**, *49*, 131–145.
16. Vieira, C.S.; Pereira, P.M.; Lopes, M.L. Feasibility of Using Selected and Mixed Recycled Aggregates from Construction and Demolition Waste in Unbound Pavement Layers. In *Advances in Transportation Geotechnics IV*; Tutumluer, E., Nazarian, S., Al-Qadi, I., Qamhia, I.I., Eds.; Lecture Notes in Civil Engineering; Springer Nature: Cham, Switzerland, 2022; Volume 164, pp. 267–279.
17. Vieira, C.S. Valorization of Fine-Grain Construction and Demolition (C&D) Waste in Geosynthetic Reinforced Structures. *Waste Biomass. Valor.* **2020**, *11*, 1615–1626.
18. Jayakody, S.; Gallage, C.; Ramanujam, J. Effects of reclaimed asphalt materials on geotechnical characteristics of recycled concrete aggregates as a pavement material. *Road Mater. Pavement Des.* **2017**, *20*, 754–772. [CrossRef]
19. Arulrajah, A.; Piratheepan, J.; Disfani, M.M. Reclaimed asphalt pavement/recycled concrete aggregate blends in pavement subbases: Laboratory and field evaluation. *J. Mater. Civil. Eng.* **2014**, *26*, 349–357. [CrossRef]
20. Saberian, M.; Li, J.; Perera, S.; Ren, G.; Tokhi, H. An experimental study on the shear behavior of recycled concrete aggregate incorporating recycled tire waste. *Constr. Build. Mater.* **2020**, *264*, 120266. [CrossRef]
21. Arulrajah, A.; Disfani, M.M.; Horpibulsuk, S.; Suksiripattanapong, C.; Prongmanee, N. Physical properties and shear strength responses of recycled construction and demolition materials in unbound pavement base/subbase applications. *Constr. Build. Mater.* **2014**, *58*, 245–257. [CrossRef]
22. Liu, W.; Zeng, Y.; Yao, Y.; Zhang, J. Experimental study and prediction model of dynamic resilient modulus of compacted subgrade soils subjected to moisture variation. *Chin. J. Geot. Eng.* **2019**, *41*, 175–183.
23. ERES Consultants Division, ARA, Inc. *Guide for Mechanistic-Empirical Design of New and Rehabilitated Pavement Structures*; ERES Consultants Division, Transportation research Board, National Research Council: Washington, DC, USA, 2004.
24. *JTG D30-2015*; Specifications for Design of Highway Subgrades. Ministry of Transport of the People's Republic of China: Beijing, China, 2015.
25. Diagne, M.; Tinjum, J.M.; Nokkaew, K. The Effects of Recycled Clay Brick Content on the Engineering Properties, Weathering Durability, and Resilient Modulus of Recycled Concrete Aggregate. *Transp. Geotech.* **2015**, *3*, 15–23. [CrossRef]
26. Arisha, A.M.; Gabr, A.R.; El-Badawy, S.M.; Shwally, S.A. Performance Evaluation of Construction and Demolition Waste Materials for Pavement Construction in Egypt. *J. Mater. Civil Eng.* **2018**, *30*, 04016064.1–04016064.18. [CrossRef]
27. Soleimanbeigi, A.; Shedivy, R.F.; Tinjum, J.M.; Edil, T.B. Climatic effect on resilient modulus of recycled unbound aggregates. *Road Mater. Pavement Des.* **2015**, *16*, 836–853. [CrossRef]

28. Wang, Y.; Song, E. Discrete Element Analysis of Influence of Rockfill Particle Shape on Bulk Density and Strength. *Rock. Soil. M.* **2019**, *40*, 2416–2426.
29. Zingg, T. Beitrag zur Schotteranalyse. Ph.D. Thesis, ETH Zurich, Zurich, Switzerland, 1935.
30. *JTG 3430-2020*; Test Methods of Soils for Highway Engineering. Ministry of Transport of the People's Republic of China: Beijing, China, 2020.
31. *JTG E42-2005*; Test Methods of Aggregate for Highway Engineering. Ministry of Transport of the People's Republic of China: Beijing, China, 2005.
32. Xiao, Y.; Tutumluer, E. Gradation and Packing Characteristics Affecting Stability of Granular Materials: Aggregate ImagingBased Discrete Element Modeling Approach. *Int. J. Geomech.* **2017**, *17*, 04016064. [CrossRef]
33. Qamhia, A., II; Chow, L.C.; Mishra, D.; Tutumluer, E. Dense-graded aggregate base gradation influencing rutting model predictions. *Transp Geotech.* **2017**, *13*, 43–51. [CrossRef]
34. Xiao, Y.J. Performance-Based Evaluation of Unbound Aggregates Affecting Mechanistic Response and Performance of Flexible Pavements. Ph.D. Thesis, University of Illinois, Urbana-Champaign, IL, USA, 2014.
35. Hardin, B.O. Crushing of soil particles. *J. Geotech. Geoenviron.* **1985**, *111*, 1177–1192. [CrossRef]
36. *AASHTO Designation T307-99*; Determining the Resilient Modulus of Soils and Aggregate Materials. American Association of State Highway and Transportation Officials: Washington, DC, USA, 2003.
37. Zhang, J.; Peng, J.; Zheng, J. Progress and Prospect of the Prediction Model of the Resilient Modulus of Subgrade Soils. *China J. Highw. Transp.* **2020**, *33*, 1–13.
38. Monismith, C.L.; Seed, H.B.; Mitry, F.G.; Chan, C.K. Predictions of pavement deflections from laboratory tests. In Proceedings of the International Conference on the Structural Design of Asphalt Pavements, Ann Arbor, MI, USA, 7–11 August 1967.
39. Moossazadeh, J.; Witczak, M.W. Prediction of subgrade moduli for soil that exhibits nonlinear behavior. *Transp. Res. Rec.* **1981**, *810*, 9–17.
40. Seed, H.B.; Mitry, F.G.; Monosmith, C.L. *Prediction of Pavement Deflection from Laboratory Repeated Load Tests*; National Cooperative Highway Research: Washington, DC, USA, 1967.
41. Pezo, R. A General Method of Reporting Resilient Modulus Tests of Soils—A Pavement Engineer's Point of View. In Proceedings of the Transportation Research Board, 72nd Annual Meeting, Washington, DC, USA, 10–14 January 1993; National Research Council: Washington, DC, USA, 1993; p. 93082.
42. Uzan, J. Characterization of Granular Material. *Transp. Res. Rec.* **1985**, *1022*, 52–59.
43. Arulrajah, A.; Piratheepan, J.; Disfani, M.M. Geotechnical and Geoenvironmental Properties of Recycled Construction and Demolition Materials in Pavement Subbase Applications. *J. Mater. Civil. Eng.* **2013**, *25*, 1077–1088. [CrossRef]
44. Arshad, M.; Ahmed, M.F. Potential use of reclaimed asphalt pavement and recycled concrete aggregate in base/subbase layers of flexible pavements. *Constr. Build. Mater.* **2017**, *151*, 83–97. [CrossRef]
45. Jie, L.; Saberian, M.; Bao, T.N. Effect of crumb rubber on the mechanical properties of crushed recycled pavement materials. *J. Environ. Manag.* **2018**, *218*, 291–299.
46. Saberian, M.; Li, J.; Boroujeni, M.; Li, C. Application of demolition wastes mixed with crushed glass and crumb rubber in pavement base/subbase. *Resour. Conserv. Recycl.* **2020**, *156*, 104722. [CrossRef]
47. Arulrajah, A.; Piratheepan, J.; Disfani, M.M. Resilient Moduli Response of Recycled Construction and Demolition Materials in Pavement Subbase Applications. *J. Mater. Civil. Eng.* **2013**, *25*, 1920–1928. [CrossRef]

materials

MDPI

Article

Aging Behavior of High-Viscosity Modified Asphalt Binder Based on Infrared Spectrum Test

Wenxuan Zhang [1], Qiang Li [1], Jiaqing Wang [1,*], Yuanpeng Meng [1] and Zhou Zhou [2]

[1] College of Civil Engineering, Nanjing Forestry University, Nanjing 210037, China; zhangwenxuan@njfu.edu.cn (W.Z.); liqiang2526@njfu.edu.cn (Q.L.); myp564335@njfu.edu.cn (Y.M.)
[2] School of Transportation, Southeast University, Nanjing 211189, China; zhouzhou1024@seu.edu.cn
* Correspondence: jiaqingw@njfu.edu.cn

Abstract: In the rapid development of sponge city construction in China, porous asphalt pavement has been widely used. The high-viscosity modified asphalt used for porous asphalt pavements is utilised in a complex aging environment. In this study, infrared spectroscopy was used to test the changes in the functional groups of high-viscosity modified asphalt under the influence of ultraviolet radiation intensity, high temperature, and water corrosion conditions. The research results showed that under the influence of several environmental factors, the high-viscosity modified asphalt has no chemical reaction but does undergo physical changes. From the perspective of the functional group index, the carbonyl index is more suitable for evaluating the degree of ultraviolet aging, and the sulfoxide group index is more suitable for evaluating the effect of temperature on aging. The high-viscosity modified asphalt aging kinetic models, established with different functional group indexes as indicators, have different activation energies. The aging kinetic model established with the carbonyl index is more suitable for simulating traditional thermal-oxidative aging. This study provides a better plan to reveal the influence of different environmental factors on the aging performance of high-viscosity modified asphalt under complex environmental conditions.

Keywords: porous asphalt pavement; high-viscosity modified asphalt; functional group index; aging kinetics

Citation: Zhang, W.; Li, Q.; Wang, J.; Meng, Y.; Zhou, Z. Aging Behavior of High-Viscosity Modified Asphalt Binder Based on Infrared Spectrum Test. *Materials* **2022**, *15*, 2778. https://doi.org/10.3390/ma15082778

Academic Editor: Francesco Canestrari

Received: 9 March 2022
Accepted: 7 April 2022
Published: 10 April 2022

Publisher's Note: MDPI stays neutral with regard to jurisdictional claims in published maps and institutional affiliations.

Copyright: © 2022 by the authors. Licensee MDPI, Basel, Switzerland. This article is an open access article distributed under the terms and conditions of the Creative Commons Attribution (CC BY) license (https://creativecommons.org/licenses/by/4.0/).

1. Introduction

Within the concept of building a green resource-saving society, "sponge cities" have attracted more and more attention from relevant government departments and practitioners. Building porous asphalt pavements is one of the effective ways to promote the construction of sponge city infrastructure [1]. Asphalt pavement with large voids has been widely used in roads due to its advantages of anti-slip ability, drainage, and noise reduction. At the same time, the main asphalt binder used in porous asphalt is high-viscosity modified asphalt (HVA), which is excessively exposed to oxygen, temperature, water, and ultraviolet radiation, thereby accelerating its aging damage, which leads to the deterioration of asphalt road performance [2–4]. The polymer degradation and polymer phase structure destruction of high-viscosity modified asphalt is distinct from that of conventional modified asphalt, which will cause some differences in performance between high-viscosity modified asphalt and conventional modified asphalt, especially in aging performance [5–7].

At present, the research focus of high-viscosity modified asphalt is mainly concentrated on preparation, modification mechanism, pavement performance, and so on. Luo [8] et al. studied the performance of the modified mixture with different high-viscosity agents, and the results showed that the most appropriate content of the high-viscosity modifier is 12% of the asphalt mass. Cai [6] et al. prepared environmental-friendly HVA using waste materials and studied the formation mechanism of its physical and mechanical properties through a series of microscopic experiments. However, the research on high-viscosity modified asphalt is still insufficient, especially that of its aging performance, which is very important to extending the durability of porous asphalt pavement.

The influence of environmental factors on the aging behavior of high-viscosity modified asphalt has also attracted the attention of researchers. Li [9] et al. studied the aging performance of HVA under complex environmental conditions. The results show that acid rain solution and ultraviolet light also have a significant impact on viscoelasticity, high and low temperature performance, and fatigue performance. It is proposed that the coupling aging effect should be considered during service. Yang [10] studied the thermal aging behavior of HVA from the perspective of cohesion, adhesion, and rheology, and the results showed that aging weakened the cohesion, adhesion, and low-temperature performance of HVA, and reduced its surface roughness. Hu [11] discussed the aging characteristics of high-viscosity modified asphalt under accelerated weathering conditions and analyzed the effects of temperature and aging time on its rheological properties and chemical changes. Li [12] studied the influence of different ultraviolet radiation intensities on asphalt aging and used DSR and FTIR tests to study the changes in rheological properties. The results showed that different ranges of ultraviolet light have different degrees of influence on asphalt aging, among which 360 nm ultraviolet light has the most significant impact on asphalt. Zou [13] analyzed the influence of different acid-base environments on asphalt aging at 60 °C, and the study showed that acid and alkaline solutions have more significant effects on asphalt aging. The current research mainly focuses on the impact of single environmental factors on the aging of high-viscosity modified asphalt. The actual service environment of high-viscosity modified asphalt is more complicated and is affected by multiple environmental factors. The aging effect is not yet clear. It is necessary to couple high-viscosity modified asphalt under different factors research on aging behavior.

Infrared spectroscopy is used as a rapid detection method to observe the changes in asphalt microstructure. It can qualitatively analyze and study the structure and chemical composition changes of asphalt cement before and after aging from a microscopic point of view, and it can also quantitatively characterize asphalt cement before and after aging [14,15]. Yang et al. [16] used FTIR to characterize the rheological properties and aging mechanism of asphalt binders mixed with high percentages of bio-adhesives. The analysis showed that the effects of aging bio-adhesives stem from three aspects: the loss of volatiles, dehydrogenation of high molecular weight compounds such as asphaltenes, and oxidation of chemical substances. Yao et al. [17] used FTIR to test different asphalt binders to obtain the spectra of chemical groups. They found that carboxylic acids and ketones are the main aging components of asphalt when exposed to air and oxygen. In addition, these functional groups and carbonyl groups are closely related to the rutting sensitivity and fatigue of asphalt mixtures. Bhupendra [18] used infrared spectroscopy to study the aging characteristics of conventional asphalt, polymer-modified asphalt (PMB), and warm-mix asphalt (WMA), and quantified the obtained infrared spectroscopy results by calculating different indicators, finally calculating the aging index of short-term and long-term aging of asphalt based on these indicators.

Infrared spectroscopy can quickly identify and mark the functional groups in asphalt to judge its chemical composition. Based on Lambert Beer law, it can carry out absolute quantitative analysis with high precision [15]. Most researchers currently use infrared spectroscopy to study and quantify the effect of aging on asphalt binders, but there are few studies on the aging characteristics of high-viscosity modified asphalt using infrared spectroscopy. As for the selection of environmental factors, some scholars choose a single UV, or UV and temperature coupled environmental conditions. However, porous asphalt pavement is always exposed to complex environments, including sunlight, ultraviolet light, and rainfall [9]. Therefore, this article is based on infrared spectroscopy experiments to study the aging characteristics of high-viscosity modified asphalt under multi-factor coupled environmental conditions, discusses the influence of different environmental factors on the chemical components of high-viscosity modified asphalt, and finally establishes the high-viscosity modified asphalt aging kinetic equation based on the functional group index.

This study shows that it is of great significance to consider the coupling effect of different environmental factors on the aging of high-viscosity modified asphalt, which is helpful to better evaluate the on-site aging behavior of porous asphalt pavement.

2. Materials and Methods

2.1. Materials

In this paper, SBS modified asphalt provided by Jiangyin Taifu Company (Jiangyin, China) was selected as the base asphalt, in which the content of SBS was 4.5%. TPS (TAFPACK-Super) high-viscosity agent was added to the SBS for preparation of high-viscosity modified asphalt (HVA), which was prepared in accordance with the Chinese standard JTGT 3350-03-2020. TPS modifier is a high-viscosity modifier specially produced for porous asphalt pavement. The full name of TPS is TAFPACK-Super. Its composition is mainly thermoplastic rubber, supplemented by adhesive resin, combined with other stabilizers to make ordinary asphalt become high-viscosity modified asphalt through mechanical mixing and processing. The TPS high-viscosity agent was in 12 wt.% of the base binder. SBS asphalt and TPS high-viscosity agent were put into a high-speed shearing machine, shearing for 30 min at 180 °C and 5000 rpm. Asphalt samples with a diameter of 140 mm and a thickness of about 0.7–1.0 mm were prepared after mixing, then settled in the asphalt aging simulation device for aging procedures. The 0.7–1.0 mm asphalt sample was prepared by pouring, and the purpose of controlling the thickness of asphalt film was achieved by controlling the mass of the pouring volume. The asphalt was heated to 180 °C, and then poured into the regular glass dish with a consistent diameter. The bottom surface of the glass dish was a circle with a radius of 7 cm, where the area was determined. After measuring the density and quality of the asphalt sample, the thickness of the asphalt sample thickness can be controlled by the pouring volume. The physical properties of base bitumen (SBS) and HVA are shown in Table 1, and the physical properties of the high-viscosity agent are shown in Table 2.

Table 1. Basic properties of two asphalt samples.

Asphalt Types	Penetration 25 °C (0.1 mm)	Softening Point (°C)	5 °C Ductility (cm)	60 °C Dynamic Viscosity (Pa·s)
SBS	64	71.2	36.5	5015
HVA	44.3	86.4	45.3	73,527

Table 2. Technical specification of TPS high-viscosity agent.

Project	Unit	Results	Standard
Particle size	mm	4.2	JT/T 860.2
Density	g/cm^3	0.8	GB/T1033
Absorption	%	0.6	-

2.2. Complex Aging Simulation Method

The ultraviolet radiation intensity in China can be divided into three gradients: 35–50 × 10^3 J/cm^2 a, 25–35 × 10^3 J/cm^2 a, and 20–25 × 10^3 J/cm^2 a; the three selected ultraviolet high-pressure mercury lamp radiation intensity levels represent the national ultraviolet distribution gradient: 10.14 × 10^{-4} w/cm^2, 13.28 × 10^{-4} w/cm^2, 17.38 × 10^{-4} w/cm^2; the water situation of acidity, neutrality, and alkalinity represents the effects to high-viscosity modified asphalt by moisture in the service process. The alkaline solution was prepared by diluting pure NaOH reagent with a mass fraction of 7%, and pH = 11 was determined by pH meter. The acid solution was made of sulfuric acid and nitric acid by selecting the ratio of C (SO_4^{2-}): C (NO_3^-) = 9:1 to simulate the acid rain solution, and determined the pH = 2.8. In the conventional aging simulation method, PAV used three temperature levels of 90 °C, 100 °C, and 110 °C. To ensure the authenticity and aging efficiency of the simulation, three temperature levels of 90 °C, 100 °C, and 110 °C were selected to simulate thermal aging.

This article adopts a self-designed coupling aging device, which can combine three environmental factors to achieve simultaneous effects of ultraviolet light, temperature, and moisture. The simulated aging equipment is composed of an environmental box, an ultraviolet high-pressure mercury lamp, and a water-immersed transparent vessel, as shown in Figure 1. The device can adjust the distance between the lamp and the test piece via the card slot to change the ultraviolet irradiance on the surface of the test piece, control the temperature through the environmental box to simulate thermal aging, and use the forming of a uniform water film on the surface of the asphalt sample to simulate water aging. It also changes the solution every five hours to ensure the integrity of the water film and the purity of the solution during the aging process.

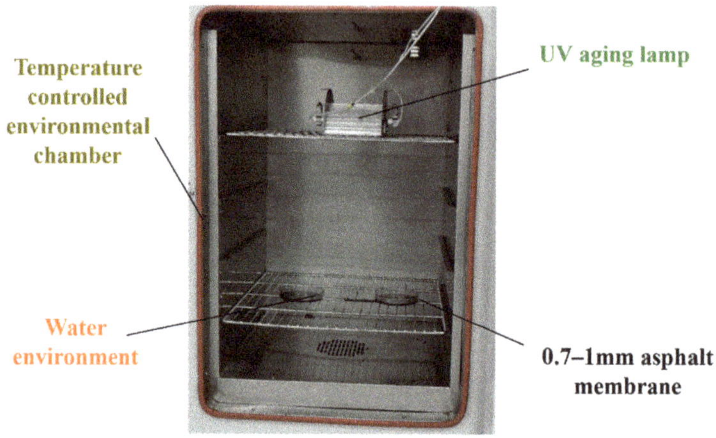

Figure 1. Complex aging simulation device.

2.3. FTIR Spectroscopy Test

A non-destructive FTIR test was performed to detect the changes in functional groups of asphalt under the influence of different environmental factors. First, a small amount of potassium bromide was ground into a powder to produce a compressed sample. Then, the heated high-viscosity pitch was evenly coated on the pressed samples of potassium bromide. Cary 630 infrared spectrometer was used to collect data in the attenuated total reflection (ATR) mode. Its wave number accuracy was greater than 0.005, the signal-to-noise ratio was greater than 5000, and 32 times per scan. The samples were heated to 70 °C, then a few samples were evenly coated on all the surfaces of ATR crystals, and their infrared spectra were measured.

2.4. Experiment Plan

The aging test plan is shown in Table 3, and the sample aging time was 40 h. "O-HVA" stands for origin HVA; "U" stands for HVA aging under the condition of ultraviolet radiation intensity; "UT" means HVA aging under the combined action of ultraviolet light and high temperature; and "UTS" means HVA aging under the combination of ultraviolet light, temperature, and different water quality solutions. At the same time, different factor levels were embedded in the coupling conditions. L, M, and H mean low, medium, and high level, respectively, A indicates acidic water conditions, and ALK indicates alkaline water condition. For example, "$U_LT_LS_N$" represents low-level ultraviolet irradiance, low-level temperature, and neutral water quality conditions.

Table 3. Aging program.

Asphalt Specimen	UV Irradiance (w/cm^{2})	Temperature (°C)	Water Environment
O-HVA	/	/	/
$U_LT_LS_N$	10.14×10^{-4}	90	neutral
$U_MT_LS_N$	13.28×10^{-4}	90	neutral
$U_HT_LS_N$	17.38×10^{-4}	90	neutral
$U_HT_MS_N$	17.38×10^{-4}	100	neutral
$U_HT_HS_N$	17.38×10^{-4}	110	neutral
$U_HT_LS_A$	17.38×10^{-4}	90	acidic
$U_HT_LS_{ALK}$	17.38×10^{-4}	90	alkaline

3. Results

3.1. The Effect of UV Intensity on Aging

The functional groups corresponding to the special absorption peaks of the infrared spectrum are shown in Table 4. The quantitative analysis of infrared spectroscopy is based on the peak area ratio related to the characteristic functional groups in the asphalt. The definition of the peak area ratio is as follows [15].

Table 4. Infrared spectrum analysis of asphalt.

Wavenumber (cm^{-1})	Spectral Peak Attribution
2924	Asymmetric stretching vibration of C-H in methylene
2852	Symmetrical stretching vibration of C-H in methylene
2729	Stretching vibration of aldehyde group
1686	C=O stretching vibration
1671	C=O stretching vibration of primary amide carbonyl
1600	Respiratory vibration of asymmetrically substituted benzene ring
1461	Scissor vibration of methylene (—CH_2—)
1377	Umbrella vibration of methyl (—CH_3—)
1031	Stretching vibration of sulfoxide group (S=O)
812\868	Stretching vibration of benzene ring
747	Bending vibration of aromatic branched chain
722	Synergistic vibration of methylene segment (CH_2) n (n $\geq$ 4)

Aliphatic functional group index:

$$I_B = A_{1377}/(A_{1462} + A_{1377} + A_{723}) \tag{1}$$

Aromatic functional group index:

$$I_{Ar} = \frac{A_{1600}}{\sum A} \tag{2}$$

Carbonyl functional group index:

$$I_{C=O} = \frac{A_{1700}}{\sum A} \tag{3}$$

Sulfoxide functional group index:

$$I_{S=O} = \frac{A_{1031}}{\sum A} \tag{4}$$

where:

A—Different peak areas; $\sum A$—The sum of different peak areas in a certain range. $\sum A = A_{1700} + A_{1600} + A_{1456} + A_{1376} + A_{1306} + A_{1162} + A_{1031} + A_{968} + A_{861} + A_{810} + A_{744} + A_{722}$.

The spectrum analysis software adopts OMNIC, takes the sulfoxide group at 1031 cm^{-1} of the original HVA as an example, and takes the tangent of the lowest point on both sides of the absorption peak as the correction baseline from which to calculate the absorption peak area A_{1031}. The schematic diagram of peak area calculation and baseline selection is shown in the Figure 2 (O_1, O_2, and O_3 are the reference points for peak area calculation). Similarly, different peak areas corresponding to other functional groups can be obtained, and the indexes of different functional groups can be calculated according to the above Formulas (1)–(4).

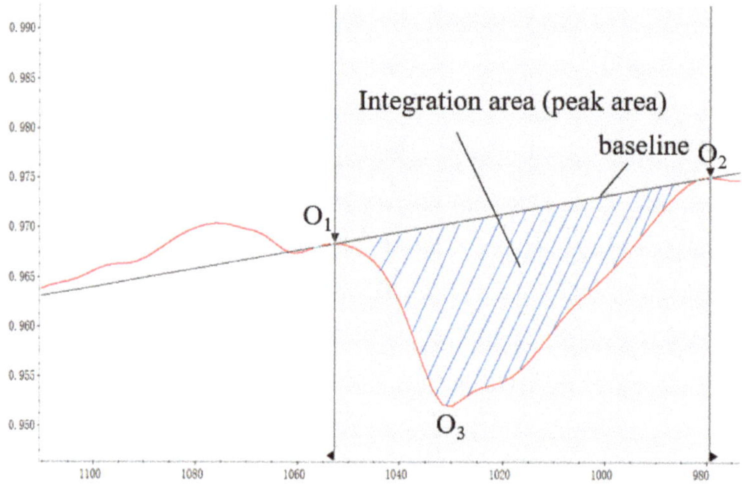

Figure 2. Peak area calculation diagram.

The infrared spectra of asphalt samples under different levels of ultraviolet radiation are shown in Figure 3. Compared with the original HVA, after three levels of ultraviolet radiation, the intensity of the carbonyl group at 1700 cm^{-1} and the sulfoxide group at 1031 cm^{-1} increased significantly. As the level of ultraviolet radiation increased, the peak intensity of C=O gradually increased, while the peak intensity of C=C gradually decreased. The change of C=O is because the asphalt molecular chain broke and formed free radicals under long-term ultraviolet irradiation, and then the formed free radicals combined with oxygen to produce several oxygen-containing groups. The absorption peak at 966 cm^{-1} is the C=C bending vibration absorption peak in SBS, and its peak intensity gradually decreases. This is because as the intensity of ultraviolet radiation increases, SBS absorbs ultraviolet light energy, and the molecular chain breaks. The absorption peaks of methyl and methylene at 1460 cm^{-1} and 1375 cm^{-1}, respectively, increased slightly, and the amplitude of the absorption spectrum increased with the increase in ultraviolet radiation level. This is mainly due to the destruction of the C=C bond of the unsaturated olefin in the high-viscosity pitch under ultraviolet irradiation, forming a C-H bond. Compared with the original HVA, the amplitude of these two bands decreased slightly after UV aging, and with the increase in UV irradiance, the amplitude decreased more.

As shown in Figure 4, when the ultraviolet irradiance reached the level of 10.14×10^{-4} w/cm^2 under the three types of ultraviolet radiation intensity, the carbonyl group started to increase. When the ultraviolet irradiance increased to 13.28×10^{-4} w/cm^2 and 17.38×10^{-4} w/cm^2, the carbonyl index increased by two times and 17 times, respectively, indicating that with the increase in ultraviolet irradiance, ultraviolet rays have a strong promotion effect on the formation of carbonyl functional groups. The sulfoxide group vibration only appeared when the ultraviolet irradiance reached the level of 13.28×10^{-4} w/cm^2, and when the ultraviolet irradiance increased to 17.38×10^{-4} w/cm^2, the sulfoxide group index only increased by 85.7%. Compared with the carbonyl index, the growth rate was much smaller, indicating that for high-viscosity modified asphalt, the carbonyl functional group is more sensitive to ultraviolet rays, and the carbonyl index is used to evaluate the effect of ultraviolet aging more obviously.

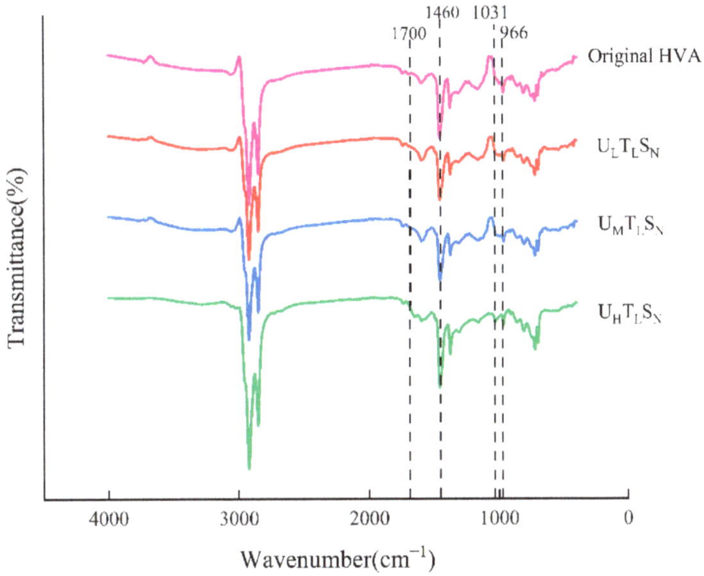

Figure 3. Variation of the spectrum of different asphalt samples under different ultraviolet irradiation intensities.

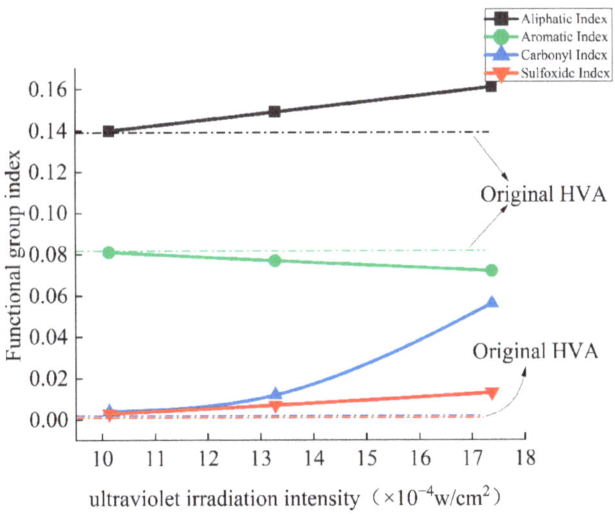

Figure 4. Variation of characteristic functional group index under different ultraviolet irradiation intensity.

3.2. The Effect of Temperature on Aging

It can be seen from Figure 5 that under different temperature levels, the absorption peak of sulfoxide group at 1031 cm^{-1} and the absorption peak of carbonyl group at

1700 cm^{-1} tended to strengthen. There was no new absorption peak during the aging process, indicating that the increase in temperature only led to the aging progresses, and no chemical reaction occurred to produce new substances. The peak of the polystyrene section near 910 cm^{-1} disappeared, which is due to the chemical reaction of the SBS polystyrene in the high-viscosity modified asphalt under thermal oxidative aging conditions [19].

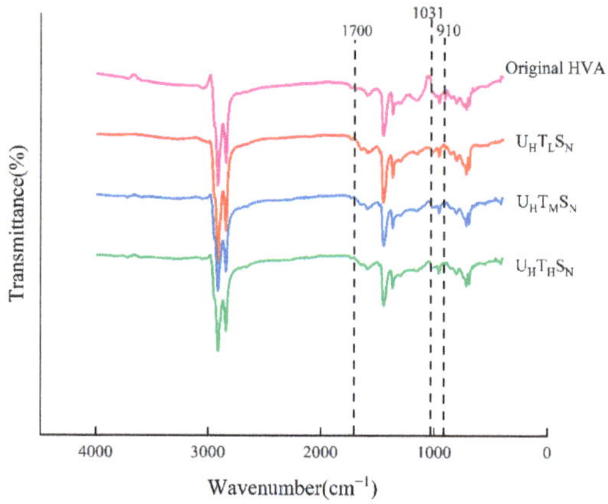

Figure 5. Variation of spectrum of different asphalt samples under different temperature levels.

In terms of characteristic functional group index, as shown in Figure 6, as the temperature increased, the aging temperature increased from 90 °C to 110 °C, the aliphatic index increased by 6.21%, the aromatic index decreased by 19.4%, and the carbonyl index increased by 14.3%. The sulfoxide base index also increased by 61.5%. This showed that in the temperature range of 90–110 °C, the changes of aliphatic, aromatic, and carbonyl functional groups were not obvious. Unlike the sensitivity to ultraviolet rays, the sulfoxide group is more sensitive to temperature. When studying the thermal aging of high-viscosity modified asphalt, it is more appropriate to use the sulfoxide index as an evaluation index.

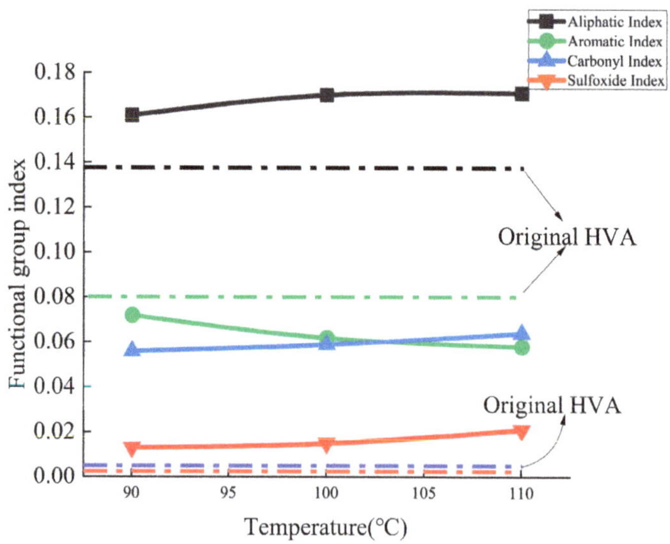

Figure 6. Change of functional group index under different temperature levels.

3.3. The Effect of Water Environment on Aging

Based on the combination of 17.38 × 10^{-4} w/cm^2, 90 °C, and a neutral water environment, the water quality was changed, and the influence of saline-alkali water and acid rain on the aging behavior of high-viscosity modified asphalt was analyzed. It can be seen from Figure 7 that the infrared spectra of high-viscosity modified asphalt under different water quality did not show different absorption peaks, indicating that SO_4^{2-} and NO_3^- in acid rain solution, and Na^+ and Cl^- in alkaline solution, did not chemically react with asphalt to generate new substances.

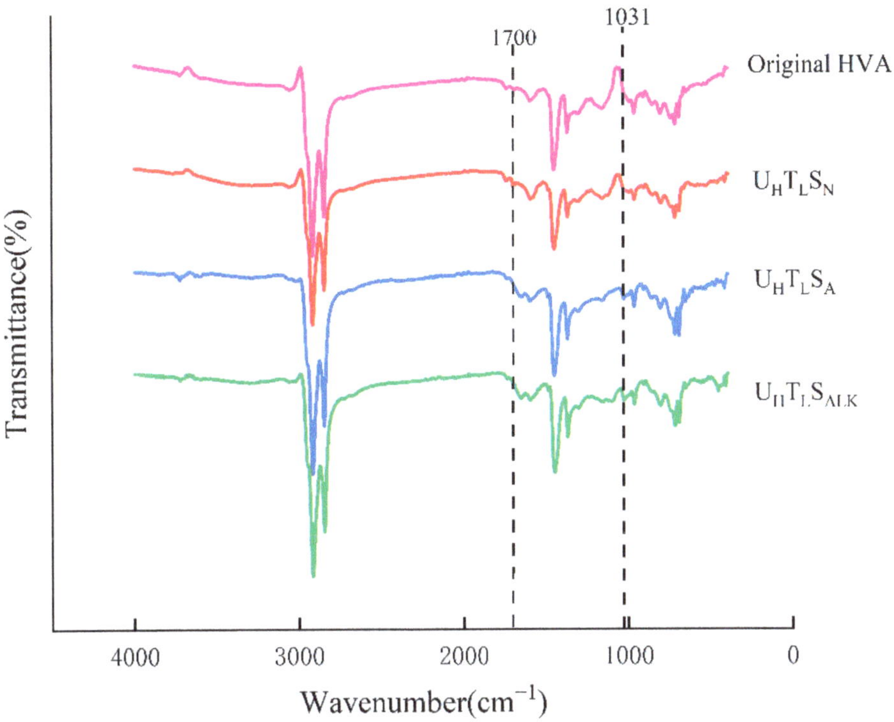

Figure 7. Variation of spectrum of different asphalt samples under different water environment.

As shown in Figure 8, compared with neutral water, the aliphatic index of the high-viscosity modified asphalt in acid rain solution increased by 7.5%, the aromatic index decreased by 4.2%, the carbonyl index increased by 41.1%, and the sulfoxide group index increased by 84.6%. In the high-viscosity pitch in alkaline solution, the aliphatic index and the aromatic index hardly changed, the carbonyl index increased by 53.6%, and the sulfoxide group index increased by 92.3%. Under the action of acidic and alkaline solutions, the range of changes of aromatic phenol and saturated phenol was very small, indicating that the light and heavy components of high-viscosity modified asphalt have not changed due to the difference in water quality, but the carbonyl index and sulfoxide index had increased. From the changes of carbonyl and sulfoxide groups, acid rain and saline-alkali groundwater will increase the aging of high-viscosity modified asphalt. This may be due to the acidity of carboxylic acids and phenols in the asphalt. The substance dissolves and ionizes in the acid rain solution to a certain extent, and the neutralization reaction occurs in the alkaline solution, which accelerates the aging of the asphalt in the water, and the sodium hydroxide and asphaltic acid in the alkaline solution can react to form organic soap compounds.

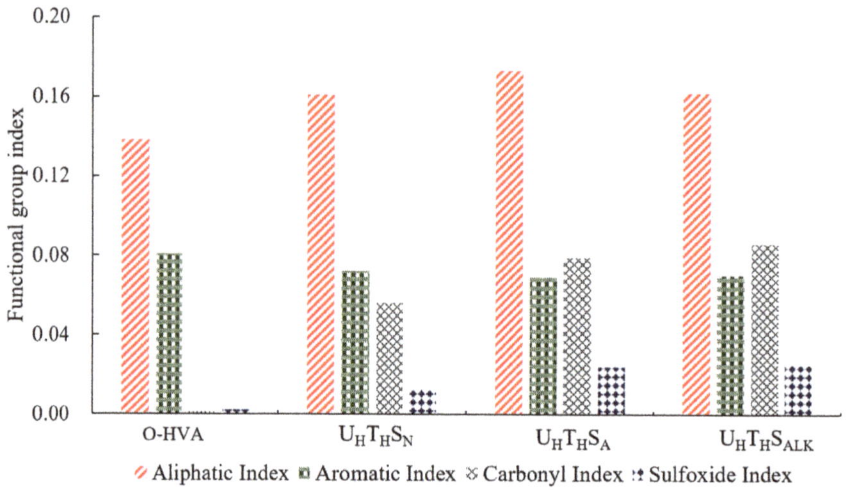

Figure 8. Change of characteristic functional group index under different water properties.

3.4. Variance Analysis

The SPSS software was used to analyze the variance of the test results at the 95% confidence level, and the significance of the influence of different factors on the aging functional group index was compared. The results are shown in Table 5. Statistical analysis shows that UV intensity, temperature, and water quality conditions all have a significant impact on the functional group index. Among them, the degree of influence of ultraviolet intensity on the carbonyl index is greater than that of temperature and water quality environment, and the influence of temperature on the sulfoxide group index is greater than that of ultraviolet intensity and water quality environment. For the aliphatic index and the aromatic index, the three have the same significance.

Table 5. Analysis of variance results.

Project	Deviation Sum of Squares	Degree of Freedom	Mean Square Error	F	p
		(1) Aliphatic index			
UV intensity	2.16×10^{-4}	3	2.16×10^{-4}	2.892	0.012
Temperature	1.29×10^{-3}	3	1.29×10^{-3}	55.099	0.011
Water environment	1.12×10^{-3}	3	1.12×10^{-3}	53.556	0.012
		(2) Aromatic Index			
UV intensity	2.81×10^{-5}	3	2.81×10^{-5}	2.770	0.013
Temperature	4.33×10^{-4}	3	4.33×10^{-4}	16.673	0.015
Water environment	1.70×10^{-4}	3	1.70×10^{-4}	46.286	0.011
		(3) Carbonyl Index			
UV intensity	7.26×10^{-4}	3	7.26×10^{-4}	3.967	0.004
Temperature	4.98×10^{-3}	3	4.98×10^{-3}	61.796	0.013
Water environment	7.70×10^{-3}	3	7.70×10^{-3}	62.551	0.013
		(4) Sulfoxide Index			
UV intensity	6.01×10^{-5}	3	6.01×10^{-5}	31.852	0.025
Temperature	4.86×10^{-4}	3	4.86×10^{-4}	15.429	0.007
Water environment	5.80×10^{-4}	3	5.80×10^{-4}	26.173	0.012

3.5. Aging Kinetic Model of High-Viscosity Modified Asphalt

Asphalt binder meets the first-order reaction under conventional thermal and oxygen aging conditions [20], and its rate equation is:

$$-\frac{dc}{dt} = kc \tag{5}$$

After integration:

$$\int -\frac{dc}{c} = \int kdt \tag{6}$$

$$-Inc = kt + B \tag{7}$$

When $t = 0$, $C = C_0$, and $B = -Inc_0$, so the general formula of first-order reaction kinetics can be obtained:

$$Inc = Inc_0 - kt \tag{8}$$

$$In(c_0/c) = kt \tag{9}$$

Arrehenius put forward the concept of activation energy through a lot of research. He believes that molecules can react without contact if they have enough energy. These high-energy molecules are called activated molecules, and inactive molecules are transformed into activated molecules. The required energy is activation energy, as shown in the following formula:

$$Ink = InA - \frac{E_a}{RT} \tag{10}$$

After obtaining the aging kinetic parameters and substituting it into Equation (10), we come to:

$$In\left(\frac{c_0}{c}\right) = Ate^{-(E_a/R)/T} \tag{11}$$

where:

E_a: activation energy;
A: Pre-reference factor;
T: Kelvin temperature;
R: Molar gas constant;
c: A functional group index;
k: Reaction rate constant at temperature T;
t: Aging time.

Take the high temperature and ultraviolet and acid rain solution environment combination (T + UV + S) as the aging condition, make the high-viscosity modified asphalt aged 0 h, 10 h, 20 h, 30 h, 40 h at the temperature level of 90 °C, 100 °C, 110 °C. The ultraviolet radiation level was 17.38×10^{-4} w/cm^2, and the infrared spectrum scanning test was performed on the aged asphalt samples. The carbonyl index I_C and the sulfoxide group index Is were used to establish the aging kinetic equation and calculate the kinetic parameters.

According to the Arrhrenius formula, the first-order aging kinetic model with I_C and I_S as parameters are obtained. The parameters are shown in Table 6.

$$In(I_c/I_{c0}) = 771.16te^{-3425.5/T} \tag{12}$$

$$In(I_s/I_{s0}) = 8.6te^{-1598.4/T} \tag{13}$$

Activation energy is an index used to evaluate the degree of difficulty of reaction. The greater the activation energy, the more difficult it is for asphalt molecules to be converted into activated molecules, the more difficult it is for the reaction to generate new free radicals, and the better the anti-aging performance. Table 6 shows that activation energy has a greater impact on the reaction rate and temperature. For a reaction with large activation energy, temperature has a greater impact on the reaction rate. Therefore, the activation energy also determines the sensitivity of the reaction rate to temperature. Compared with I_S, the aging kinetic model established with I_C as the parameter has greater activation energy. The greater the influence of temperature on the aging reaction rate, the more suitable the established model is for traditional indoor simulation of thermal oxygen aging. As Figure 9 shows, this model can be used to predict the changes of carbonyl functional groups of high-viscosity modified asphalt under different temperature environments, and provide a theoretical basis and reference for reducing the performance of high-viscosity modified asphalt due to thermal-oxidative aging during actual production, transportation, and construction.

Table 6. Arrehenius model parameters.

Kinetic Parameters	Aging Temperature (°C)	k	A	E_a (kJ·mol)
I_c	90	0.0650	771.16	28.47
	100	0.0795		
	110	0.1010		
I_s	90	0.1095	8.600	13.28
	100	0.1156		
	110	0.1342		

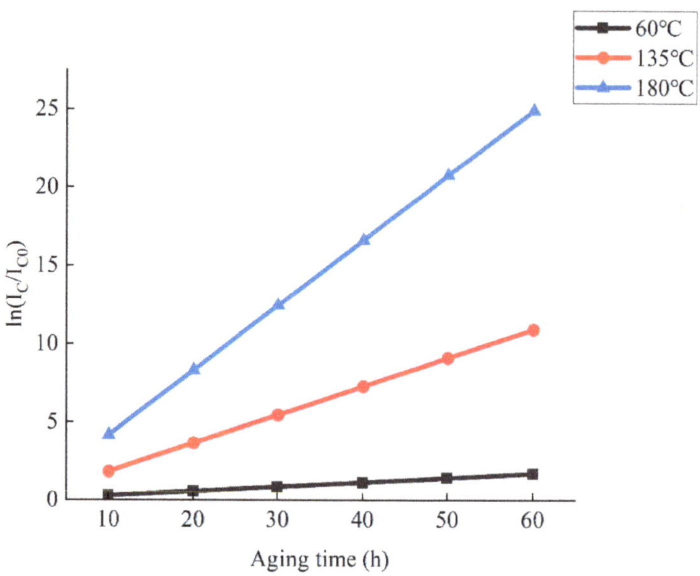

Figure 9. Changes of carbonyl functional groups at different temperatures.

4. Conclusions

This study investigated the aging behaviors of high-viscosity asphalt subjected to complex aging conditions based on the FTIR technique, and the results demonstrated that the high-viscosity asphalt was sensitive to the combined aging factors by the change of chemical functional groups. There are some main conclusions summarized as below:

(1) Compared with the unaged high-viscosity modified asphalt, the increase in UV irradiance, the increase in temperature and the change of water quality cannot make the high-viscosity modified asphalt generate new functional groups. Under the three environmental factors, the high-viscosity modified asphalt has no chemical reaction, only physical changes.

(2) When studying ultraviolet aging, it is more obvious to use the carbonyl index as the evaluation index, while the sulfoxide group is more sensitive to temperature, and it is more appropriate to use the sulfoxide index as the heat aging evaluation index.
(3) The aging kinetic model established with the carbonyl index as a parameter has greater activation energy, and the established model is more suitable for traditional indoor simulation of thermal-oxidative aging.

Author Contributions: Investigation, W.Z.; resources, Y.M.; data curation, Y.M.; writing—original draft preparation, W.Z.; writing—review and editing, Q.L. and J.W.; funding acquisition—Q.L. and W.Z.; software and methodology, Z.Z. All authors have read and agreed to the published version of the manuscript.

Funding: This research was funded by the Industry-University-Research Collaboration Project of Jiangsu Province, grant number [BY2021313]; the Postgraduate Research and Practice Innovation Program of Jiangsu Province, grant number [KYCX21_0886]; the National Natural Science Foundation of China [52108408]; the Natural Science Foundation of Jiangsu Province [BK20210617].

Institutional Review Board Statement: Not applicable.

Informed Consent Statement: Not applicable.

Data Availability Statement: All the original data of this article can be obtained by contacting the corresponding author.

Conflicts of Interest: The authors declare no conflict of interest.

References

1. Li, H.; Ding, L.; Ren, M.; Li, C.; Wang, H. Sponge City Construction in China: A Survey of the Challenges and Opportunities. *Water* **2017**, *9*, 594. [CrossRef]
2. Li, Z.; Cao, Y.; Zhang, J.; Liu, W. Urban Rainfall Characteristics and Permeable Pavement Structure Optimization for Sponge Road in North China. *Water Sci. Technol.* **2021**, *83*, 1932–1945. [CrossRef] [PubMed]
3. Wu, B.; Luo, C.; Pei, Z.; Chen, C.; Xia, J.; Xiao, P. Evaluation of the Aging of Styrene-Butadiene-Styrene Modified Asphalt Binder with Different Polymer Additives. *Materials* **2021**, *14*, 5715. [CrossRef] [PubMed]
4. Zhang, F.; Hu, C. The Composition and Ageing of High-Viscosity and Elasticity Asphalts. *Polym. Compos.* **2017**, *38*, 2509–2517. [CrossRef]
5. Ma, X.; Li, Q.; Cui, Y.; Ni, A. Performance of Porous Asphalt Mixture with Various Additives. *Int. J. Pavement Eng.* **2018**, *19*, 355–361. [CrossRef]
6. Cai, J.; Song, C.; Zhou, B. Investigation on high-viscosity asphalt binder for permeable asphalt concrete with waste materials. *J. Clean. Prod.* **2019**, *228*, 40–51. [CrossRef]
7. Geng, L.; Xu, Q.; Ren, R.; Wang, L.; Yang, X.; Wang, X. Performance Research of High-Viscosity Asphalt Mixture as Deck-Paving Materials for Steel Bridges. *Road Mater. Pavement* **2017**, *18*, 208–220. [CrossRef]
8. Luo, Y.; Zhang, K.; Li, P.; Yang, J.; Xie, X. Performance Evaluation of Stone Mastic Asphalt Mixture with Different High Viscosity Modified Asphalt Based on Laboratory Tests. *Constr. Build. Mater.* **2019**, *225*, 214–222. [CrossRef]
9. Li, Q.; Zeng, X.; Wang, J.; Luo, S.; Meng, Y.; Gao, L. Aging performance of high viscosity modified asphalt under complex heat-light-water coupled conditions. *Constr. Build. Mater.* **2022**, *325*, 126314. [CrossRef]
10. Yang, J.; Zhang, Z.; Shi, J.; Yang, X.; Fang, Y. Comparative Analysis of Thermal Aging Behavior and Comprehensive Performance of High Viscosity Asphalt (HVA) from Cohesion, Adhesion and Rheology Perspectives. *Constr. Build. Mater.* **2022**, *317*, 125982. [CrossRef]
11. Hu, M.; Sun, G.; Sun, D.; Lu, T.; Ma, J.; Deng, Y. Accelerated Weathering Simulation on Rheological Properties and Chemical Structure of High Viscosity Modified Asphalt: A Temperature Acceleration Effect Analysis. *Constr. Build. Mater.* **2021**, *268*, 121120. [CrossRef]
12. Li, Y.; Wu, S.; Liu, Q.; Dai, Y.; Li, C.; Li, H.; Nie, S.; Song, W. Aging Degradation of Asphalt Binder by Narrow-Band UV Radiations with a Range of Dominant Wavelengths. *Constr. Build. Mater.* **2019**, *220*, 637–650. [CrossRef]
13. Zou, Y.; Amirkhanian, S.; Xu, S.; Li, Y.; Wang, Y.; Zhang, J. Effect of Different Aqueous Solutions on Physicochemical Properties of Asphalt Binder. *Constr. Build. Mater.* **2021**, *286*, 122810. [CrossRef]
14. Hou, X.; Lu, S.; Chen, Z.; Xiao, F. Applications of Fourier Transform Infrared Spectroscopy Technologies on Asphalt Materials. *Measurement* **2018**, *121*, 304–316. [CrossRef]
15. Singh, B.; Saboo, N.; Kumar, P. Use of Fourier transform infrared spectroscopy to study ageing characteristics of asphalt binders. *Petrol. Sci. Technol.* **2017**, *35*, 1648–1654. [CrossRef]

16. Yang, X.; You, Z.; Mills-Beale, J. Asphalt binders blended with a high percentage of bio-binders: Aging mechanism using FTIR and rheology. *J. Mater. Civ. Eng.* **2015**, *27*, 401–415. [CrossRef]
17. Yao, H.; Dai, Q.; You, Z. Fourier Transform Infrared Spectroscopy Characterization of Aging-Related Properties of Original and Nano-Modified Asphalt Binders. *Constr. Build. Mater.* **2015**, *101*, 1078–1087. [CrossRef]
18. Singh, B.; Kumar, P. Effect of Polymer Modification on the Ageing Properties of Asphalt Binders: Chemical and Morphological Investigation. *Constr. Build. Mater.* **2019**, *205*, 633–641. [CrossRef]
19. Zhang, F.; Hu, C.; Zhuang, W. The Research for Low-Temperature Rheological Properties and Structural Characteristics of High-Viscosity Modified Asphalt. *J. Therm. Anal. Calorim.* **2018**, *131*, 1025–1034. [CrossRef]
20. Yang, B.; Wei, J.; Xie, J.; Zhao, Z. Aging Kinetic Equation of Pure Asphalt and Trinidad-Lake-Asphalt-Modified Asphalt. *Sci. Adv. Mater.* **2017**, *9*, 1081–1086. [CrossRef]

materials

MDPI

Article

Mechanical Performance of 3D Printed Concrete in Steam Curing Conditions

Bolin Wang [1], Xiaofei Yao [1], Min Yang [1,*], Runhong Zhang [2,3], Jizhuo Huang [4,*], Xiangyu Wang [5], Zhejun Dong [6] and Hongyu Zhao [2,7,*]

[1] CCCC First Highway Consultants Co., Ltd., Xi'an 710075, China; wbl1993lang@163.com (B.W.); slbyxf@163.com (X.Y.)
[2] Institute for Smart City of Chongqing University in Liyang, Chongqing University, Liyang 213300, China; zhangrh@cqu.edu.cn
[3] College of Aerospace Engineering, Chongqing University, Chongqing 400045, China
[4] College of Civil Engineering, Fuzhou University, 2 Xue Yuan Rd., University Town, Fuzhou 350116, China
[5] School of Design and Built Environment, Curtin University, Perth 6102, Australia; xiangyu.wang@curtin.edu.au
[6] School of Civil Engineering, Suzhou University of Science and Technology, Suzhou 215009, China; knight.cc.ccc@gmail.com
[7] School of Civil Engineering, Chongqing University, Chongqing 400045, China
* Correspondence: yangminccccc@sina.com (M.Y.); jzhuang@fzu.edu.cn (J.H.); 20211601069@cqu.edu.cn (H.Z.)

Abstract: Three-dimensional (3D) concrete printing (3DCP) technology attracts significant attention from research and industry. Moreover, adequate mechanical performance is one of the primary properties for materials, meeting the demand of structural safety using 3DCP technology. However, research on curing conditions as the significant influence factor of mechanical capacity is required to accelerate the practical application of 3DCP technology. This study aims to explore the impact of various steam curing conditions (heating rate, constant temperature time, and constant temperature) on the mechanical performance of printed concrete containing solid wastes. Moreover, the optimal steam curing conditions are obtained for compressive, tensile, and flexural properties in different directions. Subsequently, anisotropies in the mechanical properties of printed composites and interlayer bonding behaviors are investigated when various curing conditions are employed. The result shows that steam curing conditions and solid waste incorporation improves the interlayer bond for 3D printed cement-based composites.

Keywords: 3D concrete printing; curing conditions; mechanical capacity; solid waste; anisotropy; sustainability

Citation: Wang, B.; Yao, X.; Yang, M.; Zhang, R.; Huang, J.; Wang, X.; Dong, Z.; Zhao, H. Mechanical Performance of 3D Printed Concrete in Steam Curing Conditions. *Materials* **2022**, *15*, 2864. https://doi.org/10.3390/ma15082864

Academic Editor: Francesco Fabbrocino

Received: 1 March 2022
Accepted: 10 April 2022
Published: 13 April 2022

Publisher's Note: MDPI stays neutral with regard to jurisdictional claims in published maps and institutional affiliations.

Copyright: © 2022 by the authors. Licensee MDPI, Basel, Switzerland. This article is an open access article distributed under the terms and conditions of the Creative Commons Attribution (CC BY) license (https://creativecommons.org/licenses/by/4.0/).

1. Introduction

Additive manufacturing is a method exhibiting the characteristics of saving material, being economical, having high construction speeds, and enabling flexible design, and has been employed in various fields [1–3]. In the last decade, the interest in additive manufacturing from the building industry has significantly increased [4–6]. The rapid growth and elaborate research in this field indicate that additive manufactured construction structures are not a dream but will be a reality in the near future, although they are still in their nascent stage [7–9]. Meanwhile, additive manufacturing in construction has developed varying technologies such as contour crafting, D-shape, and extrusion-based 3DCP [10–12]. 3DCP technology is the cementitious composites deposited in layer-by-layer shapes by the coordination of a computer design and the help of a printer [2,13]. Advantages of 3DCP include no requirement for formwork, waste reduction, and significant saving in costs and labor [14–18]. Consequently, 3DCP has attracted the most attention from both academia and industry in the additive manufacturing area [19–21].

To ensure structural safety using 3DCP technology, desirable materials using the 3DCP method demand adequate mechanical properties for load bearing [1,19,22]. Additionally, high early strength is essential to ensure the maintenance of printed shapes and printed layers without collapse [23–25]. The curing conditions and material mix proportions are the main influencing factors on the mechanical properties of a material [26–28]. Moreover, the steam curing method improves the hydration and early strength of materials to benefit 3DCP [29–31]. However, improper steam curing technology used in printed concrete creates an interface transition zone between the cementitious material and aggregate resulting in thermal expansion deformation which produces microcracks [32,33]. As the microcracks grow, the long-term performance of steam-cured concrete may be reduced. Furthermore, conventional concrete by 3DCP methodology generally uses fine aggregates, leading to more cement usage, CO_2 emissions, and energy consumption than traditional casting concrete due to the higher cement content [34–36]. Compared to cast concrete, higher evaporation rates, surface drying, and cement content for printed concrete can induce shrinkage cracking due to the lack of formwork [37–39]. Disposing of anisotropy (the distinct property of 3D printing materials) is crucial to promoting the use of 3DCP technology in large-scale construction [40–42]. Therefore, research on the optimum steam curing conditions of mechanical capacity and anisotropy of 3DCP is required.

The rapid development of industrial production and demolition has led to a huge amount of various solid wastes [43–45]. Meanwhile, more than 95% of urban solid wastes are directed to landfill, thus impacting the health of humans and the environment [46–48]. As a result, effectively improving the reuse of solid wastes is urgent to obtain sustainability [49–51]. FA, SF, and GGBS are solid wastes in the by-product of the energy and smelting industry, which need to be resolved at a considerable cost [52–54]. However, cementitious composites using 3DCP technology employ the combination of FA, SF, and GGBS to replace cement, improving sustainability, shrinkage resistance, steam curing adaptability, durability, and the mechanical properties of materials [25,55,56]. Printed materials with waste minerals incorporated improve the internal pore structure and the particle fineness gradation, based on the theory of dense packing [57–59].

This study aims to investigate the influence of mechanical performance when various steam curing conditions are employed on the concrete using 3DCP technology. Orthogonal experiments are conducted to reduce the experimental workload and obtain a high-sensitivity evaluation indicator. Additionally, a coefficient is used to quantitatively assess the anisotropy of the materials. Subsequently, a further analysis of printed interlayer bonding performance is conducted. Thereby, the optimum printed concrete steam curing conditions for various mechanical performance and anisotropy are acquired.

2. Materials and Methods

2.1. Raw Materials

The precursor materials of the samples in this study were 42.5 R Portland cement (produced by China Anhui Conch Group Company Limited, Wuhu, China), FA (Class F, obtained from Shenglong Technology Industry Co., Ltd., Weinan, China), GGBS (produced by Delong Powder Engineering Material Co., Ltd., Xi'an, China), and SF (provided by Linyuan micro silica powder Co., Ltd., Xi'an, China). The chemical compositions of the precursor materials are shown in Table 1. Quartz sand with a particle grading range of between 0.21 mm and 0.55 mm wa used because it has less mud content than natural sand, as shown in Figure 1. Additionally, the good gradation of quartz sand guaranteed that the pipe and nozzle of the printer were not blocked. The physical and mechanical properties of PVA fibers utilized in this study are shown in Table 2. Additionally, the high-efficiency polycarboxylic acid type water reducer and hydroxypropyl methylcellulose (viscosity at 40,000 Cp) were employed to improve the working performance of 3D printed concrete.

Table 1. Chemical compositions, specific surface area, fineness of 45μm sieve residue, and density of the cement, FA, SF, and GGBS (wt.%).

Precursors	Content (wt.%)										Specific Surface Area (m^2/kg)	Fineness of 45 μm Sieve Residue (%)	Density (kg/m^3)
	SiO_2	Al_2O_3	Fe_2O_3	CaO	Ignition Loss	MgO	SO_3	K_2O	Na_2O	Cl			
Cement	19.4–21.5	4.1–4.9	2.8–2.9	61.9–64.2	1.9–2.0	1.1–1.2	3.0–3.2	0.6–0.7	0.1–0.2	0.01–0.02	350–400	-	2800–3200
FA	43–54	28–34	8–13	0.4–0.5	1.6–4.7	1.1–2.3	0.5–1.2	2–4	0.8–1.5	0.01–0.02	-	6.7–7.3	2400–2500
SF	93–97	-	-	0.26–0.28	1.0–1.1	-	-	-	-	0.01–0.02	20,000–27,000	2.2–4.1	320–380
GGBS	34.7–38.2	9.1–10.2	0.5–0.7	38.8–40.5	0.6–0.8	9.9–11.1	0.1–1.8	0.12–0.14	0.24–0.29	0.01–0.04	420–480	5.8–7.5	2800–2900

Figure 1. (**a**) FA, (**b**) SF, (**c**) GGBS, and (**d**) quartz sand used in specimens of this experiment [60].

Table 2. Main properties of the PVA fibers.

Diameter (μm)	Length (mm)	Density (kg/m^3)	Elastic Modulus (GPa)	Tensile Strength (MPa)	Elongation (%)
39	18	1.2	76.5	1950	6

2.2. *Orthogonal Experimental Design of Mix Proportion*

The mixing proportions of 3D printed concrete mainly include a water binder ratio, sand binder ratio, polycarboxylate superplasticizer, PVA fiber, hydroxypropyl methylcellulose, etc. The water binder ratio of this experiment was 0.36 based on a trial test considering the balance between printable fluidity and mechanical strength. The sand binder ratio at 1.3 was verified to provide smooth extrusion and reasonable cost. The optimal material mix proportion was determined by previous tests of mechanical and printable capacity [60]. Therefore, the components of FA, SF, and GGBS were 20 wt.%, 15 wt.%, and 10 wt.%, respectively, in the cementitious materials which were calculated when the total percentage of binder materials was 100 wt.%, as shown in Table 3. PVA fiber, hydroxypropyl methylcellulose, and polycarboxylate superplasticizer content in the mortar were 0.21 wt.%, 0.23 wt.%, and 0.016 wt.%, respectively. Hydroxypropyl methylcellulose promotes the dispersion of mortar and fiber, thus enhancing the water-retaining property of the material. Moreover, the addition of hydroxypropyl methylcellulose improved viscosity.

Table 3. Control mix proportion of this test.

Components (wt.%)								
Cement	FA	SF	GGBS	Sand	Water	Polycarboxylate Superplasticizer	PVA Fiber	Hydroxypropyl Methylcellulose
55	20	15	10	130	36	0.23	0.21	0.016

The orthogonal test aimed to explore the mechanical capacity of the printed concrete influenced by various steam curing conditions. With reference to Chinese criteria [61] and previous research [62], three factors (heating rate, holding time, holding temperature) and three levels were designed in the orthogonal experimental shown in Table 4. The study utilized range analysis because of higher intelligibility and more convenient calculation compared to analysis of variance. The range analysis method utilizes the average value calculation of each level and obtains the range value, and subtracts the minimum level average value from the maximum level average value. The primary and secondary relationships affecting the index value are determined by the range value.

Table 4. The curing conditions of samples in this test.

Test Group	Curing Conditions		
	e (°C/h)	E (h)	f (°C)
1	10	6	50
2	10	8	70
3	10	10	60
4	15	6	70
5	15	8	60
6	15	10	50
7	20	6	60
8	20	8	50
9	20	10	70
C_c C_p	Curing in natural condition		

Note: e represents heating-up rate; E represents thermostatic period; f represents thermostatic temperature, C_c represents control cast sample, C_p represents control printed sample, natural condition is 20 °C average temperature and 60% humidity.

2.3. Load Direction Definition

Load direction was determined to explore the anisotropy of printed composites. Figure 2a shows that the three directions of X, Y, and Z are orthogonal. The X, Y, and Z directions are the outer points perpendicular to the centre of the cross-section of the printed layer, the profile center of the printed layer, and the center of the plane section of the print layer, respectively. F_{xy} and F_{xz} are the X-direction line loads, which extend in Y direction and Z direction, respectively, as shown in Figure 2b. F_{yx} and F_{yz} represent the line loads in the Y direction extending in the X direction and Z direction, respectively. F_{zx} and F_{zy} denote the Z-direction line loads extending in the X direction and Y direction, respectively.

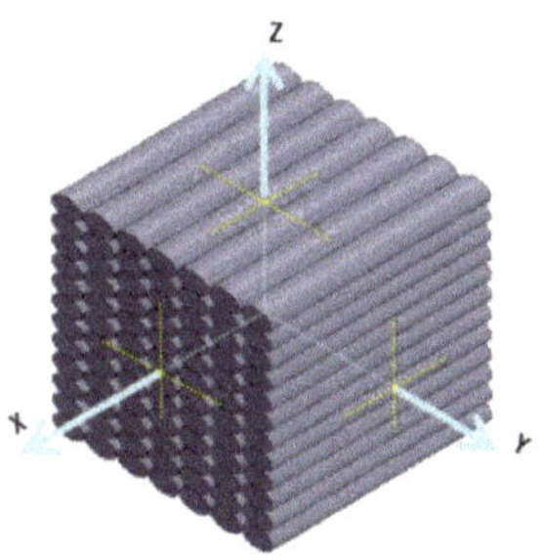

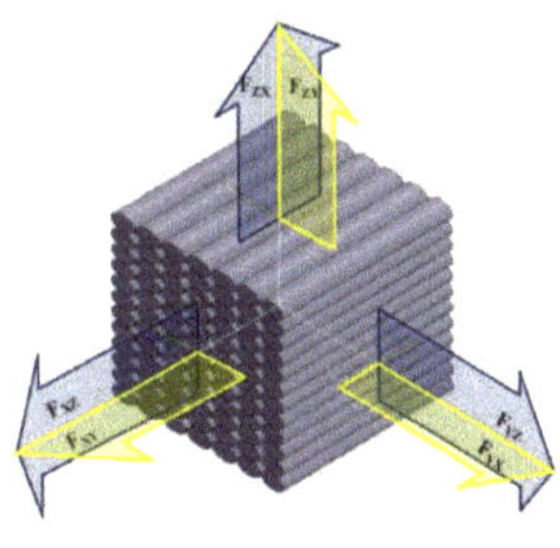

Figure 2. (**a**) Coordinate system, (**b**) loading direction.

2.4. Mechanical Performance Test

The mix proportion of printable mortar referred to in Table 3 was used for the mechanical performance test. The structure was extruded and deposited by 3DCP equipment according to the designative printing path, as shown in Figure 3. Additionally, the printing direction of each layer was Y. The height, width of printing nozzle, and nozzle moving velocity were 38 mm, 14 mm, and 12 cm/s, respectively. To improve printing efficiency and structure quality, a large-sized printing nozzle was used for the structure fabrication. Moreover, it enabled a more stable deposition of the printing filaments compared to the circular nozzle, preventing collapse. Each layer was manufactured with 12 round-trip continuous paths and the number of vertical deposition layers was 20. The printing path interval was 40 mm and the height of each layer was set to 13 mm. The structure used a bucket (0.05 m^3) of printing material and its printing time was 15 min. The open time of the mortar was 40 min. The printing and equipment parameters were based on previous experimental results of buildability and extrudability [60]. The round-trip continuous paths ensured that the mutual extrusion of printing filament was not improved.

(a) (b)

Figure 3. (**a**) Printing path model, (**b**) printing process [60].

By adopting the Chinese national standard GB/T 50081-2002 [63,64], compressive, flexural, and tensile tests were employed to evaluate the mechanical performance influenced by various steam curing conditions. The cube samples (100 mm × 100 mm × 100 mm) and prismatic specimens (50 mm × 50 mm × 200 mm), used in different direction tests, were obtained by structural cutting after 8 h of natural maintenance, as shown in Figure 4a,b. A printed structure was cut into three cube compression test samples; and three cube specimens were used in the X, Y, and Z directions splitting tension test, as shown in Figure 4a. The splitting tension test specimens in the X direction were loaded in an XY or XZ direction, specimens in the Y directions were loaded in an YX or YZ direction, and specimens in the Z directions were loaded in an ZX or ZY direction. Six prismatic flexure tests specimens loaded in various directions (XY, XZ, YX, YZ, ZX, and ZY) were acquired from the printed structure. To ensure the accuracy of the mechanical properties experiment, the cube or prismatic specimens, which were employed in the same direction test, were cut from the same position in the printed structures and had the same interlayer number. After an 8-h cut time, the test samples were placed into a zky-400b steam box to cure one circle, as shown in Figure 4c. The steam curing conditions of the printed samples refer to the orthogonal experimental design (Table 4). Subsequently, the test samples were cured for 28 days in natural conditions. The experiments of 28-day compressive, flexural, and tensile performance utilized 9 group samples with various steam curing conditions and 2 control groups cured in natural conditions. Subsequently, uniaxial compressive strength in the X, Y, and Z directions was determined by compression tests in which each group contained three cube specimens, as shown in Figure 5a. Each group of flexural and splitting tensile capacity tests contained six prismatic specimens to be under the line flexural and tensile load of XY, XZ, YX, YZ, ZX, and ZY directions, as shown in Figure 5b,c. Moreover, the average and coefficient of variation of mechanical performance were obtained by measuring three group samples at each curing condition from Table 4.

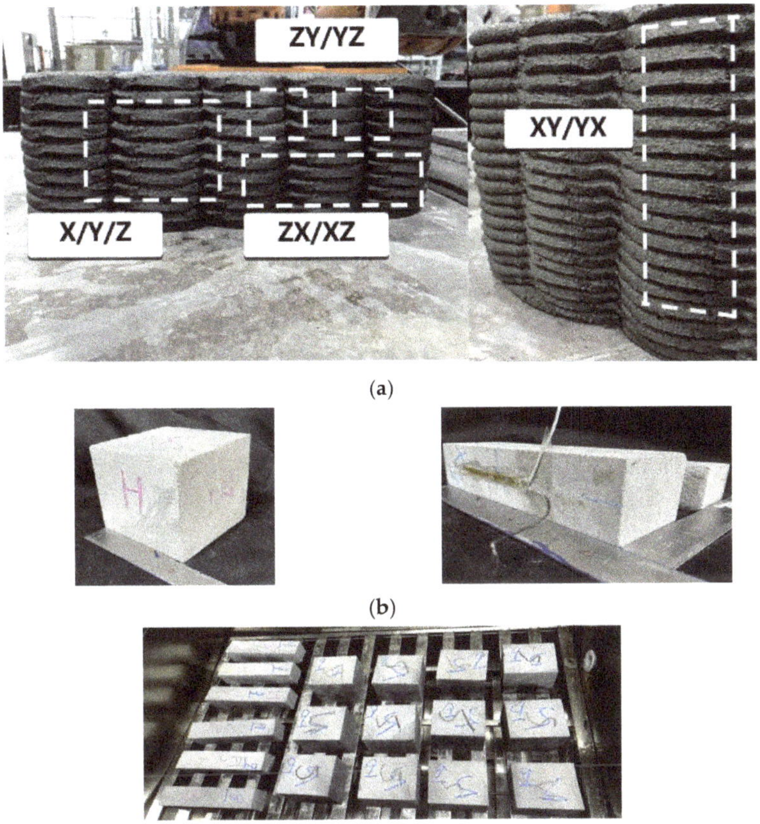

(**a**)

(**b**)

(**c**)

Figure 4. (**a**) Cutting position, (**b**) specimen size, and (**c**) steam curing condition [60].

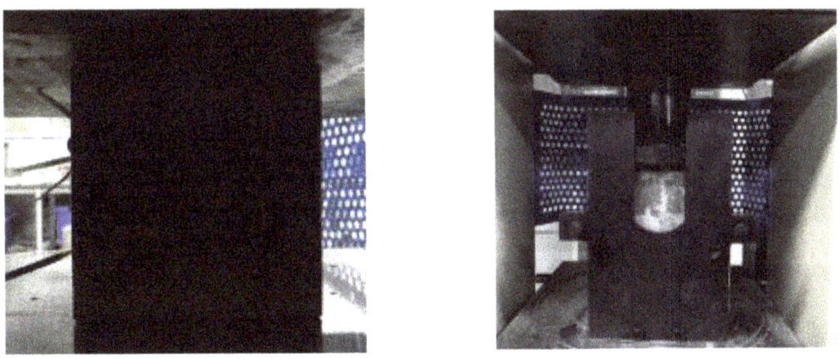

(**a**) (**b**)

Figure 5. *Cont.*

(c)

Figure 5. (**a**) Compressive, (**b**) splitting tensile capacity, and (**c**) flexural tests.

2.5. Anisotropy Assessment

This study applied an anisotropy coefficient to represent the influence of the printing process on the mechanical behavior of the materials, as described in Equations (1) and (2).

$$f_{avg} = \frac{\sum_{n=1}^{i} f_{xn} + \sum_{n=1}^{i} f_{yn} + \sum_{n=1}^{i} f_{zn}}{3i} \quad (1)$$

$$I_a = \frac{\sqrt{(f_{x1} - f_{avg})^2 + \ldots + (f_{xi} - f_{avg})^2 + (f_{y1} - f_{avg})^2 + \ldots + (f_{yi} - f_{avg})^2 + (f_{z1} - f_{avg})^2 + \ldots + (f_{zi} - f_{avg})^2}}{f_{avg}} \quad (2)$$

where i is the number of load directions in the direction of the main load, f_{xi}, f_{yi}, and $f_{zi,}$ are the average strength of i-th load direction in the X, Y, and Z direction, respectively, f_{avg} is the average strength of all loads, and I_a is the anisotropy coefficient. Thereby, the value of I_a is the positive correlation with anisotropy of the printed material. The smaller value of I_a ought to be employed to improve the feasibility of printed structures when the requirement for mechanical property directions of the structure are not determined.

3. Results and Discussion

3.1. Analysis of Mechanical Performance in Various Curing Conditions

3.1.1. Compressive Performance

The compressive strength (CS) test data and variation coefficient of the samples in the various curing conditions were summarized in the L9 (3^3) orthogonal experimental table (Table 5). The average values of CS_x, CS_y, and CS_z were obtained by the results and calculations of the orthogonal test, as shown in Figure 6. The CS_x range values in the various curing conditions of heating-up rate, thermostatic period, and thermostatic temperature were 7.70 MPa, 1.90 MPa, and 3.83 MPa, respectively, and determined the degree of influence of the three steam curing conditions (heating-up rate > thermostatic temperature > thermostatic period). Meanwhile, the CS_y range values in the varying conditions of heating-up rate, thermostatic period, and thermostatic temperature were 4.23 MPa, 7.50 MPa, and 5.47 MPa, respectively, which determined the influence degree of thermostatic period > thermostatic temperature > heating-up rate. Furthermore, the CS_z range values in the various conditions of heating-up rate, thermostatic period, and thermostatic temperature were 4.37 MPa, 8.23 MPa, and 7.30 MPa, respectively, which determined the influence degree of thermostatic period > thermostatic temperature > heating-up rate.

Table 5. Orthogonal experimental table of various test group designs, curing conditions, compressive strength test data, and coefficients of variation.

Test Group	Curing Conditions			Compressive Strength (MPa)			Coefficient of Variation (%)		
	e (°C/h)	E (h)	f (°C)	CS_x	CS_y	CS_z	CS_x	CS_y	CS_z
1	10	6	50	48.0	55.0	36.0	5.12	3.75	9.13
2	10	8	70	52.0	45.8	46.9	7.20	8.86	9.51
3	10	10	60	48.3	50.3	43.3	7.66	6.81	5.58
4	15	6	70	46.1	56.7	39.4	8.60	5.72	9.27
5	15	8	60	42.3	42.7	40.8	4.89	7.21	8.66
6	15	10	50	47.5	45.7	32.9	5.59	4.98	9.40
7	20	6	60	39.6	43.6	44.9	6.43	5.61	5.08
8	20	8	50	42.0	44.3	44.9	5.99	6.45	7.37
9	20	10	70	43.6	50.5	31.7	7.02	4.41	8.81
C_c	Curing in natural condition				42.7			7.53	
C_p				48.1	56.3	35.8	4.37	3.80	7.96

Note: CS_x represents compressive capacity in the X direction; CS_y represents compressive capacity in the Y direction; CS_z represents compressive capacity in the Z direction; C_c means control cast sample; C_p means control printed sample.

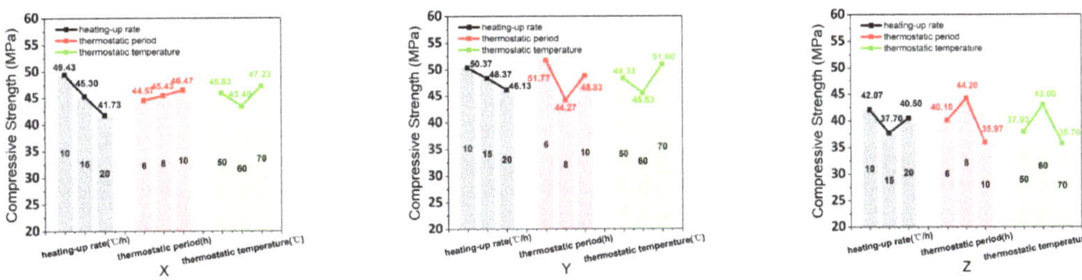

(a) (b) (c)

Figure 6. (**a**) CS_x, (**b**) CS_y, and (**c**) CS_z affected by various heating-up rates, thermostatic periods, and thermostatic temperatures.

Figure 6a shows that the CS_x had a negative correlation, positive correlation, and parabolic correlation with the curing conditions of heating-up rate, thermostatic period, and thermostatic temperature, respectively. Additionally, Figure 6b shows that negative correlation, parabolic correlation, parabolic correlation were achieved by the CS_y with the curing conditions of heating-up rate, thermostatic period, and thermostatic temperature, respectively. Figure 6c shows that CS_z had a parabolic correlation with the curing conditions of heating-up rate, thermostatic period, and thermostatic temperature. Moreover, the relation between CS_z and heating-up rate was a concave-shape curve, which differed from thermostatic period and thermostatic temperature. Thereby, the optimal parameter of steam curing conditions for CS_x and CS_y are 10 °C/h heating-up rate, 10h thermostatic period, and 70 °C thermostatic temperature based on the correlation with the various curing conditions. Moreover, the optimal parametersfor steam curing conditions for CS_z were 10 °C/h heating-up rate, 8h thermostatic period, and 60 °C thermostatic temperature.

3.1.2. Splitting Tensile Performance

The test data and variation coefficient of the splitting tensile strength (TS) for this study are summarized in the L_9 (3^3) orthogonal experimental tables (Tables 6 and 7). TS of the printed specimen was influenced by the steam curing technology and interlayer effect, as shown in Table 6. Applying steam curing technology to cast samples has some negative effects on the mechanical properties due to a heterogeneous distribution of air voids in the specimens. The cast concrete demonstrates uneven expansion under heating-up conditions due to the thermal expansion of air and water. However, the steam curing technology can improve the TS of printed samples because of an improvement in the bond strength between layers. Additionally, the air voids in the printed samples are relatively homogeneous due to the printing path impact. Small dimensions of the air bubbles and a tight internal structure are produced as a result of the deposition effects of the 3DCP technology. The addition of various solid wastes can improve the hydration of the products and the pore structure, thereby enhancing the steam curing adaptability of 3D printed cement-based materials.

Table 6. Orthogonal experimental table of various test group designs, curing conditions, and splitting tensile strength test data.

Test Group	Curing Conditions			Splitting Tensile Strength (MPa)					
	e (°C/h)	E (h)	f (°C)	TS_{xy}	TS_{xz}	TS_{yx}	TS_{yz}	TS_{zx}	TS_{zy}
1	10	6	50	2.680	4.201	5.468	3.081	2.788	4.253
2	10	8	70	5.380	3.488	4.113	4.520	3.807	4.087
3	10	10	60	4.654	2.756	3.813	4.591	2.896	3.693
4	15	6	70	4.024	3.164	4.348	4.316	2.891	4.902
5	15	8	60	5.367	2.865	4.278	4.546	3.336	3.801
6	15	10	50	4.628	2.228	4.138	4.049	2.547	2.947
7	20	6	60	4.182	2.222	5.284	3.081	2.999	4.208
8	20	8	50	4.113	2.699	4.145	2.489	2.355	4.074
9	20	10	70	4.456	2.611	3.501	3.145	3.145	3.126
C_c	Curing in natural condition			4.43					
C_p				4.533	2.018	2.814	2.588	2.426	2.909

Note: TS_{xy} represents splitting tensile strength in the XY direction; TS_{xz} represents splitting tensile strength in the XZ direction; TS_{yx} represents splitting tensile strength in the YX direction, TS_{yz} represents splitting tensile strength in the YZ direction, TS_{zx} represents splitting tensile strength in the ZX direction, TS_{zy} represents splitting tensile strength in the ZY direction.

Table 7. Orthogonal experimental table of various test group designs, curing conditions, and variation coefficient of splitting tensile strength.

Test Group	Curing Conditions			Coefficient of Variation (%)					
	e (°C/h)	E (h)	f (°C)	TS_{xy}	TS_{xz}	TS_{yx}	TS_{yz}	TS_{zx}	TS_{zy}
1	10	6	50	2.34	5.27	5.90	9.27	1.93	3.99
2	10	8	70	9.46	7.40	10.10	7.56	5.17	5.60
3	10	10	60	4.43	8.34	7.93	2.45	6.50	2.39
4	15	6	70	8.60	9.12	9.25	6.06	9.78	7.08
5	15	8	60	10.03	5.59	9.74	4.64	4.37	2.65
6	15	10	50	5.43	3.45	8.65	5.73	8.08	3.93
7	20	6	60	9.81	7.06	7.51	1.48	5.59	2.64
8	20	8	50	7.77	2.68	5.15	4.84	6.83	8.89
9	20	10	70	9.91	3.93	10.78	9.07	8.66	6.75
C_c	Curing in natural condition			4.15					
C_p				5.34	3.78	9.71	5.39	7.13	1.85

The average values of TS_{xy}, TS_{xz}, TS_{yx}, TS_{yz}, TS_{zx}, and TS_{zy} are obtained by the results and calculations of the orthogonal test, as shown in Figure 7. The relation between TS_{xy} and the three curing conditions (thermostatic period, thermostatic temperature, heating-up rate) was a parabolic correlation as presented in Figure 7a. Meanwhile, the optimal curing conditions were 15 °C/h (heating-up rate), 8 h (thermostatic period), and 60 °C (thermostatic temperature) for TS_{xy}. Figure 7b demonstrates that TS_{xz} had a negative correlation, negative correlation, and parabolic correlation with heating-up rate, thermostatic period, and thermostatic temperature, respectively. A heating-up rate of 10 °C/h, 6h thermostatic period, and 70 °C thermostatic temperature were the optimal curing conditions for TS_{xz}. Figure 7c shows that the optimal curing conditions of heating-up rate, thermostatic period, and thermostatic temperature for TS_{yx} were 10 °C/h, 6 h, 50 °C, respectively. Parabolic correlation, negative correlation, and negative correlation, respectively, are the relation between TS_{yx} and the three curing conditions (heating-up rate, thermostatic period, and thermostatic temperature). Figure 7d shows that TS_{yz} had a negative correlation with heating-up rate, negative correlation with thermostatic period, and parabolic correlation with thermostatic temperature. A 15 °C/h heating-up rate, 10h thermostatic period, and 60 °C thermostatic temperature were the optimal curing conditions for TS_{yz}. A negative correlation, parabolic correlation, and positive correlation were the relation of TS_{yx} with the heating-up rate, thermostatic period, and thermostatic temperature, respectively, as shown in Figure 7e. For TS_{yx} 10 °C/h, 8 h, and 70 °C were the optimal heating-up rate, thermostatic period, and thermostatic temperature for curing conditions. Figure 7f demonstrates that heating-up rate, thermostatic period, and thermostatic temperature had a negative correlation, parabolic correlation, and positive correlation, respectively, with TS_{zy}. Furthermore, the optimal curing conditions were 10 °C/h (heating-up rate), 8 h (thermostatic period), and 70 °C (thermostatic temperature) for TS_{zy}.

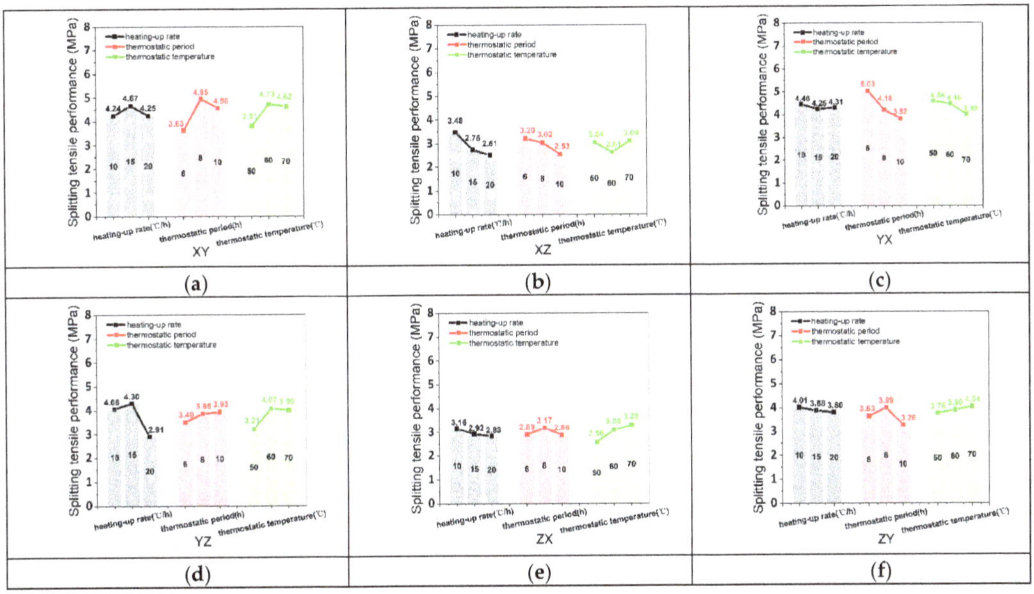

Figure 7. (**a**) TS_{xy}, (**b**) TS_{xz}, (**c**) TS_{yx}, (**d**) TS_{yz}, (**e**) TS_{zx}, and (**f**) TS_{zy} affected by various heating-up rates, thermostatic periods, and thermostatic temperatures.

3.1.3. Flexural Performance

The L_9 (3^3) orthogonal experimental tables (Tables 8 and 9) summarize the test data and variation coefficients of flexural strength (FS). Figure 8 shows that the average values of FS_{xy}, FS_{xz}, FS_{yx}, FS_{yz}, FS_{zx}, and FS_{zy} were obtained by the results and calculations of the orthogonal test. Compared to the control cast, the flexural strength of the printed samples was commonly decreased because of the weak surface of the printing interlayer. Besides, the practical structure ought to utilize ZY and YZ directions to resist the flexural load.

The optimal curing conditions were 15 °C/h (heating-up rate), 10 h (thermostatic period), and 60 °C (thermostatic temperature) for FS_{xy}, as presented in Figure 8a. Figure 8b demonstrates the optimal steam curing conditions (15 °C/h heating-up rate, 10 h thermostatic period, and 50 °C thermostatic temperature) for FS_{xz}. Figure 8c shows that the optimal curing conditions of heating-up rate, thermostatic period, and thermostatic temperature for FS_{yx} were 10 °C/h, 10 h, and 70 °C, respectively. Figure 8d exhibits that 20 °C/h heating-up rate, 8h thermostatic period, 50 °C thermostatic temperature were the optimal curing conditions for FS_{yz}. Dor FS_{yx}, 20 °C/h, 8 h, and 60 °C were the optimal heating-up rate, thermostatic period, and thermostatic temperature of curing condition, as shown in Figure 8e. Figure 8f demonstrates that the optimal curing conditions were 10 °C/h (heating-up rate), 8 h (thermostatic period), and 70 °C (thermostatic temperature) for FS_{zy}.

Table 8. Orthogonal experimental table of various test group designs, curing conditions, and flexural strength test data.

Test Group	Curing Conditions			Flexural Strength (MPa)					
	e (°C/h)	E (h)	f (°C)	TS_{xy}	FS_{xz}	FS_{yx}	FS_{yz}	FS_{zx}	FS_{zy}
1	10	6	50	5.94	8.10	6.12	12.06	8.28	11.70
2	10	8	70	5.65	7.56	6.74	10.29	8.64	11.65
3	10	10	60	6.14	8.31	5.94	10.64	8.98	10.60
4	15	6	70	5.55	8.98	5.64	9.59	7.17	9.46
5	15	8	60	7.00	7.57	6.43	11.08	9.91	10.38
6	15	10	50	7.42	9.85	6.25	11.09	7.94	11.74
7	20	6	60	6.87	8.11	5.25	10.37	10.29	11.16
8	20	8	50	5.68	9.53	6.10	12.45	8.29	12.37
9	20	10	70	5.34	8.31	7.36	11.29	7.57	9.05
C_c	Curing in natural condition					11.50			
C_p				5.63	9.57	6.13	12.72	8.97	12.20

Note: FS_{xy} represents flexural strength in the XY direction, FS_{xz} represents flexural strength in the XZ direction; FS_{yx} represents flexural strength in the YX direction, FS_{yz} represents flexural strength in the YZ direction, FS_{zx} represents flexural strength in the ZX direction, FS_{zy} represents flexural strength in the ZY direction.

Table 9. Orthogonal experimental table of various test group designs, curing conditions, and variation coefficient of flexural strength.

Test Group	Curing Conditions			Coefficient of Variation (%)					
	e (°C/h)	E (h)	f (°C)	FS_{xy}	FS_{xz}	FS_{yx}	FS_{yz}	FS_{zx}	FS_{zy}
1	10	6	50	5.33	3.58	8.68	6.14	7.85	8.41
2	10	8	70	7.21	5.47	9.33	5.49	8.29	5.21
3	10	10	60	10.3	4.18	6.75	6.35	8.89	5.47
4	15	6	70	6.81	3.93	10.26	5.37	9.37	5.26
5	15	8	60	4.5	6.11	5.37	2.99	5.06	4.47
6	15	10	50	5.16	4.51	7.29	5.1	7.11	7.2
7	20	6	60	8.48	6.4	9.2	6.8	10.03	1.8
8	20	8	50	8.28	1.65	8.93	7.2	6.13	8.39
9	20	10	70	7.08	9.51	9.19	6.23	8.55	6.19
C_c	Curing in natural condition					3.8			
C_p				2.15	6.07	5.68	3.74	1.15	5.14

3.2. Anisotropy Assessment in Various Curing Conditions

The test data of compressive strength, splitting tensile strength, and flexural strength employed Equations (1) and (2) to obtain the results of various anisotropy coefficients, as demonstrated in Table 10. The table shows that the anisotropy coefficient of splitting tensile performance was similar to the anisotropy coefficient of flexural performance and two-time compressive performance. The improvement of the weak interlayer surface of the printed concrete with added 0.21 wt% PVA fiber was limited by the brittleness of cementitious material. However, the steam curing condition improved the hydration of the materials, thus accelerating the development of composite strengths.

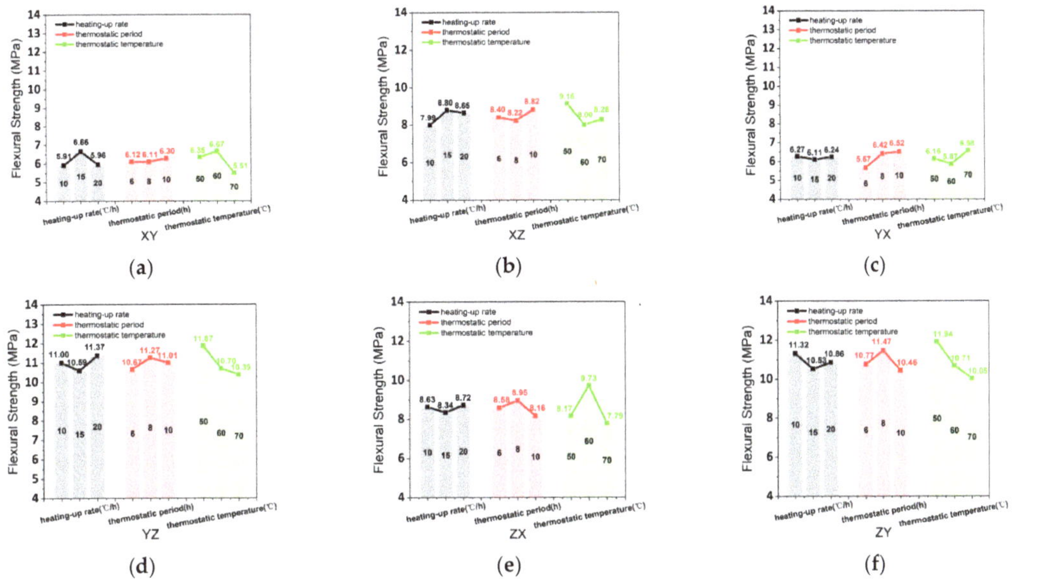

Figure 8. (**a**) FS_{xy}, (**b**) FS_{xz}, (**c**) FS_{yx}, (**d**) FS_{yz}, (**e**) FS_{zx}, and (**f**) FS_{zy} affected by various heating-up rates, thermostatic periods, and thermostatic temperatures.

Table 10. Orthogonal experimental table of various test group designs, curing conditions, and anisotropy coefficient of mechanical performance.

Test Group	Curing Conditions			Compressive Performance	Splitting Tensile Performance	Flexural Performance
	e (°C/h)	E (h)	f (°C)	I_{ac28} (%)	I_{at} (%)	I_{af} (%)
1	10	6	50	29.33	65.00	68.10
2	10	8	70	9.70	34.80	59.62
3	10	10	60	10.78	48.25	54.78
4	15	6	70	26.03	43.55	54.03
5	15	8	60	3.38	49.64	50.29
6	15	10	50	26.78	63.85	53.88
7	20	6	60	9.15	67.14	59.86
8	20	8	50	4.95	59.49	72.67
9	20	10	70	32.07	41.67	54.23
C_p	Curing in natural condition			31.22	67.41	71.87

Note: I_{ac28} represents anisotropy coefficient of compressive performance, I_{ap} represents anisotropy coefficient of splitting tensile performance, and I_{az} represents anisotropy coefficient of flexural performance.

3.2.1. Anisotropy Assessment of Compressive Performance

The average anisotropy coefficient of compressive strength is shown in Table 10 and calculated with various curing conditions (heating-up rate, thermostatic period, and thermostatic temperature), as shown in Figure 9. The compressive strength range values of various heating-up rates, thermostatic periods, and thermostatic temperatures were 3.34%, 17.20%, and 14.83%, respectively, obtaining the degree of influence of the three curing conditions on I_{ac} (thermostatic period > thermostatic temperature > heating-up rate). Figure 9 shows the relations between compressive strength and the three curing conditions (heating-up rate, thermostatic period, and thermostatic temperature) had a parabolic correlation. The influence of thermostatic period and thermostatic temperature exhibited the same variation trend. Moreover, the I_{ac} variation values with the influence of the thermostatic period and thermostatic temperature were similar. With the curing time increased, the influence degrees of the three curing conditions were reduced for I_{ac}. Thereby, the printed samples possessed the lowest anisotropy of compressive performance when heating-up rate, thermostatic period, and thermostatic temperature were 20 °C/h, 8 h, and 60 °C, respectively.

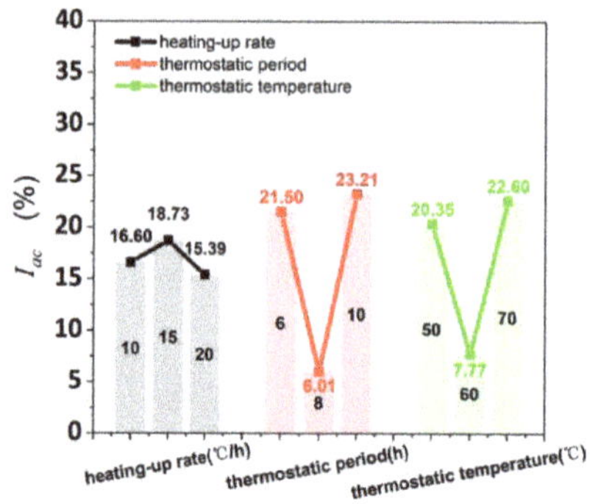

Figure 9. I_{ac} affected by various heating-up rates, thermostatic periods, and thermostatic temperatures.

3.2.2. Anisotropy Assessment of Splitting Tensile Performance

Figure 10 shows that the average anisotropy coefficient of the splitting tensile strength was calculated by the data of the orthogonal test (Table 10). Furthermore, the I_{ap} range values of varying curing conditions of heating-up rate, thermostatic period, and thermostatic temperature were 6.75%, 10.59%, and 22.77%, respectively, obtaining the influence degree of thermostatic temperature > thermostatic period > heating-up rate. The I_{ap} had the positive correlation, parabola correlation, and negative correlation with the curing conditions of heating-up rate, thermostatic period, and thermostatic temperature, respectively. Hence, 20 °C/h heating-up rate, 8h thermostatic period, and 70 °C thermostatic temperature were the optimum steam curing condition to reduce the anisotropy of splitting tensile performance.

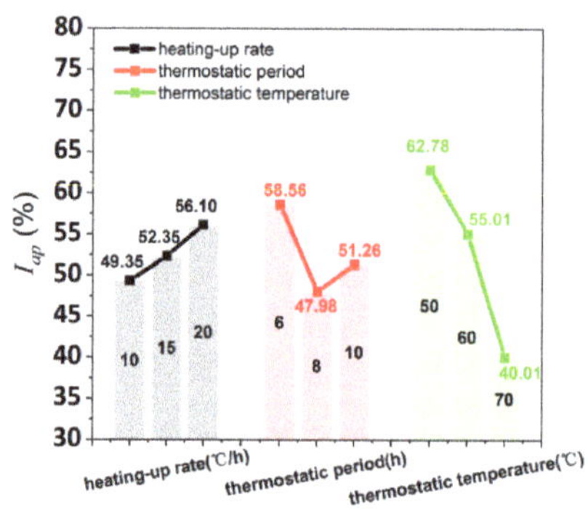

Figure 10. I_{ap} affected by various heating-up rates, thermostatic periods, and thermostatic temperatures.

3.2.3. Anisotropy Assessment of Flexural Performance

The data of the orthogonal test in Table 10 was used to calculate the average anisotropy coefficient of flexural strength, as shown in Figure 11. The I_{az} strength range values of the various curing conditions of heating-up rate, thermostatic period, and thermostatic temperature were 9.52%, 6.56%, and 9.91%, respectively, obtaining the influence degree of three solid wastes on I_{az} (thermostatic temperature > heating-up rate > thermostatic period). Simultaneously, the I_{az} had a parabola correlation with the three steam curing conditions (heating-up rate, thermostatic period, and thermostatic temperature). Therefore, the lowest anisotropy of flexural performance for printed concrete was acquired when heating-up rate, thermostatic period, and thermostatic temperature were 15 °C/h, 10 h, and 60 °C, respectively.

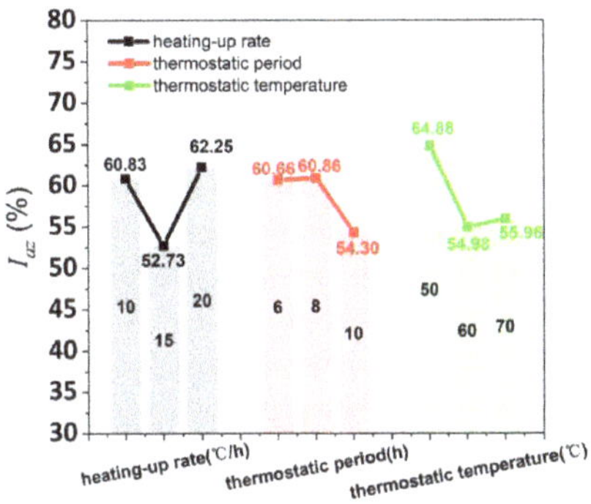

Figure 11. I_{az} affected by various heating-up rates, thermostatic periods, and thermostatic temperatures.

3.3. *Analysis of Interlayer Bonding Capacity*

Figure 12a shows that the sequence of compressive performance averages were $CS_y > CS_x > CS_z$. The compressive performance averages of the printed samples were higher than the control cast. Figure 12b,c demonstrates that the splitting tensile and flexural performance averages of the printed samples were lower than the cast. However, the splitting tensile performance averages of the printed specimens in the XZ and ZX directions were close to cast. For the flexural capacity, the averages of the printed specimens in the YZ and ZY directions were 0.52 MPa and 0.6 MPa, slightly lower than cast, respectively.

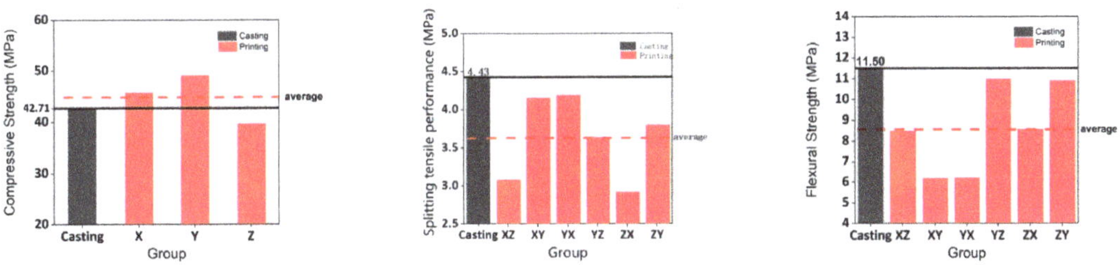

(a) (b) (c)

Figure 12. The mechanical performance average comparison of cast and printed samples in the varying directions: (**a**) compressive performance, (**b**) splitting tensile performance, (**c**) flexural performance.

The interlayer bonding capacity influences the mechanical performance of printed concrete and is related to the steam curing method, solid waste incorporation, and the printing parameters. The impact of steam curing conditions on mechanical performance is discussed in Sections 3.1 and 3.2. For solid waste, printed material incorporates various solid wastes to improve the microcrystalline nuclear effect, pozzolan effect, and micro aggregate effect leading to a dense structure. The microcrystalline nuclear effect accelerates the hydration reaction and improves the homogenization of the hydration product distribution. Furthermore, solid waste absorbing the calcium hydroxide of hydration enhances the hydration of the cement to produce more C-S-H gels, thereby improving the microstructure of the material. Meanwhile, the void ratio reduction of the material enables the cohesion of the aggregate interface up and increases the weak interface of printed concrete.

The printable filament height and weight were 38mm and 13mm, respectively. The layer height of printable mortar was lower than the nozzle height of the printer due to the additional pressure of the upper printed layer. Figure 13a,b exhibit the diagrammatic sketches of horizontal interlayer and vertical interlayer, respectively, for the printed materials using the 3DCP method. Figures 13c and 14 show that the main crack angle generated on the C plane was oblique 45°, which was different from cast (20°–30°) when the X-direction compression was loaded on the samples. Additionally, some vertical cracks were generated on the C plane because the weak interlayer influenced the slenderness ratio and shear span ratio of the printed material to result in shearing failure.

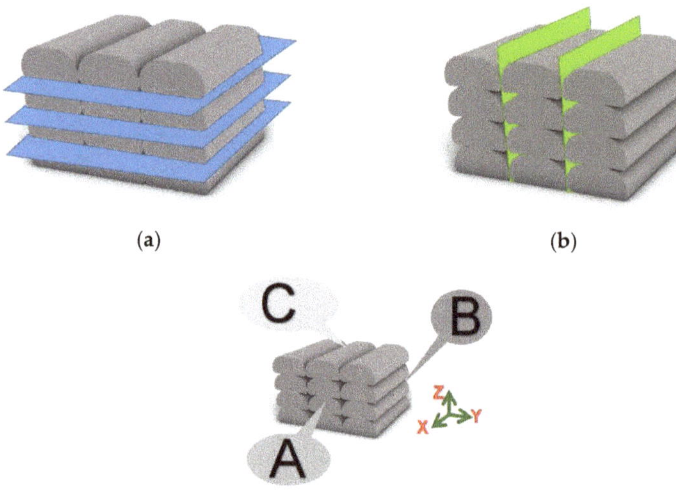

Figure 13. The horizontal interlayer diagrammatic sketch between printing layers (**a**), the vertical interlayer diagrammatic sketch between printing filaments (**b**), and the diagrammatic sketch of printed cube sample faces (**c**).

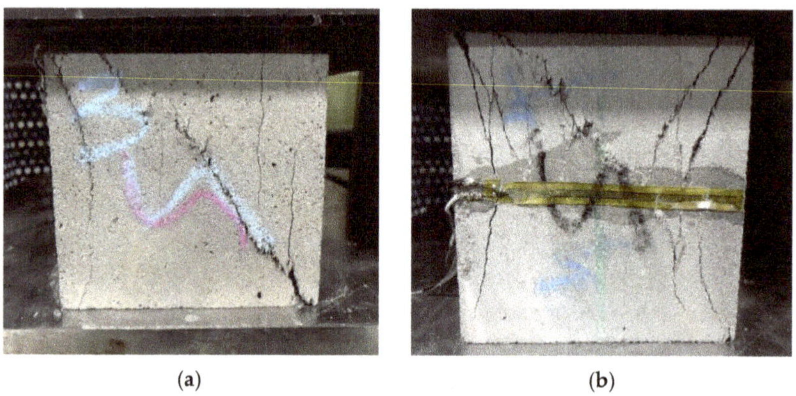

Figure 14. The printed cube sample cracks of (**a**) C and (**b**) reverse plane on the X-direction compression.

Figure 15 demonstrates that a cross-shaped main crack and a horizontal main crack were generated on the C plane and the reverse plane, respectively, when the Y-direction compression was loaded on the samples. Moreover, a slide was created on the weakest vertical interlayer of the printed specimen when the compression was loaded, hence proving that the horizontal interlayer bond was higher than the vertical interlayer bond. Two main vertical cracks and shear oblique cracks were generated on the forward and reverse plane respectively of the A plane when the Z-direction compression was loaded on the samples, as shown in Figure 16. Due to the rectangular printable filament, some controllable deformation may improve the vertical interlayer bond and dense microstructure at the deposition effect, thereby reducing anisotropy for material. Hence, the improvement of fluidity is beneficial to the horizontal interlayer bond of materials without horizontal additional pressure.

(**a**) (**b**)

Figure 15. The printed cube sample cracks of (**a**) C plane and (**b**) reverse plane.

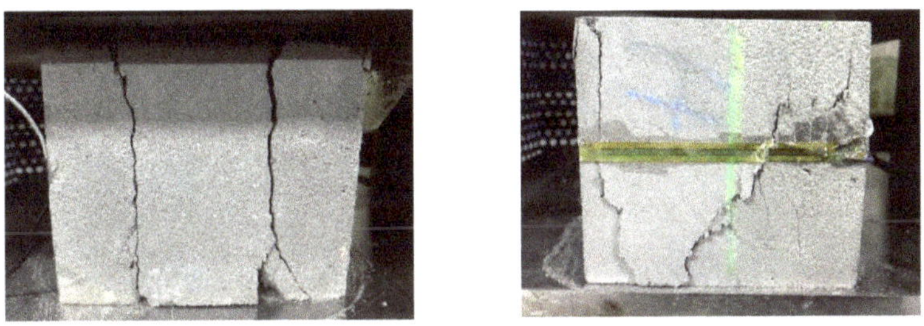

(**a**) (**b**)

Figure 16. The printed cube sample cracks of (**a**) A and (**b**) reverse plane.

4. Conclusions

This study proposed that the anisotropy coefficient and orthogonal experiment be applied to evaluate the mechanical performance and anisotropy of concrete using 3DCP technology when various steam curing conditions are used for printed materials. Subsequently, the optimum steam curing conditions and interlayer bonding were investigated. The following conclusions can be drawn:

1. The most influential factors for solid wastes for CS_x, CS_y, and CS_z are heating-up rate, thermostatic period, and thermostatic period, respectively. The optimal steam curing conditions of CS_x and CS_y are 10 °C/h heating-up rate, 10 h thermostatic period, and 70 °C for the thermostatic temperature. The optimal F_z curing conditions are 10 °C/h heating-up rate, 8h thermostatic period, and 60 °C thermostatic temperature.
2. The optimal curing conditions (thermostatic period, thermostatic temperature, heating-up rate) are 15 °C/h, 8 h, and 60 °C for TS_{xy}; 10 °C/h, 6 h, and 70 °C for TS_{xz}; 10 °C/h, 6 h, and 50 °C for TS_{yx}; 15 °C/h, 10 h, and 60 °C for TS_{yz}; 10 °C/h, 8 h, and 70 °C for TS_{zx}; 10 °C/h, 8 h, and 70 °C for TS_{zy}; respectively.
3. The optimal flexural strengths are obtained when the thermostatic period, thermostatic temperature, heating-up rate are 15 °C/h, 10 h, 60 °C for FS_{xy}; 15 °C/h, 10 h, 50 °C for FS_{xz}; 10 °C/h, 10 h, 70 °C for FS_{yx}; 20 °C/h, 8 h, 50 °C for FS_{yz}; 20 °C/h, 8 h, 60 °C for FS_{zx}; and 10 °C/h, 8 h, 70 °C for FS_{zy}; respectively.

4. The optimal heating-up rate, thermostatic period, and thermostatic temperature are 20 °C/h, 8 h, 60 °C for compressive performance; and 20 °C/h, 8 h, 70 °C for splitting tensile performance; and 15 °C/h, 10 h, 60 °C for flexural performance; respectively, to reduce the anisotropy.
5. The interlayer bonding capacity is influenced by steam curing conditions, solid waste incorporation, and printing parameters. Solid waste incorporation can improve the microstructure and interface bond of the printed concrete. Furthermore, some controllable deformation may improve the vertical interlayer bond and dense microstructure at the deposition effect, thereby reducing anisotropy for the material. Meanwhile, the bond of the horizontal interlayer mainly depends on the fluidity of materials without horizontal additional pressure.

Author Contributions: Conceptualization, B.W.; methodology, B.W.; software, X.Y.; writing—review and editing, M.Y.; project administration, X.Y.; formal analysis, R.Z.; investigation, J.H.; funding acquisition, M.Y.; visualization, Z.D.; validation, X.W.; writing—original draft preparation, H.Z.; visualization, H.Z. All authors have read and agreed to the published version of the manuscript.

Funding: The research is supported by the key industry science and technology project of Ministry of Transport in 2020 "Development of Concrete 3D Printing Materials, Equipment and Systems" (2020-MS1-062) and the key technology research of the China Communications Construction Company on "3D Printing of Concrete Structure of Transportation Infrastructure" (2020-ZJKJ-07).

Institutional Review Board Statement: Not applicable.

Informed Consent Statement: Not applicable.

Data Availability Statement: The data presented in this study are openly available.

Conflicts of Interest: The authors declare no conflict of interest.

References

1. Sun, J.; Aslani, F.; Lu, J.; Wang, L.; Huang, Y.; Ma, G. Fresh and mechanical behaviour of developed fibre-reinforced lightweight engineered cementitious composites for 3D concrete printing containing hollow glass microspheres. *Ceram. Int.* **2021**, *47*, 27107–27121. [CrossRef]
2. Sun, J.; Huang, Y.; Aslani, F.; Wang, X.; Ma, G. Mechanical enhancement for EMW-absorbing cementitious material using 3D concrete printing. *J. Build. Eng.* **2021**, *41*, 102763. [CrossRef]
3. Zhang, J.; Sun, Y.; Li, G.; Wang, Y.; Sun, J.; Li, J. Machine-learning-assisted shear strength prediction of reinforced concrete beams with and without stirrups. *Eng. Comput.* **2020**, *38*, 1293–1307. [CrossRef]
4. Chen, Y.; Chaves Figueiredo, S.; Yalçinkaya, Ç.; Çopuroğlu, O.; Veer, F.; Schlangen, E. The effect of viscosity-modifying admixture on the extrudability of limestone and calcined clay-based cementitious material for extrusion-based 3D concrete printing. *Materials* **2019**, *12*, 1374. [CrossRef]
5. Mou, B.; Bai, Y. Experimental investigation on shear behavior of steel beam-to-CFST column connections with irregular panel zone. *Eng. Struct.* **2018**, *168*, 487–504. [CrossRef]
6. Sun, Y.; Li, G.; Zhang, J.; Sun, J.; Xu, J. Development of an ensemble intelligent model for assessing the strength of cemented paste backfill. *Adv. Civ. Eng.* **2020**, *2020*, 1–6. [CrossRef]
7. Huang, H.; Huang, M.; Zhang, W.; Yang, S. Experimental study of predamaged columns strengthened by HPFL and BSP under combined load cases. *Struct. Infrastruct. Eng.* **2021**, *17*, 1210–1227. [CrossRef]
8. Zhang, W.; Tang, Z. Numerical modeling of response of CFRP–Concrete interfaces subjected to fatigue loading. *J. Compos. Constr.* **2021**, *25*, 04021043. [CrossRef]
9. Zhang, C.; Ali, A. The advancement of seismic isolation and energy dissipation mechanisms based on friction. *Soil Dyn. Earthq. Eng.* **2021**, *146*, 106746. [CrossRef]
10. Asprone, D.; Auricchio, F.; Menna, C.; Mercuri, V. 3D printing of reinforced concrete elements: Technology and design approach. *Constr. Build. Mater.* **2018**, *165*, 218–231. [CrossRef]
11. Tsai, Y.-H.; Wang, J.; Chien, W.-T.; Wei, C.-Y.; Wang, X.; Hsieh, S.-H. A BIM-based approach for predicting corrosion under insulation. *Autom. Constr.* **2019**, *107*, 102923. [CrossRef]
12. Xu, S.; Wang, J.; Shou, W.; Ngo, T.; Sadick, A.-M.; Wang, X. Computer vision techniques in construction: A critical review. *Arch. Comput. Methods Eng.* **2021**, *28*, 3383–3397. [CrossRef]
13. Hou, L.; Wu, S.; Zhang, G.K.; Tan, Y.; Wang, X. Literature review of digital twins applications in construction workforce safety. *Appl. Sci.* **2021**, *11*, 339. [CrossRef]

14. Ma, G.; Wang, L.; Ju, Y. State-of-the-art of 3D printing technology of cementitious material—An emerging technique for construction. *Sci. China Technol. Sci.* **2018**, *61*, 475–495. [CrossRef]
15. Wu, C.; Wang, X.; Chen, M.; Kim, M.J. Differential received signal strength based RFID positioning for construction equipment tracking. *Adv. Eng. Inform.* **2019**, *42*, 100960. [CrossRef]
16. Wang, L.; Yuan, J.; Wu, C.; Wang, X. Practical algorithm for stochastic optimal control problem about microbial fermentation in batch culture. *Optim. Lett.* **2019**, *13*, 527–541. [CrossRef]
17. Skibicki, S.; Kaszyńska, M.; Techman, M. Maturity testing of 3D printing concrete with inert microfiller. *Proc. MATEC Web Conf.* **2018**, *219*, 03008. [CrossRef]
18. Hoffmann, M.; Skibicki, S.; Pankratow, P.; Zieliński, A.; Pajor, M.; Techman, M. Automation in the Construction of a 3D-Printed Concrete Wall with the Use of a Lintel Gripper. *Materials* **2020**, *13*, 1800. [CrossRef]
19. Sun, J.; Aslani, F.; Wei, J.; Wang, X. Electromagnetic absorption of copper fiber oriented composite using 3D printing. *Constr. Build. Mater.* **2021**, *300*, 124026. [CrossRef]
20. Liu, J.; Wu, C.; Wu, G.; Wang, X. A novel differential search algorithm and applications for structure design. *Appl. Math. Comput.* **2015**, *268*, 246–269. [CrossRef]
21. Sun, Y.; Li, G.; Zhang, J.; Sun, J.; Huang, J.; Taherdangkoo, R. New Insights of Grouting in Coal Mass: From Small-Scale Experiments to Microstructures. *Sustainability* **2021**, *13*, 9315. [CrossRef]
22. Singh, V.; Gu, N.; Wang, X. A theoretical framework of a BIM-based multi-disciplinary collaboration platform. *Autom. Constr.* **2011**, *20*, 134–144. [CrossRef]
23. Sun, J.; Wang, J.; Zhu, Z.; He, R.; Peng, C.; Zhang, C.; Huang, J.; Wang, Y.; Wang, X. Mechanical Performance Prediction for Sustainable High-Strength Concrete Using Bio-Inspired Neural Network. *Buildings* **2022**, *12*, 65. [CrossRef]
24. Wang, J.; Dai, Q.; Si, R.; Guo, S. Investigation of properties and performances of Polyvinyl Alcohol (PVA) fiber-reinforced rubber concrete. *Constr. Build. Mater.* **2018**, *193*, 631–642. [CrossRef]
25. Sun, J.; Wang, Y.; Yao, X.; Ren, Z.; Zhang, G.; Zhang, C.; Chen, X.; Ma, W.; Wang, X. Machine-Learning-Aided Prediction of Flexural Strength and ASR Expansion for Waste Glass Cementitious Composite. *Appl. Sci.* **2021**, *11*, 6686. [CrossRef]
26. Shen, P.; Lu, L.; He, Y.; Wang, F.; Hu, S. The effect of curing regimes on the mechanical properties, nano-mechanical properties and microstructure of ultra-high performance concrete. *Cem. Concr. Res.* **2019**, *118*, 1–13. [CrossRef]
27. Sun, J.; Ma, Y.; Li, J.; Zhang, J.; Ren, Z.; Wang, X. Machine learning-aided design and prediction of cementitious composites containing graphite and slag powder. *J. Build. Eng.* **2021**, *43*, 102544. [CrossRef]
28. Aslani, F.; Sun, J.; Bromley, D.; Ma, G. Fiber-reinforced lightweight self-compacting concrete incorporating scoria aggregates at elevated temperatures. *Struct. Concr.* **2019**, *20*, 1022–1035. [CrossRef]
29. Wang, J.; Dai, Q.; Si, R.; Guo, S. Mechanical, durability, and microstructural properties of macro synthetic polypropylene (PP) fiber-reinforced rubber concrete. *J. Clean. Prod.* **2019**, *234*, 1351–1364. [CrossRef]
30. Sun, J.; Huang, Y.; Aslani, F.; Ma, G. Electromagnetic wave absorbing performance of 3D printed wave-shape copper solid cementitious element. *Cem. Concr. Compos.* **2020**, *114*, 103789. [CrossRef]
31. Aslani, F.; Hou, L.; Nejadi, S.; Sun, J.; Abbasi, S. Experimental analysis of fiber-reinforced recycled aggregate self-compacting concrete using waste recycled concrete aggregates, polypropylene, and steel fibers. *Struct. Concr.* **2019**, *20*, 1670–1683. [CrossRef]
32. Sun, J.; Wang, Y.; Liu, S.; Dehghani, A.; Xiang, X.; Wei, J.; Wang, X. Mechanical, chemical and hydrothermal activation for waste glass reinforced cement. *Constr. Build. Mater.* **2021**, *301*, 124361. [CrossRef]
33. Wang, J.; Dai, Q.; Si, R. Experimental and Numerical Investigation of Fracture Behaviors of Steel Fiber–Reinforced Rubber Self-Compacting Concrete. *J. Mater. Civ. Eng.* **2022**, *34*, 04021379. [CrossRef]
34. Mehta, P.K.; Monteiro, P.J. *Concrete: Microstructure, Properties, and Materials*; McGraw-Hill Education: New York, NY, USA, 2014.
35. Ma, W.; Wang, X.; Wang, J.; Xiang, X.; Sun, J. Generative Design in Building Information Modelling (BIM): Approaches and Requirements. *Sensors* **2021**, *21*, 5439. [CrossRef]
36. Huang, H.; Xue, C.; Zhang, W.; Guo, M. Torsion design of CFRP-CFST columns using a data-driven optimization approach. *Eng. Struct.* **2022**, *251*, 113479. [CrossRef]
37. Zhang, W.; Li, H.; Li, Y.; Liu, H.; Ding, X. Application of deep learning algorithms in geotechnical engineering: A short critical review. *Artif. Intell. Rev.* **2021**, *54*, 5633–5673. [CrossRef]
38. Zhang, J.; Wang, J.; Dong, S.; Yu, X.; Han, B. A review of the current progress and application of 3D printed concrete. *Compos. Part A Appl. Sci. Manuf.* **2019**, *125*, 105533. [CrossRef]
39. Ali, A.; Zhang, C.; Bibi, T.; Zhu, L.; Cao, L.; Li, C.; Hsiao, P.-C. Investigation of five different low-cost locally available isolation layer materials used in sliding base isolation systems. *Soil Dyn. Earthq. Eng.* **2022**, *154*, 107127. [CrossRef]
40. Li, J.; Qin, Q.; Sun, J.; Ma, Y.; Li, Q. Mechanical and conductive performance of electrically conductive cementitious composite using graphite, steel slag, and GGBS. *Struct. Concr.* **2020**, *23*, 533–547. [CrossRef]
41. Feng, W.; Wang, Y.; Sun, J.; Tang, Y.; Wu, D.; Jiang, Z.; Wang, J.; Wang, X. Prediction of thermo-mechanical properties of rubber-modified recycled aggregate concrete. *Constr. Build. Mater.* **2022**, *318*, 125970. [CrossRef]
42. Wolfs, R.; Bos, F.; Salet, T. Early age mechanical behaviour of 3D printed concrete: Numerical modelling and experimental testing. *Cem. Concr. Res.* **2018**, *106*, 103–116. [CrossRef]
43. Sun, J.; Lin, S.; Zhang, G.; Sun, Y.; Zhang, J.; Chen, C.; Morsy, A.M.; Wang, X. The effect of graphite and slag on electrical and mechanical properties of electrically conductive cementitious composites. *Constr. Build. Mater.* **2021**, *281*, 122606. [CrossRef]

44. Sun, J.; Wang, X.; Zhang, J.; Xiao, F.; Sun, Y.; Ren, Z.; Zhang, G.; Liu, S.; Wang, Y. Multi-objective optimisation of a graphite-slag conductive composite applying a BAS-SVR based model. *J. Build. Eng.* **2021**, *44*, 103223. [CrossRef]
45. Lu, N.; Wang, H.; Wang, K.; Liu, Y. Maximum probabilistic and dynamic traffic load effects on short-to-medium span bridges. *Comput. Modeling Eng. Sci.* **2021**, *127*, 345–360. [CrossRef]
46. Wu, C.; Wu, P.; Wang, J.; Jiang, R.; Chen, M.; Wang, X. Critical review of data-driven decision-making in bridge operation and maintenance. *Struct. Infrastruct. Eng.* **2020**, *18*, 47–70. [CrossRef]
47. Zhao, R.; Zhang, L.; Guo, B.; Chen, Y.; Fan, G.; Jin, Z.; Gua, X.; Zhu, J. Unveiling substitution preference of chromium ions in sulphoaluminate cement clinker phases. *Compos. Part B Eng.* **2021**, *222*, 109092. [CrossRef]
48. Tang, Y.; Feng, W.; Chen, Z.; Nong, Y.; Guan, S.; Sun, J. Fracture behavior of a sustainable material: Recycled concrete with waste crumb rubber subjected to elevated temperatures. *J. Clean. Prod.* **2021**, *318*, 128553. [CrossRef]
49. Zhang, G.; Chen, C.; Li, K.; Xiao, F.; Sun, J.; Wang, Y.; Wang, X. Multi-objective optimisation design for GFRP tendon reinforced cemented soil. *Constr. Build. Mater.* **2022**, *320*, 126297. [CrossRef]
50. Sun, J.; Tang, Y.; Wang, J.; Wang, X.; Wang, J.; Yu, Z.; Cheng, Q.; Wang, Y. A Multi-objective Optimisation Approach for Activity Excitation of Waste Glass Mortar. *J. Mater. Res. Technol.* **2022**, *17*, 2280–2304. [CrossRef]
51. Liu, W.; Guo, Z.; Wang, C.; Niu, S. Physico-mechanical and microstructure properties of cemented coal Gangue-Fly ash backfill: Effects of curing temperature. *Constr. Build. Mater.* **2021**, *299*, 124011. [CrossRef]
52. González-Fonteboa, B.; Martínez-Abella, F.; Martínez-Lage, I.; Eiras-López, J. Structural shear behaviour of recycled concrete with silica fume. *Constr. Build. Mater.* **2009**, *23*, 3406–3410. [CrossRef]
53. Xu, D.; Liu, Q.; Qin, Y.; Chen, B. Analytical approach for crack identification of glass fiber reinforced polymer–sea sand concrete composite structures based on strain dissipations. *Struct. Health Monit.* **2020**, *10*, 1475921720974290. [CrossRef]
54. Wang, M.; Yang, X.; Wang, W. Establishing a 3D aggregates database from X-ray CT scans of bulk concrete. *Constr. Build. Mater.* **2022**, *315*, 125740. [CrossRef]
55. Shi, T.; Lan, Y.; Hu, Z.; Wang, H.; Xu, J.; Zheng, B. Tensile and Fracture Properties of Silicon Carbide Whisker-Modified Cement-Based Materials. *Int. J. Concr. Struct. Mater.* **2022**, *16*, 2. [CrossRef]
56. Li, Y.; Che, P.; Liu, C.; Wu, D.; Du, Y. Cross-scene pavement distress detection by a novel transfer learning framework. *Comput.-Aided Civ. Infrastruct. Eng.* **2021**, *36*, 1398–1415. [CrossRef]
57. Sanjayan, J.G.; Nematollahi, B.; Xia, M.; Marchment, T. Effect of surface moisture on inter-layer strength of 3D printed concrete. *Constr. Build. Mater.* **2018**, *172*, 468–475. [CrossRef]
58. Zhang, W.; Zhang, R.; Wu, C.; Goh, A.T.C.; Lacasse, S.; Liu, Z.; Liu, H. State-of-the-art review of soft computing applications in underground excavations. *Geosci. Front.* **2020**, *11*, 1095–1106. [CrossRef]
59. Zheng, J. Application of 3D Printing Technology by Continuous Laser Curing in Space Optimization of Ocean Park. *J. Coast. Res.* **2020**, *103*, 966–969. [CrossRef]
60. Wang, B.; Zhai, M.; Yao, X.; Wu, Q.; Yang, M.; Wang, X.; Huang, J.; Zhao, H. Printable and Mechanical Performance of 3D Printed Concrete Employing Multiple Industrial Wastes. *Buildings* **2022**, *12*, 374. [CrossRef]
61. *GB/T 51231-2016*; Technical Standard for Assembled Buildings with Concrete Structure. China National Standard: Beijing, China, 2016.
62. Shi, J.; Liu, B.; He, Z.; Wu, X.; Tan, J.; Chen, J.; Jiang, J. Properties evolution of high-early-strength cement paste and interfacial transition zone during steam curing process. *Constr. Build. Mater.* **2020**, *252*, 119095. [CrossRef]
63. *GB/T 50082-2009*; Standard for Test Methods of Long-Term Performance and Durability of Ordinary Concrete. China National Standard: Beijing, China, 2009.
64. Luo, Y.; Zheng, H.; Zhang, H.; Liu, Y. Fatigue reliability evaluation of aging prestressed concrete bridge accounting for stochastic traffic loading and resistance degradation. *Adv. Struct. Eng.* **2021**, *24*, 3021–3029. [CrossRef]

materials

MDPI

Article

Investigation on Water Transformation and Pore Structure of Cement-Stabilized Dredged Sediment Based on NMR Technology

Shiquan Wang [1], Xingxing He [2,*], Guanghua Cai [3,*], Lei Lang [4], Hongrui Ma [1], Shunmei Gong [1] and Zhiyong Niu [1]

1 National Engineering Research Center of Coal Mine Water Hazard Controlling, School of Resources and Civil Engineering, Suzhou University, Suzhou 234000, China; wsq@ahszu.edu.cn (S.W.); wsq_cersm@163.com (H.M.); hxx_cersm@163.com (S.G.); nzy1232022@163.com (Z.N.)
2 State Key Laboratory of Hydroscience and Engineering, Tsinghua University, Beijing 100084, China
3 School of Civil Engineering, Nanjing Forestry University, Nanjing 210037, China
4 State Key Laboratory of Geomechanics and Geotechnical Engineering, Institute of Rock and Soil Mechanics, Chinese Academy of Sciences, Wuhan 430071, China; budameiqian@163.com
* Correspondence: hexingxing16@mails.ucas.ac.cn (X.H.); ghcai@njfu.edu.cn (G.C.)

Abstract: Cement-stabilized dredged sediment (CDS) when used as a new road construction material cannot only solve the problem of abandoned sediment disposal, but also effectively save natural soil resources. This study aimed to evaluate the strength and permeability of CDS and establish corresponding prediction models from the perspective of a stabilization mechanism. The soil–water composition and pore size distribution were investigated by the nuclear magnetic resonance (NMR) technique. The results demonstrated that more liquid pore water inside the CDS specimen transformed into combined water with cement hydration. The amount of combined water, which essentially characterized the hydration process of cement, presented a linear relationship with log (t). The cementation and filling action of hydrates resulted in the transformation of large pores into smaller ones, hence the optimal pore size decreased with an increasing curing period and cement content. The stress–strain curves and hydraulic conductivity were determined based on unconfined compression and flexible wall penetration tests, respectively. The unconfined compressive strength increased exponentially with the amount of combined water, and the functional correlations of hydraulic conductivity and micropore parameters were established. The reliability of the NMR technique as a new method to study the microscopic evolution mechanism of the strength and permeability of CDS was further verified by scanning electron microscopy and mercury intrusion porosimetry tests.

Keywords: dredged sediment; cement-stabilized; NMR; permeability; strength; pore structure; water transformation

Citation: Wang, S.; He, X.; Cai, G.; Lang, L.; Ma, H.; Gong, S.; Niu, Z. Investigation on Water Transformation and Pore Structure of Cement-Stabilized Dredged Sediment Based on NMR Technology. *Materials* **2022**, *15*, 3178. https://doi.org/10.3390/ma15093178

Academic Editor: F. Pacheco Torgal

Received: 6 March 2022
Accepted: 24 April 2022
Published: 28 April 2022

Publisher's Note: MDPI stays neutral with regard to jurisdictional claims in published maps and institutional affiliations.

Copyright: © 2022 by the authors. Licensee MDPI, Basel, Switzerland. This article is an open access article distributed under the terms and conditions of the Creative Commons Attribution (CC BY) license (https://creativecommons.org/licenses/by/4.0/).

1. Introduction

Large quantities of sediments are dredged annually from harbors, waterways, and lakes to build port facilities, maintain shipping capacity, or protect water environment ecology [1–3]. The dredged sediments are considered waste due to their high-water content, high compressibility and low bearing capacity, and are difficult to be directly utilized without any treatment [4–6]. Cement-stabilized dredged sediment (CDS) when used as road construction material cannot only solve the problem of abandoned sediment disposal, but also effectively save natural soil resources [7–10]. Achour et al. [11] designed the method of road construction using solidified sludge and analyzed the road deflection and the tensile strength of borehole samples. Wang et al. [12] studied the effect of high curing temperature on the engineering performance of lime/cement-treated marine silt as an alternative material for road construction. Dubois et al. [13] developed a road construction material based on fine dredged sediments in combination with sand and binders, and the designed mixture met the threshold specified in the leaching toxicity test. The geomechanical properties of CDS have been extensively studied based on the cement content or soil–water/cement ratio in the past [14–17]. However, there are few studies that evaluate the strength and permeability of CDS and establish corresponding prediction models from the perspective of a stabilization mechanism.

Cement stabilization of dredged sediment involves following three aspects [18–20]: (1) the direct hydration reaction between cement and pore water in dredged sediment; (2) the secondary pozzolanic reaction between calcium hydroxide (cement hydration product) and active silicon aluminum dissolved from the clay minerals; and (3) the interaction between the soil aggregates and formed hydrates. When the cement is mixed into the dredged sediment, a large number of hydrates such as ettringite (AFt), calcium aluminate hydrate (CAH), calcium silicate hydrate (CSH), and calcium hydroxide (CH) are produced to bond the soil aggregates and fill the pores. Consequently, the internal structure of CDS becomes denser and firmer, resulting in an increase in strength and a decrease in hydraulic conductivity. Thus, it can be seen that the geomechanical properties of CDS are governed primarily by the number of hydrates. At the same time, the water in the CDS can be divided into liquid pore water (PW) and combined water (CW). The PW can be converted into the CW with cement hydration, resulting in the redistribution of water inside the CDS. Furthermore, the CW is an integral part of hydrates that can indirectly characterize the number of hydrates generated inside the CDS. Therefore, the real-time monitoring of the moisture state inside CDS can track the hydration reaction progress and reveal the evolution mechanism of the geomechanical properties (strength and permeability).

Recently, nuclear magnetic resonance (NMR) technology has been applied to detect water distribution and pore structure characteristics in the field of geotechnical engineering. Yu et al. [21] confirmed that the macrodynamic characteristics of soil observed in triaxial tests are closely related to the micro-pore size of soil obtained in the NMR test. Tian et al. [22] studied freezing–thawing characteristics of three soils by NMR technology and calculated the threshold and content of bound water in the soils. Yao et al. [23] investigated the pore–water status (including adsorbed water and capillary water) of expansive clay with different dry densities under the action of temperature and salt solution based on the NMR technique. NMR is a technique to investigate the content and distribution of protons (i.e., hydrogen nuclei) in a unit volume. Under the action of a magnetic field, the macroscopic magnetization vector of the proton group will be biased and out of balance. When the radio frequency stops, the proton group will recover from the non-equilibrium state to the equilibrium state. In this process, the nuclear magnetic signal starts to decay freely, which is referred to as the free induction decay (FID) curve. The peak point on the FID curve is proportional to the number of protons in the sample, so it can be used to determine the water content. At the same time, the shape of the FID curve is related to the transverse relaxation time (T_2) of the proton, and the T_2 distribution curve of pore water in the samples can be obtained by Fourier transformation [24]. In a uniform magnetic field, the T_2 of liquid water was calculated as Equation (1),

$$\frac{1}{T_2} = \rho_2 \left(\frac{S}{V} \right)_{\text{pore}} \tag{1}$$

where ρ_2 is the transverse relaxation rate, a parameter characterizing the property of porous material, and S and V represent the surface area and volume of the pores where liquid water resided, respectively. When considering the pore as a cylindrical shape, Equation (1) can also be abbreviated as,

$$\frac{1}{T_2} = \rho_2 \frac{2}{R} \tag{2}$$

where R was the pore radius. Equations (1) and (2) show that the T_2 value is proportional to the pore radius R, that is, a higher value of T_2 represents a larger pore size [25]. Therefore, the T_2 distribution curve can reflect the distribution of pore water and pore size in the porous media, and the peak area below the curve represents the corresponding water content [26].

Strength and permeability are the key performance parameters to evaluate the use of CDS as road building materials. This study aims to investigate the mechanism of water transformation and pore structure evolution during cement stabilization by the NMR technique and to establish a quantitative relationship between macro-geomechanical properties and microstructure. A series of laboratory specimens were prepared for different cement contents and curing periods. Furthermore, the stress–strain curves were determined based on an unconfined compression test, and the effect of the water content parameters on the strength development was quantitatively evaluated from the perspective of the stabilization mechanism. Similarly, mathematical models for evaluating the relationship between hydraulic conductivity, which are measured by a flexible wall permeameter and the representative parameters for pore size distribution, were proposed. Finally, the reliability of the NMR technique as a new method to study the microscopic evolution mechanism of strength and permeability of CDS was further verified by scanning electron microscopy (SEM) and mercury intrusion porosimetry (MIP) tests.

2. Materials and Methods

2.1. Raw Materials

The dredged sediment used in this study was taken from the bottom of a river in Suzhou, China. The sediment collected from the site was first sieved to remove impurities with a large particle size. The basic physical properties of dredged sediment are shown in Table 1. According to the X-ray diffraction (XRD) test (Figure 1), the sediment was mainly composed of quartz, illite, kaolinite, and montmorillonite. According to the Unified Soil Classification System, the dredged sediment is classified as clay of high plasticity (CH). Ordinary Portland cement (P.O. 42.5) was used as the curing agent in this study, and the physical properties and chemical compositions are listed in Table 2.

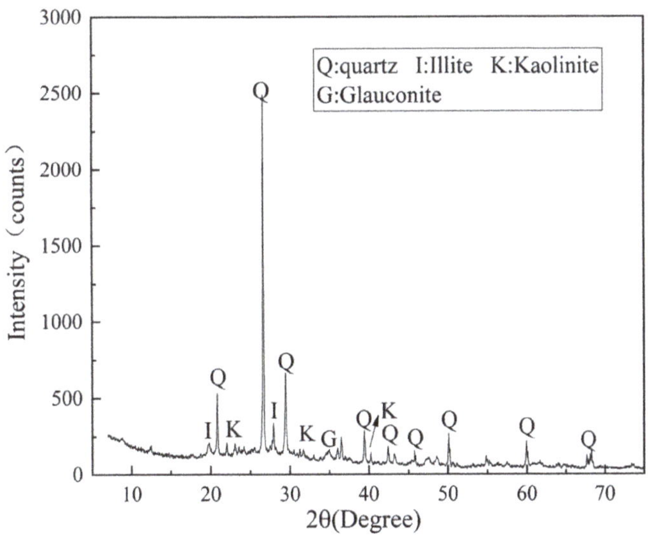

Figure 1. The XRD pattern of dredged sediment.

Table 1. Physical characteristics of dredged sediment.

Parameters	Values
Specific gravity	2.71
Liquid limit, %	50.0
Plastic limit, %	25.0
Plastic index (I_P)	25.2
Clay fraction (d < 0.005 mm), %	21.0
Silt fraction (0.005 mm < d < 0.075 mm), %	64.0
Sand fraction (d > 0.075 mm), %	15.0

Table 2. Properties of cement utilized for this experiment.

Physical Property	Value	Chemical Composition (%)	Value	Mineral Composition	Value
Ignition loss, %	3.76	Silica, SiO_2	21.3	C_3S, %	56.54
Specific gravity	3.13	Calcium oxide, CaO	64.8	C_2S, %	22.56
Fineness, m^2/kg	354	Alumina, Al_2O_3	5.2	C_3A, %	8.32
Initial setting time, min	208	Ferric oxide, Fe_2O_3	3.3	C_4AF, %	10.32
Final setting time, min	258	Magnesium oxide, MgO	2.47		
UCS [a] in 3 d, MPa	30.3	Chloride, Cl^-	0.021		
UCS in 28 d, MPa	43.2	Sulfur oxide, SO_3	2.83		
		Sodium oxide, Na_2O	0.08		

Note: [a] means unconfined compression strength.

2.2. Testing Methods

2.2.1. NMR Test

The FID curves of the CDS specimen were determined using the Carr-Purcell-Meiboom-Gill technique with a 23 MHz NMR setup [24]. The schematic diagram of the NMR experimental system is shown in Figure 2. The NMR equipment consists of a sample tube (Φ60 mm × H60 mm), data acquisition–analysis system, radio frequency (RF) system, and magnet unit. At the predetermined curing period, the sealed CDS specimens were put into the tube for NMR detection. Each NMR test can be completed in less than three minutes. After finishing all the detections, these sealed CDS specimens were put back into the cabinet for further curing. Then, the FID data were inverted and the T_2 distribution curves were drawn. Based on the NMR theory: (1) the integral area under the T_2 distribution curve (termed as the "peak area") represents the total population of pore water molecules in the T_2 range; (2) the horizontal coordinate of the curve (transverse relaxation time T_2) represents the size of interstitial pores saturated with water; and (3) the "T_2 at peak (T_{2P})" refers to the T_2 value corresponding to the maximum NMR signal, thereby representing the optimal aperture [23].

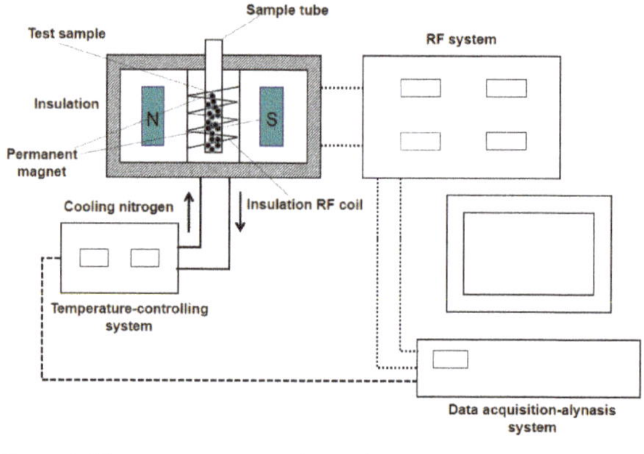

Figure 2. The structural principles of the NMR setup.

2.2.2. Geomechanical Tests

The unconfined compression test (UCT) was carried out by a hydraulic servo testing machine at a vertical loading rate of 1 mm/min, which was terminated when the peak strength of the specimen was attained or 8% of its axial strain was reached. The average value of three specimens was taken as the final unconfined compressive strength (UCS). The hydraulic conductivity testing (HCT) was carried out with the flexible wall permeameter. The specimen was vacuumized for 30 min and then saturated with deionized water for 24 h before the test. During the test, the confining pressure was kept at 200 kPa, and the osmotic pressure difference was kept at 100 kPa between the upper and lower ends of the specimens.

2.2.3. Microscopic Experiment

The micromorphology of CDS was observed by using scanning electron microscopy (SEM, ZEISS MERLIN Compact, Muenchen, Germany). The pore structure was characterized by performing a mercury intrusion porosimetry (MIP) test, which can achieve a maximum pressure of 227 MPa. Sample pieces for microscopic analysis were immediately pre-treated with liquid nitrogen freezing and vacuum drying for 48 h. Furthermore, the blocks used for MIP and SEM analysis were taken from the undisturbed parallel CDS specimen to exclude the influence of micro-damage caused by external factors.

2.3. Mixed Design and Specimen Preparation

Table 3 shows the mixed design and corresponding test items. Based on the mass ratios of dry sediment, three different cement contents (5%, 10%, 15%) were added. Firstly, the retrieved sediment was prepared into a uniform slurry with an initial water content of 60% (1.2 times the liquid limit). Then, the prepared sediment paste was homogeneously mixed with a predetermined amount of cement inside a mechanical mixer as fast as possible. Afterward, the freshly stabilized CDS slurry was subsequently poured into cylindrical PVC molds (50 mm in diameter and 50 mm in height) and vibrated to dissipate the bubbles. The upper and lower ends of the mold were covered with plastic film to prevent water evaporation. The prepared specimens, together with the molds, were then transferred into a standard curing room (20 ± 2 °C, relative humidity ≥95%). Finally, the demolded specimens were sealed with a plastic membrane and cured until the predetermined period (i.e., 1, 3, 7, 14, 28 days). Six parallel specimens were prepared for each mixture.

Table 3. Mix design and testing program.

Test Items	Curing Period (Days)	Cement Content (%)
NMR	1, 3, 7, 14, 28	5%, 10%, 15%
UCS	1, 3, 7, 14, 28	5%, 10%, 15%
HCT	7, 14, 28	5%, 10%, 15%
SEM	7, 28	5%, 15%
MIP	28	5%, 15%

3. Results and Discussion

3.1. NMR Test Results

3.1.1. Characteristics of the T_2 Distribution Curve

The T_2 distribution curves of three groups of specimens with different cement contents during the curing period were depicted in Figure 3a–c. It can be seen that the T_2 distribution curves of all the specimens showed a bimodal distribution (primary peak and secondary peak) and spanned several orders of magnitude (0.01–10 ms). Since the pore size is proportional to relaxation time T_2, the bimodal distributions indicated the existence of macro- and micro-pores within the CDS matrix. The T_2 curves of the three groups of CDS specimens all shifted to the lower left with an increasing curing period. That is, the total peak area, the T_2 at peak (T_{2P}), and the maximum NMR signal gradually decreased, which denoted that the liquid PW inside the CDS specimen transformed into CW with cement hydration. At the same time, the proportion of large pores decreased and that of the small pores increased, and the distribution of internal pores became relatively uniform.

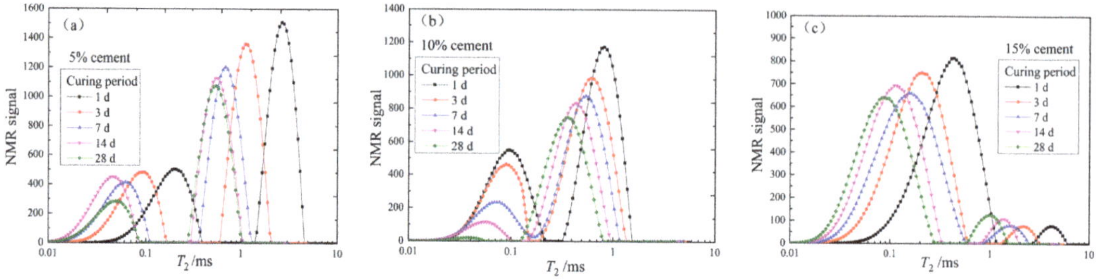

Figure 3. The T_2 distribution curves of three groups of specimens: (**a**) 5% cement content, (**b**) 10% cement content, (**c**) 15% cement content.

To highlight the influence of different cement contents more clearly, the T_2 distribution curves of three groups of CDS specimens at the 28-day curing period are plotted in Figure 4. Similarly, the total peak area and the T_{2P} decreased gradually with increasing cement content. However, the main peak of the T_2 distribution curves was on the left for the specimen with 5% and 10% cement, while that of the specimen with 15% cement was opposite, implying that the internal pore structure of CDS specimens changed significantly. Therefore, with the increase of cement content, more liquid PW was involved in cement hydration and transformed into CW, and the optimal pore size decreased while the proportion of small pores increased.

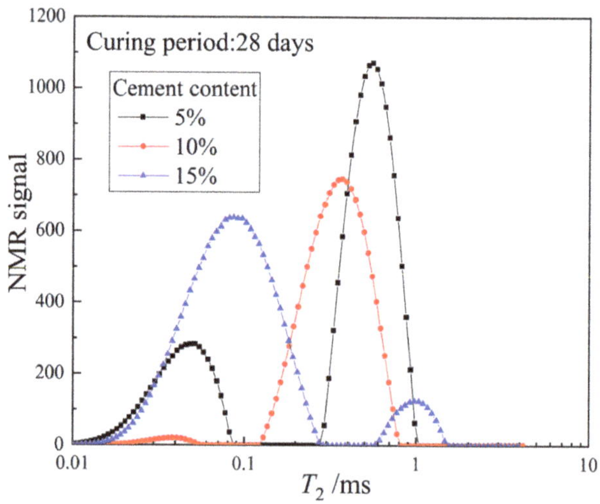

Figure 4. The 28-day T_2 distribution curves of specimens with different cement content.

3.1.2. Water Transformation Mechanism during Cement Stabilization

Since the CDS specimens were in sealed condition, the reduced liquid PW mass was exactly equal to the mass of generated CW. Given this, the corresponding water parameters were calculated as follows:

(I) The untreated dredged sediment with a known initial water content was taken as the standard sample, and its PW mass (m_{pws}) and the corresponding total peak area of the

T_2 distribution curve (A_s) were taken as the reference. Then, the PW mass corresponding to the unit peak area (u_{pw}) was calculated:

$$u_{pw} = \frac{m_{pws}}{A_s} \tag{3}$$

(II) The T_2 distribution curve of each CDS specimen (shown in Figure 3) was integrated to calculate the peak area (A_i), then the PW mass of each CDS (m_{pwi}) was calculated:

$$m_{pwi} = u_{pw} \times A_i = \frac{m_{pws}}{A_s} \times A_i \tag{4}$$

(III) The difference between the initial PW mass (m_{pw0}) and the PW mass (m_{pwi}) of each CDS specimen was the CW mass generated by the hydration reaction (m_{cw}):

$$m_{cw} = m_{pw0} - m_{pwi} \tag{5}$$

Figure 5 shows the functional relationship between the CW mass and curing period of three groups of specimens. Figure 6 depicts the cement hydration model established by Wang and Dong, respectively [27,28], where hydration degree $\alpha = Q_{(t)}/Q_{(max)}$, $Q_{(t)}$ was the hydration heat at time t, and $Q_{(max)}$ was the complete hydration heat. The model in this study indirectly represented the hydrates production in the CDS specimen from the perspective of water transformation, while Figure 6 shows the test results of the cement hydration heat measured by the heat of solution method. It is worth noting that m_{cw}, α, or $Q_{(t)}$ were all essential parameters that characterized cement hydration progress and all the presented linear relationships with log (t), which effectively proved the reliability of NMR technology in exploring the mechanism of water transformation in CDS. Moreover, it can be found by fitting the data in Figures 3 and 5 that the variation of T_{2P} with m_{cw} also follows a power function form of $y = ax^b$, which indicated the optimal pore size decreased with cement hydration, shown in Figure 7.

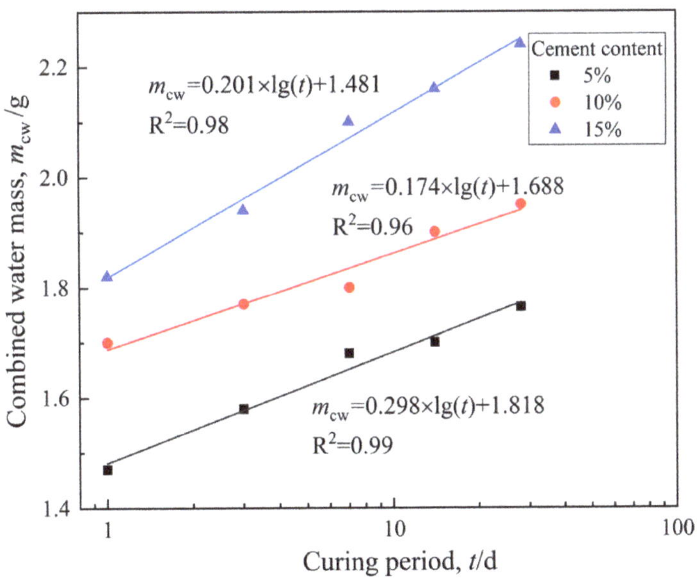

Figure 5. Relationship between CW mass and curing period.

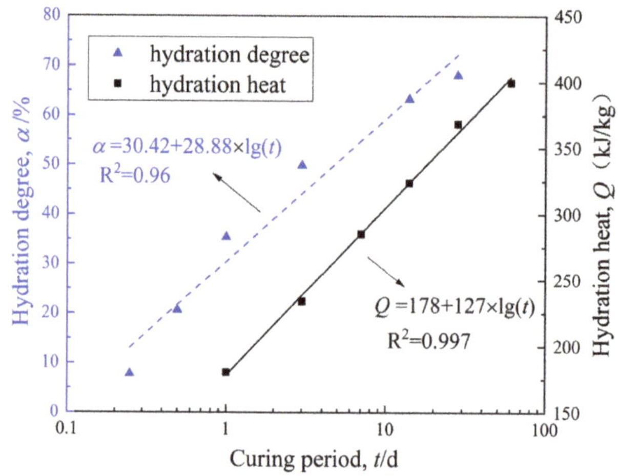

Figure 6. Cement hydration model established by Wang [27] and Dong [28].

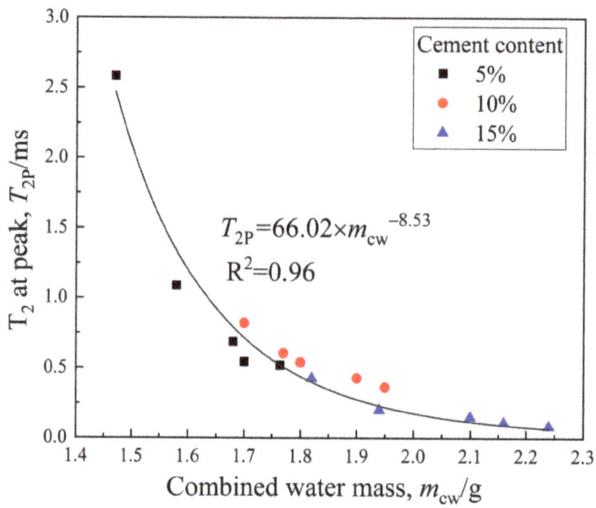

Figure 7. The T_2 at peak versus the combined water.

3.1.3. Pore Structure Characteristics

Combined with the research results of Tian et al. [24], the transverse relaxation rate, ρ_2, of CDS is chosen as 0.22. Then, the cumulative pore size distribution curve can be obtained by calculating the T_2 distribution data in Figure 3 with Equation (2). Taking the specimen with 15% cement as an example, the calculation results were plotted in Figure 8 (the ordinate indicates the percentage of pores smaller than a certain pore size). D_{50} is the median entrance throat diameter when the cumulative total pores reach 50%, which can be selected as a representative indicator to characterize the pore structure characteristics of the CDS matrix [29]. The D_{50} of specimens with 15% cement was 0.29, 0.14, 0.13, 0.10, and 0.09 μm for 1, 3, 7, 14, and 28 days, respectively, thereby indicating that the average pore size gradually decreased with the curing period. Furthermore, based on the feature and shape of the pore size distribution curve (Figure 8), the pore can be divided into three different grades by the pore diameter, namely "micropores" (<0.1 μm), "mesopores" (0.1~1.0 μm), and "macropores" (>1.0 μm) [30].

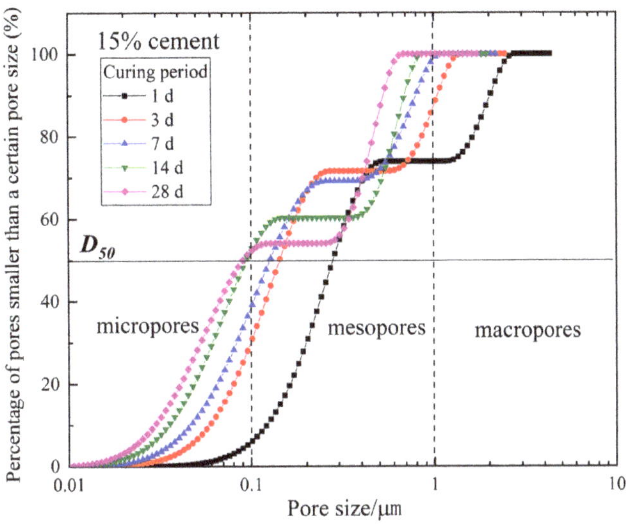

Figure 8. The cumulative pore size distribution curve of the specimen with 15% cement.

The pore structure evolution of the specimen with 15% cement was shown in Figure 9. It can be observed that within 7 days of curing, the mesopores ranging from 0.1~1 μm in all specimens constituted the majority of the pore space, accounting for more than 50% of the total pore volume. However, the proportion of mesopores and macropores gradually decreased with the curing period. Moreover, the percentage of micropores (with a diameter between 0.01 and 0.1 μm) increased significantly to more than 50% (53% and 59% for 14 and 28 days of curing, respectively), while the macropores (>1.0 μm) disappeared. This evolution of pore structure can be attributed to the cement stabilization effect, that is, gels such as calcium silicate hydrate (CSH) bound soil aggregates more tightly and expansive minerals such as ettringite (AFt) filled intergranular spaces, resulting in the transformation of large pores into smaller ones.

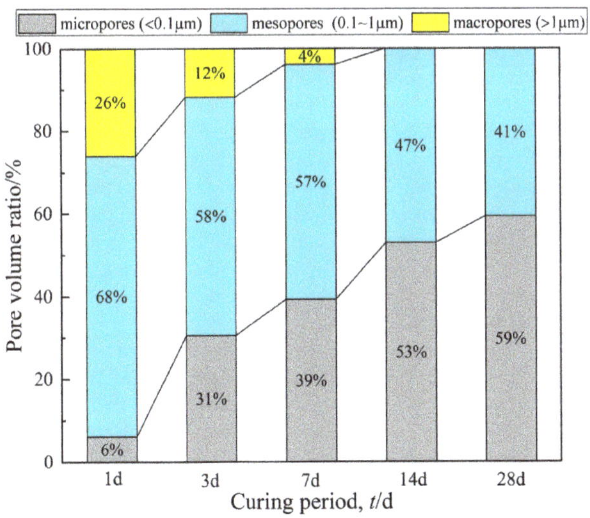

Figure 9. The pore structure evolution of the specimen with 15% cement.

MIP is a mature method to determine the pore structure of porous materials. In this study, the MIP results (differential intrusion volume curves) of specimens with 5% and 15% cement after 28 days of curing were compared with the pore size distribution determined by NMR, which is shown in Figure 10. Furthermore, the following conclusions can be drawn: the characteristics of the pore size distribution curves determined by the MIP and NMR were highly consistent, and the optimal pore sizes were also very close to each other. However, the pore sizes measured by MIP were generally slightly larger than those determined by NMR, especially in the range of pore diameters greater than 0.5 μm, and the pore size distribution curve determined by MIP was not smooth. The reason may be that in the process of the mercury injection test, the larger mercury pressure destroyed the skeleton structure of CDS and caused internal crack propagation. In general, this proved the reliability and accuracy of the NMR technique as a rapid and non-destructive method for detecting the pore structure of geotechnical media.

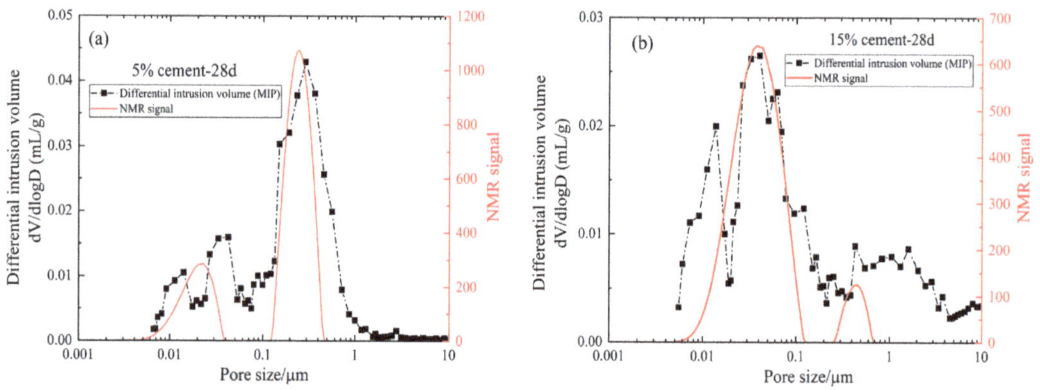

Figure 10. Comparison of MIP and NMR experimental results of specimens cured for 28 days: (**a**) 5% cement content, (**b**) 15% cement content.

3.2. *Geomechanical Properties (Strength and Permeability)*

3.2.1. Stress–Strain Relationship and Strength Development

Figure 11 shows the stress–strain curves of CDS specimens at 1, 3, 7, 14, and 28 days of curing. The CDS specimens cured for one day all showed significant ductility, regardless of the cement content. When the curing period exceeded seven days, the cured specimens became more brittle with obvious peak strength. The increase in the brittleness and strength of CDS was caused by the cementation structure of soil aggregates. That is, the more hydrates (characterized by m_{cw}, as shown in Figure 5) produced by cement hydration resulted in a denser and firmer microstructure of the solidified matrix, and the reduced pore water also exerted a weaker lubrication effect on the specimens under external compression. It can be seen that m_{cw} may be a suitable parameter to reflect the structure formed by the interaction between hydrates and soil aggregates in CDS. Therefore, it is necessary to further explore the quantitative relationship between the strength and water distribution of CDS.

The strength development with curing time is shown in Figure 12. It is noted that there is a linear relationship between the UCS (MPa) and curing period (days). The increase of strength with the curing period was due to the fact that hydration is a time-dependent process. On the other hand, the difference in the strength of 1-day specimens and 28-day specimens tended to increase for CDS with a higher cement content. Furthermore, the normalized strength (labeled as q_u/q_{u-28d}) can be obtained by dividing the strength of a particular curing time by the 28-day strength. In addition, the relationship between the normalized strength and curing period can be expressed by the following formula for specimens with different cement contents in this study (vide Figure 13).

$$q_u/q_{u-28d} = 0.252 + 0.028 \times t \tag{6}$$

where q_u is the strength to be estimated at t days of curing, q_{u-28d} is the 28-day curing strength, and t is the curing time (days). The fitting results are of very good quality as indicated by the high coefficient of determination ($R^2 > 0.96$). The inevitable scatter of the normalized q_u data may be due to the variability of UCT specimens at different curing ages.

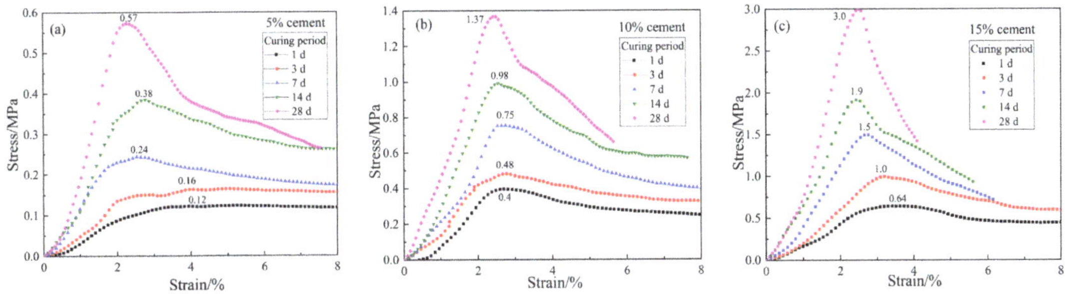

Figure 11. Stress–strain curves of CDS: (**a**) 5% cement content, (**b**) 10% cement content, (**c**) 15% cement content.

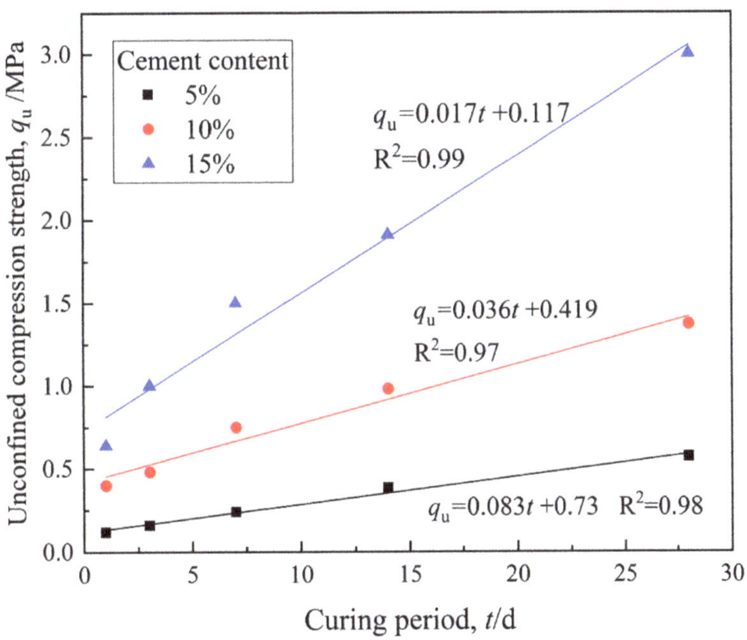

Figure 12. Unconfined compressive strength development versus curing period.

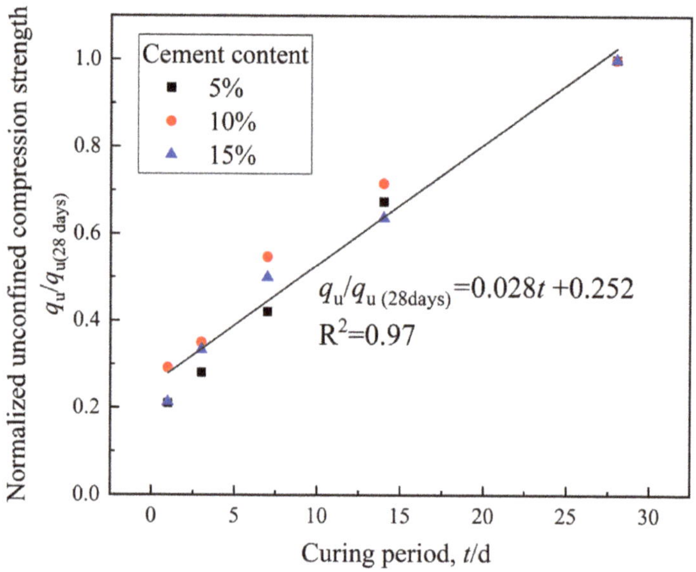

Figure 13. Variation of normalized strength with curing period.

3.2.2. Hydraulic Conductivity

Figure 14 presents the hydraulic conductivity change of CDS and shows that prolonging the curing period reduced the permeability. Furthermore, the decreased rate of hydraulic conductivity k from 7 days to 28 days (i.e., $(k_{7d} - k_{28d})/k_{7d} \times 100\%$) tended to increase for CDS with higher cement content, indicating that the effect of the curing period was related to the cement content. It is worth mentioning that the increase in cement content also reduced the hydraulic conductivity, which was consistent with the experimental results of the cement-stabilized marine clay with metakaolin by Deng [29]. This can be attributed to the decrease of porosity and average pore size with the increase of hydrates, which led to increased resistance to water flow through the CDS [31].

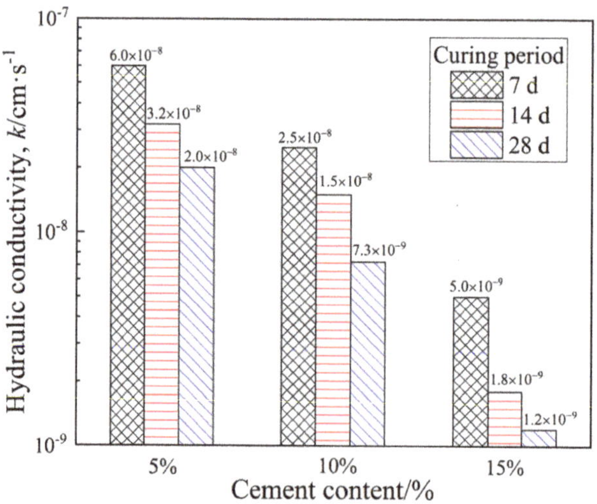

Figure 14. The hydraulic conductivity of CDS.

3.3. *Quantitative Relationship between Macro-Geomechanical Properties and Microstructure*

3.3.1. Relationship between Pore Structure Characteristics and Hydraulic Conductivity

According to Poiseuille's theory of laminar flow in porous media, the permeability of water depends on the pore size of the materials [32]. Existing studies have shown the effectiveness of D_{50} as a representative parameter to characterize pore size distribution. Through detailed analysis of the above data, the permeability of CDS is intrinsically related to its pore size characteristics. Consequently, the relationship between hydraulic conductivity k and D_{50} in this study is presented in Figure 15, which further confirmed that hydraulic conductivity is the macro-representation reflection of micropore structures [33–35].

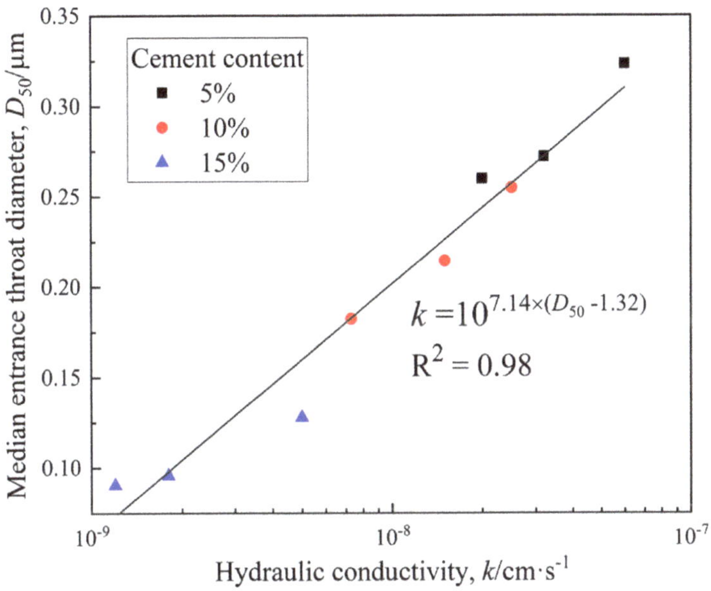

Figure 15. Relationship between k and D_{50}.

3.3.2. Relationship between Strength and Water Distribution

The correlation of the unconfined compressive strength q_u with combined water m_{cw} can be obtained by fitting the data from Figures 5 and 12, as shown in Figure 16. The q_u of CDS can be well normalized by the parameter m_{cw} irrespective of the curing period and cement content. This observation is highly consistent with the results presented in the reference from Zhu et al. [20], in which three kinds of sediments were solidified with cement and the exponential function relationship between q_u and m_{cw} was established by a centrifugal method. It can be observed that q_u increased exponentially with m_{cw}, which represented the number of hydrates and was attributed to the fact that, during the later curing period, the three-dimensional skeleton of the CDS matrix had been formed and the cementation and filling effect of the regenerated hydrates were more significant. In addition, most of the existing empirical models for strength development accounted for cement content and curing age generally at a certain initial water content. However, in this study, the parameter m_{cw} can comprehensively characterize the influence of the two factors mentioned above on the strength development of CDS from the view of a stabilization mechanism.

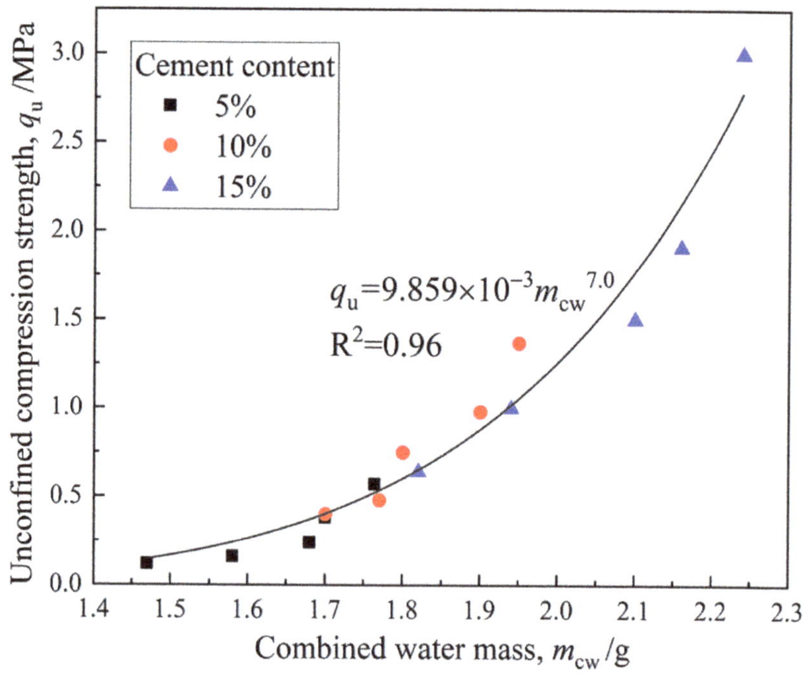

Figure 16. Relationship between q_u and m_{cw}.

3.4. SEM Characterization

The micromorphology of the stabilized specimens with 5% and 15% cement at 7 and 28 days of curing is shown in Figure 17a–d. For the specimens with 5% cement cured for seven days (Figure 17a), a certain amount of acicular ettringite (AFt) can be observed, while a small amount of gel hydrate resulted in an obvious loose skeleton of solidified matrix with macropores and transfixion cracks, thus showing lower strength and higher hydraulic conductivity. After curing for 28 days (Figure 17b), although the fabric became relatively denser, the intergranular pores larger than 5 microns still existed. For the specimens with 15% cement cured for 7 days (Figure 17c), the soil particles were encapsulated and cemented by a large number of gel substances such as C-S-H, and the large pores disappeared. Furthermore, ettringite appeared again between the gap of aggregates within specimens cured for 28 days, and the coupling effect of ettringite filling and gel cementation led to an increase in the proportion of micropores and non-transfixable pores, which is presented in Figure 17d. Consequently, this denser and stronger structure of the CDS matrix showed superior strength and poor permeability (Figures 12 and 14). Therefore, the microstructure characteristics measured by SEM intuitively reveal the evolution mechanism of strength and permeability, which were also highly consistent with NMR results. It should be pointed out that the pore diameter shown by SEM was slightly larger than that measured by the NMR technique because the pore size was expanded by the frost–heave force generated during the freeze-drying process.

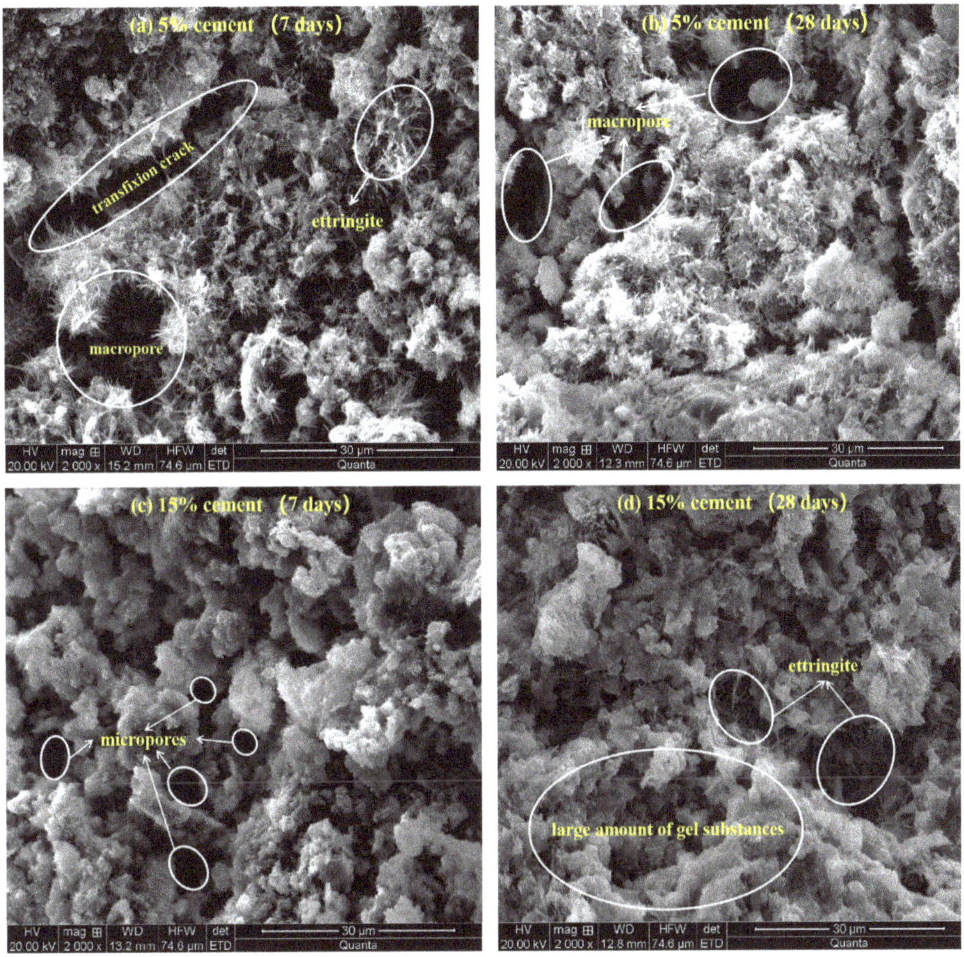

Figure 17. The micromorphology of stabilized specimens: (**a**) 5% cement-7 days, (**b**) 5% cement-28 days, (**c**) 15% cement-7 days, (**d**) 15% cement-28 days.

4. Discussion

The microscopic testing methods, such as SEM, MIP, XRD, and TGA (thermogravimetric analysis), mainly explore the cement stabilization mechanism from the perspectives of micromorphology, pore structure, and mineral composition. However, the deficiencies of these methods are as follows: (1) The integrity and initial structure of the original specimen are inevitably damaged by the complex pretreatment, such as the freeze-drying process in the SEM and MIP tests; (2) a small amount of test sample taken locally (tens of micrograms or a few grams) may not be adequately representated, such as the powder sample used in XRD and TGA experiments; (3) the hydration process of the test specimen is forced to stop, making it difficult to carry out continuous tests on the same specimen throughout the curing period; and (4) the number of hydrates generated inside a specimen cannot be measured accurately, hence the essential relationship between the geomechanical behaviors and hydration reaction cannot be constructed quantitatively. In contrast, the NMR technology is used to test a full-size initial specimen directly, without any pretreatment that may cause microdamage inside the specimen. Furthermore, the NMR test can be carried out for the whole curing period of the same specimen, and each measurement only lasts for several minutes. The hydration progress can be accurately tracked by monitoring the conversion of pore water and combined water, and the pore structure characteristics can be deduced from the T_2 distribution curve. Therefore, compared with the above traditional microscopic testing methods, the NMR technique is a non-destructive, continuous, straightforward, and simple method to study the stabilization mechanism of CDS, while providing convincing results.

5. Conclusions

(1) The total peak area and the T_2 at peak gradually decreased with an increasing curing period and cement content, indicating that more liquid pore water inside the CDS specimen transformed into combined water with cement hydration. The amount of combined water, m_{cw}, which essentially characterized the hydration process of cement, presented a linear relationship with log (t).

(2) The cementation and filling action of hydrates resulted in the transformation of large pores into smaller ones, then the proportion of mesopores and macropores gradually decreased with the curing period. In addition, the characteristics of the pore size distribution curves determined by the MIP and NMR were highly consistent. Given that there was an essential relationship between permeability and pore size distribution, the functional correlations of hydraulic conductivity k and micropore parameters D_{50} were established.

(3) With the increase in the curing period, more hydrates contributed to the denser and firmer microstructure within the solidified matrix, and the stress–strain characteristic of CDS specimens changed from ductility to brittleness. Furthermore, the relationship between the normalized strength q_u/q_{u-28d} and curing period t can be characterized by a linear function.

(4) The parameter m_{cw} can comprehensively characterize the influence of the curing period and cement content on the strength development of CDS from the view of a stabilization mechanism and q_u increased exponentially with m_{cw}. The microstructure characteristics measured by SEM revealed the evolution mechanism of strength and permeability more intuitively.

(5) In the actual project of filling subgrade with CDS, increasing cement content can improve the strength and reduce hydraulic conductivity in a short term, which is helpful to promote the progress of some special projects with tight construction periods. In general, the curing period should be extended to at least 28 days, which is conducive to the full hydration and stabilization effect of cement.

Author Contributions: Conceptualization, X.H. and G.C.; methodology, S.W.; software, Z.N. and L.L.; validation, S.G., H.M. and L.L.; formal analysis, H.M.; investigation, S.W.; resources, S.G.; data curation, S.W.; writing—original draft preparation, S.W.; writing—review and editing, X.H.; visualization, Z.N.; supervision, L.L.; project administration, X.H.; funding acquisition, G.C. All authors have read and agreed to the published version of the manuscript.

Funding: This research was supported by the Start-up Fund for Doctoral Research of Suzhou University (No. 2019jb18); Key Projects of Natural Science Research in Colleges and Universities of Anhui Province (No. KJ2021A1112); Key Research Project of Suzhou University (No. 2021yzd10).

Institutional Review Board Statement: Not applicable.

Informed Consent Statement: Not applicable.

Data Availability Statement: The data that support the findings of this study are available on request from the corresponding authors.

Acknowledgments: The authors are sincerely thankful for the funding support.

Conflicts of Interest: The authors declare no conflict of interest.

References

1. He, X.; Chen, Y.; Tan, X.; Wang, S.; Liu, L.J. Determining the water content and void ratio of cement-treated dredged soil from the hydration degree of cement. *Eng. Geol.* **2020**, *279*, 105892. [CrossRef]
2. Lang, L.; Chen, B.; Li, N. Utilization of lime/carbide slag-activated ground granulated blast-furnace slag for dredged sludge stabilization. *Mar. Georesources Geotechnol.* **2020**, *39*, 659–669. [CrossRef]
3. Wang, D.; Xiao, J.; Gao, X. Strength gain and microstructure of carbonated reactive MgO-fly ash solidified sludge from East Lake, China. *Eng. Geol.* **2019**, *251*, 37–47. [CrossRef]
4. Bi, J.; Chian, S.C. Modelling of three-phase strength development of ordinary Portland cement- and Portland blast-furnace cement-stabilised clay. *Geotechnique* **2020**, *70*, 80–89. [CrossRef]
5. Lang, L.; Chen, B.; Duan, H. Modification of nanoparticles for the strength enhancing of cement-stabilized dredged sludge. *J. Rock Mech. Geotech. Eng.* **2021**, *13*, 694–704. [CrossRef]
6. Wang, D.; Gao, X.; Liu, X. Strength, durability and microstructure of granulated blast furnace slag-modified magnesium oxychloride cement solidified waste sludge. *J. Clean. Prod.* **2021**, *292*, 126072. [CrossRef]
7. Zentar, R.; Dubois, V.; Abriak, N.E. Mechanical behaviour and environmental impacts of a test road built with marine dredged sediments. *Resour. Conserv. Recycl.* **2008**, *52*, 947–954. [CrossRef]
8. Miraoui, M.; Zentar, R.; Abriak, N.E. Road material basis in dredged sediment and basic oxygen furnace steel slag. *Constr. Build. Mater.* **2012**, *30*, 309–319. [CrossRef]
9. Siham, K.; Fabrice, B.; Edine, A.N. Marine dredged sediments as new materials resource for road construction. *Waste Manag.* **2008**, *28*, 919–928. [CrossRef] [PubMed]
10. Wang, D.; Abriak, N.E.; Zentar, R. Dredged marine sediments used as novel supply of filling materials for road construction. *Mar. Geotechnol.* **2017**, *35*, 472–480. [CrossRef]
11. Achour, R.; Abriak, N.E.; Zentar, R. Valorization of unauthorized sea disposal dredged sediments as a road foundation material. *Environ. Technol.* **2014**, *35*, 1997–2007. [CrossRef]
12. Wang, D.; Zentar, R.; Abriak, N.E. Temperature-accelerated strength development in stabilized marine soils as road construction materials. *J. Mater. Civ. Eng.* **2017**, *29*, 04016281. [CrossRef]
13. Dubois, V.; Abriak, N.E.; Zentar, R. The use of marine sediments as a pavement base material. *Waste Manag.* **2009**, *29*, 774–782. [CrossRef]
14. Horpibulsuk, S.; Nagaraj, T.S.; Miura, N. Assessment of strength development in cement-admixed high water content clays with Abrams' law as a basis. *Géotechnique* **2003**, *53*, 439–444. [CrossRef]
15. Horpibulsuk, S.; Miura, N.; Nagaraj, T.S. Clay–Water/Cement Ratio Identity for Cement Admixed Soft Clays. *J. Geotech. Geoenviron. Eng.* **2005**, *131*, 187–192. [CrossRef]
16. Yao, K.; Pan, Y.; Jia, L. Strength evaluation of marine clay stabilized by cementitious binder. *Mar. Georesources Geotechnol.* **2020**, *38*, 730–743. [CrossRef]
17. Zhang, R.J.; Santoso, A.M.; Tan, T.S. Strength of high water-content marine clay stabilized by low amount of cement. *J. Geotech. Geoenviron. Eng.* **2013**, *139*, 2170–2181. [CrossRef]
18. Zentar, R.; Abriak, N.E.; Dubois, V. Effects of salts and organic matter on Atterberg limits of dredged marine sediments. *Appl. Clay Sci.* **2009**, *42*, 391–397. [CrossRef]
19. Zhu, W.; Chiu, C.F. Role of Soil Water in Cement-Based Treatment of Dredged Materials. In *Advances in Environmental Geotechnics*; Springer: Berlin/Heidelberg, Germany, 2010; pp. 395–404.
20. Zhu, W.; Zhang, C.L.; Chiu, A. Soil–Water Transfer Mechanism for Solidified Dredged Materials. *J. Geotech. Geoenviron. Eng.* **2007**, *133*, 588–598. [CrossRef]
21. Yu, J.; Sun, M.; He, S. Accumulative Deformation Characteristics and Microstructure of Saturated Soft Clay under Cross-River Subway Loading. *Materials* **2021**, *14*, 537. [CrossRef]
22. Yao, C.; Wei, C.; Ma, T. Experimental Investigation on the Influence of Thermochemical Effect on the Pore–Water Status in Expansive Soil. *Int. J. Geomech.* **2021**, *21*, 04021080. [CrossRef]
23. Tian, H.; Wei, C.; Wei, H. Freezing and thawing characteristics of frozen soils: Bound water content and hysteresis phenomenon. *Cold Reg. Sci. Technol.* **2014**, *103*, 74–81. [CrossRef]
24. Tian, H.; Wei, C.; Wei, H. An NMR-based analysis of soil–water characteristics. *Appl. Magn. Reson.* **2014**, *45*, 49–61. [CrossRef]
25. Rao, B.; Su, X.; Qiu, S. Meso-mechanism of mechanical dewatering of municipal sludge based on low-field nuclear magnetic resonance. *Water Res.* **2019**, *162*, 161–169. [CrossRef]
26. Bird, N.; Preston, A.R.; Randall, E.W. Measurement of the size distribution of water-filled pores at different matric potentials by stray field nuclear magnetic resonance. *Eur. J. Soil Sci.* **2005**, *56*, 135–143. [CrossRef]
27. Wang, Y.-W. *Study on Hydration Model and Application of Mineral Admixture Cement*; Zhejiang University: Zhejiang, China, 2014. (In Chinese)
28. Dong, J.; Li, Z. Exothermic Model of Fly Ash-Cement System Hydration. *Coal Ash China* **2010**, *6*, 18–20. (In Chinese)
29. Deng, Y.; Yue, X.; Liu, S. Hydraulic conductivity of cement-stabilized marine clay with metakaolin and its correlation with pore size distribution. *Eng. Geol.* **2015**, *193*, 146–152. [CrossRef]
30. Wang, L.; Chen, L.; Tsang, D.W.C.W. Recycling dredged sediment into fill materials, partition blocks, and paving blocks: Technical and economic assessment. *J. Clean. Prod.* **2018**, *199*, 69–76. [CrossRef]

31. Fan, X.; Zhu, W.; Qian, Y.; Wu, S.; Lin, N. Increasing the hydraulic conductivity of solidified sewage sludge for use as temporary landfill cover. *Adv. Civ. Eng.* **2019**, *2019*, 8163563. [CrossRef]
32. Childs, E.C.; Collis-George, N. The permeability of porous materials. *Proc. R. Soc. Lond. Ser. A Math. Phys. Sci.* **1950**, *201*, 392–405.
33. Xu, W.; Li, K.; Chen, L.; Kong, W.; Liu, C. The impacts of freeze-thaw cycles on saturated hydraulic conductivity and microstructure of saline-alkali soils. *Sci. Rep.* **2021**, *11*, 18655. [CrossRef] [PubMed]
34. Yuan, S.Y.; Liu, X.F.; Buzzi, O. Effects of soil structure on the permeability of saturated Maryland clay. *Geotechnique* **2019**, *69*, 72–78. [CrossRef]
35. Spagnoli, G.; Romero, E.; Fraccica, A.; Arroyo, M.; Gomez, R. The effect of curing conditions on the hydromechanical properties of a metakaolin-based soilcrete. *Geotechnique* **2020**, *72*, 455–469. [CrossRef]

materials
Article

Performance Prediction of Cement Stabilized Soil Incorporating Solid Waste and Propylene Fiber

Genbao Zhang [1], Zhiqing Ding [2], Yufei Wang [3,*], Guihai Fu [1], Yan Wang [4], Chenfeng Xie [5], Yu Zhang [6], Xiangming Zhao [4], Xinyuan Lu [4] and Xiangyu Wang [3,*]

1 College of Civil Engineering, Hunan City University, Yiyang 413000, China; genbao@hncu.edu.cn (G.Z.); fuguihai@hncu.edu.cn (G.F.)
2 Institute for Smart City of Chongqing University in Liyang, Chongqing University, Changzhou 213300, China; 22390016@student.uwa.edu.au
3 School of Design and Built Environment, Curtin University, Perth, WA 6102, Australia
4 School of Architectural Engineering, Nanjing Institute of Technology, Nanjing 211167, China; wy1823572519@126.com (Y.W.); m17327386130@163.com (X.Z.); luxinyuan6688@163.com (X.L.)
5 Urban and Rural Construction and Investment Group Limited, Putian 351100, China; 13599889911@163.com
6 General Contracting Company of CCFED, Changsha 410000, China; zy15802629415@126.com
* Correspondence: wangyf0113_suz@163.com (Y.W.); xiangyu.wang@curtin.edu.au (X.W.)

Abstract: Cement stabilized soil (CSS) yields wide application as a routine cementitious material due to cost-effectiveness. However, the mechanical strength of CSS impedes development. This research assesses the feasible combined enhancement of unconfined compressive strength (UCS) and flexural strength (FS) of construction and demolition (C&D) waste, polypropylene fiber, and sodium sulfate. Moreover, machine learning (ML) techniques including Back Propagation Neural Network (BPNN) and Random Forest (FR) were applied to estimate UCS and FS based on the comprehensive dataset. The laboratory tests were conducted at 7-, 14-, and 28-day curing age, indicating the positive effect of cement, C&D waste, and sodium sulfate. The improvement caused by polypropylene fiber on FS was also evaluated from the 81 experimental results. In addition, the beetle antennae search (BAS) approach and 10-fold cross-validation were employed to automatically tune the hyperparameters, avoiding tedious effort. The consequent correlation coefficients (R) ranged from 0.9295 to 0.9717 for BPNN, and 0.9262 to 0.9877 for RF, respectively, indicating the accuracy and reliability of the prediction. K-Nearest Neighbor (KNN), logistic regression (LR), and multiple linear regression (MLR) were conducted to validate the BPNN and RF algorithms. Furthermore, box and Taylor diagrams proved the BAS-BPNN and BAS-RF as the best-performed model for UCS and FS prediction, respectively. The optimal mixture design was proposed as 30% cement, 20% C&D waste, 4% fiber, and 0.8% sodium sulfate based on the importance score for each variable.

Keywords: cement stabilized soil; fiber-reinforced soil; mechanical strength; waste utilization; Back Propagation Neural Network; Random Forest; beetle antennae search

Citation: Zhang, G.; Ding, Z.; Wang, Y.; Fu, G.; Wang, Y.; Xie, C.; Zhang, Y.; Zhao, X.; Lu, X.; Wang, X. Performance Prediction of Cement Stabilized Soil Incorporating Solid Waste and Propylene Fiber. *Materials* **2022**, *15*, 4250. https://doi.org/10.3390/ma15124250

Academic Editor: Didier Snoeck

Received: 9 April 2022
Accepted: 7 June 2022
Published: 15 June 2022

Publisher's Note: MDPI stays neutral with regard to jurisdictional claims in published maps and institutional affiliations.

Copyright: © 2022 by the authors. Licensee MDPI, Basel, Switzerland. This article is an open access article distributed under the terms and conditions of the Creative Commons Attribution (CC BY) license (https://creativecommons.org/licenses/by/4.0/).

1. Introduction

Cement stabilized soil (CSS) is a routine cementitious material that yields wide applications including leakage-stopping, slope reinforcement, and foundation treatments [1]. However, weak strength and large deformation impede its extensive development. Construction and demolition (C&D) waste resolve the imperfection by physical and bonding strength enhancement. C&D waste particles evolve mechanical support in the CSS matrix due to higher hardness, resulting in better unconfined compressive strength (UCS) performance. The CSS mechanical property is further improved by grinding-incineration treated C&D waste, which represents a positive effect on mortar bonding strength [2,3]. Moreover, C&D waste demonstrates stronger enhancement potential under the excitation of saline solution. Previous literature explores ion activation, proving SO_4^{2-} ions as outstanding catalysts in cement hydration which reduce the initial setting time by 81.1% [4,5]. In addition, Na^+ provides an alkali situation, leading to compact hydration products. Furthermore, C&D waste endows CSS with corrosion resistance. Sulfate attack remarkably declines the CSS durability as evidenced by structure failure and sulfate heaving. C&D waste mitigates the deterioration by introducing corrosion-resistant particles [6,7]. On the other hand, the incorporation of C&D waste concurrently renders CSS more prone to crack. The non-homogeneous surface forms weaken the interfacial transition zone (ITZ) compared to that between cement, resulting in undesirable mechanical properties [8–11]. Increasing cement dosage has been proven to be an enhancer, while the specimens conduct brittle destruction [3]. Therefore, the combinatorial strategies that best improve the mechanical performance of CSS have been thoroughly investigated.

Polypropylene fiber yields the widest application range given its potent improvement in mechanical performance and durability. First, the distributed fiber generates random intersection in the specimen, which represents characteristics of preventing crack generation and suppressing brittle destruction [12–14]. Mechanical strengths, therefore, enhance significantly as evidenced by the 115% increment of UCS [15,16]. Previous studies have proven likewise promotion with various fibrous materials [17,18]. Second, polypropylene fiber mitigates the sulfate attack. Dobrovolski et al. [19] summarized that fiber bridge entraps more air, where the voids compensate for the formation of the hydrate phase and expansion. However, previous literature has not extensively explored the coupling effect of cement, C&D waste, and polypropylene fiber on CSS performance as theoretically prediction is always feeble against many variables.

Machine learning (ML) models have therefore been proposed as a potential contender to tackle the hindrance [20–22]. ML techniques yield the ability of information extraction and pattern generation based on the given dataset and consequently output the predicted value [23,24]. The prediction demonstrates outstanding accuracy with the multi-variable dataset, which renders the ML approach being extensively explored in recent decades [25–27]. Moreover, a conventional laboratory test costs a large amount of time and money to achieve accuracy, while ML techniques forecast the results without laborious work and explicit programming. The baseline models including K-Nearest Neighbor (KNN), logistic regression (LR), and multiple linear regression (MLR) are the most widely used ML models owing to their ease of understanding. However, the simplicity impedes their wider application in the cementitious material domain. Specifically, baseline models emerge with insignificant computing power to address the non-linear relationship between multiple variables (treatment, proportion, etc.) and mechanical properties (UCS, FS, etc.) [28]. Complex ML models are therefore introduced to adequately predict the strength. Artificial Neural Network (ANN), Random Forest (RF), Gaussian process regression (GPR), and Support Vector Machines (SVM) are all ML models that yield wide application in engineering material science. Among these, ANN and RF possess imperative potential as standalone and ensemble models, respectively, due to their successful prediction of concrete compressive strength and conductivity [29–32].

Hyperparameters on the other hand, hinder the more extensive application of ML approaches. ML models represent reliable workability depending on several hyperparameters. Appropriate ML models consume a large amount of time and effort to conduct trial-and-error methods, locating the accurate parameters. Consequently, swarm intelligence (SI)-based metaheuristic algorithms, including firefly algorithm (FA) [33,34] and particle swarm optimization (PSO) [35,36], are proposed to avoid tedious tuning. However, SI concurrently brings undesired computation intensity [37,38]. Therefore, the beetle antennae search (BAS) algorithm is applied to mitigate the problem. BAS demonstrates benefits comprising easy implementation and convergence, yielding the capability of automatically tuning the hyperparameters [39]. The construction of the algorithm originated from the beetle foraging behavior, leading to the order movement of the group and the optimal hyperparameter combination [40,41].

The purpose of this paper is to experimentally investigate the CSS performance enhancement by C&D waste, polypropylene fiber, and sodium sulfate. C&D waste was incorporated by 10%, 20%, and 30% to substitute cement. The dosing level of polypropylene fiber and sodium sulfate were 1%, 2%, 4% and 0.2%, 0.4%, 0.8%, respectively. UCS test, flexural strength (FS) test, and direct shear test were conducted to examine the coupling enhancement on CSS mechanical properties. The Back Propagation Neural Network (BPNN) and FR with BAS algorithm tuning hyperparameters were employed to predict the UCS and FS performance of CSS.

2. Experimental Programs

2.1. Materials

The soil and C&D waste within this research were sourced from the construction site of Zhushan Road metro station in Nanjing. All particles were pre-dried and ground to sizes less than 5 mm. The physical and mechanical properties of soil are summarized in Table 1. The Portland cement 42.5 was utilized as a stabilizer and cementitious binder. Polypropylene fiber with a length and density of 10 mm and 0.91 g/cm^3, respectively, was employed to enhance mechanical performance. Table 2 listed the detailed mechanical properties. Moreover, the air gun was applied to refrain fibrous material from agglomeration. Sodium sulfate was used to provide alkali catalysis.

Table 1. Physical and mechanical properties of soil sample.

Soil Properties	Value
Specific gravity	2.69
Liquid limit (%)	38.87
Plastic limit (%)	21.55
Plasticity index	17.32
Maximum dry unit weight (kN/m^3)	1.51
Optimum moisture content (%)	25.37

Table 2. Physical and mechanical properties of polypropylene fiber.

Polypropylene Fiber Properties	Value
Diameter (μm)	10
Cut length (mm)	10
Density (g/cm^3)	0.91
Tensile strength (MPa)	486
Stretching limit (%)	15
Acid resistance	Excellent
Alkali resistance	Excellent

2.2. Mixture Design

The variables in this research were the content of Portland cement, C&D waste, polypropylene fiber, and sodium sulfate. Each dosing level was determined based on its weight ratio to the pre-dried soil. Particularly, Portland cement and C&D waste to soil ratios were defined as 10%, 20%, and 30%. Polypropylene fiber was incorporated, accounting for 1%, 2%, and 4% of the soil weight. The dosing proportions of sodium sulfate were 0.2%, 0.4%, and 0.8%. As for water, the weight ratio was maintained constantly at 80%. The consequent 81 combinations along with 3 control groups (conventional CSS) were cast for experiments.

2.3. Mechanical Tests

UCS, FS, and direct shear tests were conducted to estimate the CSS mechanical performance. The procedure for all mechanical tests was prepared strictly in accordance with GB/T50123-1999 [42]. Soil samples for direct shear tests were shaped as 61.8 mm × 20 mm (diameter × height) particularly. The normal stress was applied vertically at $\sigma_n = 50$, 100,150, and 200 kPa to examine the shear strength parameters. The single-doped specimens used for the direct shear test contained C&D waste at 10%, 20%, and 30% content, while the variable dosage design for UCS and FS specimens was consistent with the aforementioned. Cubic (50 mm × 50 mm × 50 mm) and cuboid (40 mm × 40 mm × 160 mm) specimens were cast, respectively, for UCS and FS tests. After vibrating, mortar samples were wrapped in a thin membrane and cured in the standard curing condition (20 ± 2 °C temperature and 95% relative humidity) until tests were conducted at 7-, 14-, and 28-days of the curing period. On the day before the tests, specimens were shifted from the curing chamber and soaked in 24 °C water for 24 h to pattern the humid working situation. YAW-4206 and DY-208JX automatic pressure testers were employed to conduct the UCS and FS tests with a 0.04 MPa/s loading rate. The average data of three replicated specimens after eliminating the error was recorded as the ultimate test result, listed in Appendix A.

2.4. Machine Learning Models

2.4.1. Baseline Models

Baseline models including LR, MLR, and KNN were selected, in contrast to BPNN and RF to assess the prediction accuracy. Regression models (LR and MLR) identify the relationship between predictor and output, possessing the benefits of minimum computation and easy implementation. Equations (1) and (2) display the principles of LR and MLR models.

$$\mathrm{Ln}\frac{p}{1-p} = b_0 + \sum_{k=1}^{n} b_k x_k \quad (1)$$

$$Y = \beta_0 + \beta_1 x_1 + \beta_2 x_2 + \cdots + \beta_n x_n \quad (2)$$

Within the proposed equations above, x_k and p represent independent and dependent variables; b_0 and b_k stand for constant coefficients; Y is the predicted strength of CSS; x_i and β_i (where $i = 1, 2, 3, \ldots, n$) denote the considered variables in laboratory test design and regression coefficient, respectively.

KNN algorithm estimates mechanical performance through similitude between inputting values. Specifically, KNN models detect the most similar observations in the dataset and output the average value as the ultimate prediction [43]. The pre-defined function calculates the distance between neighbors with Euclidean distances, assigning all neighbors with the same weight (Equation (3)) [44,45]. KNN models, therefore, possess the superiority of effective prediction among large datasets [29].

$$d(i,j) = \sqrt{(y_{i1} - y_{j1})^2 + (y_{i2} - y_{j2})^2 + \cdots + (y_{in} - y_{jn})^2} \quad (3)$$

where i and j represent the detected points and d is the abbreviation of Euclidean distance.

2.4.2. Back Propagation Neural Network (BPNN)

BPNN, as one type of ANN algorithm trained by the Back Propagation (BP) technique, has been employed to successfully develop the prediction pattern for the mechanical strength of cementitious materials. The ANN model is essentially a neural network, consisting of an input layer, output layer, and hidden layer(s). As illustrated in Figure 1, each neuron yields the ability of a processing unit, merging information from the former layer to transport the combination to the subsequent nodes [46]. The following equation presents the neuron connection between upper and lower layers in the mathematical version.

$$y = \max\left(0, \sum_i w_i x_i + b\right) \quad (4)$$

where y represents the output value from the lower layer; w_i stands for the weight; x_i is the received data from the former layer, and b denotes the bias between neurons. The neuron loop iterates until the mean squared error (*MSE*) reaches the pre-set value, ending the training process [47].

$$MSE = \frac{1}{n}\sum_{i=1}^{n}(y_i - \hat{y}_i)^2 \quad (5)$$

where y_i denotes the model prediction, and $\hat{y}_i$ is the result estimated by labels.

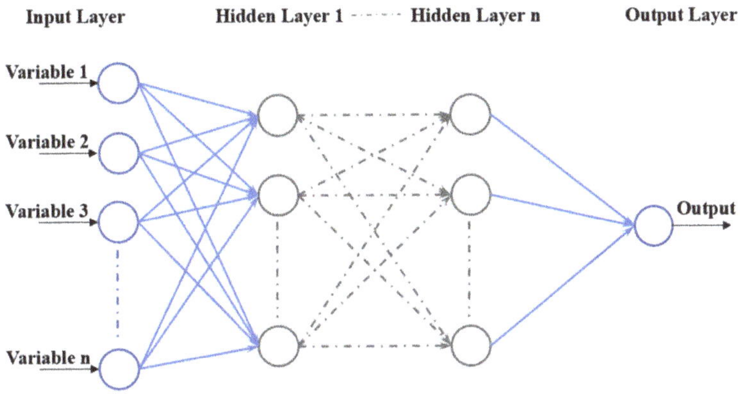

Figure 1. Structure of ANN network.

Figure 2 represents the BP flow, which is the research approach to updating the bias and weights in the neuron network by calculating the difference between the predicted output and the actual strength from the dataset [48,49]. The BP technique endows the BPNN models with sensitivity to hyperparameters which affect the ultimate accuracy.

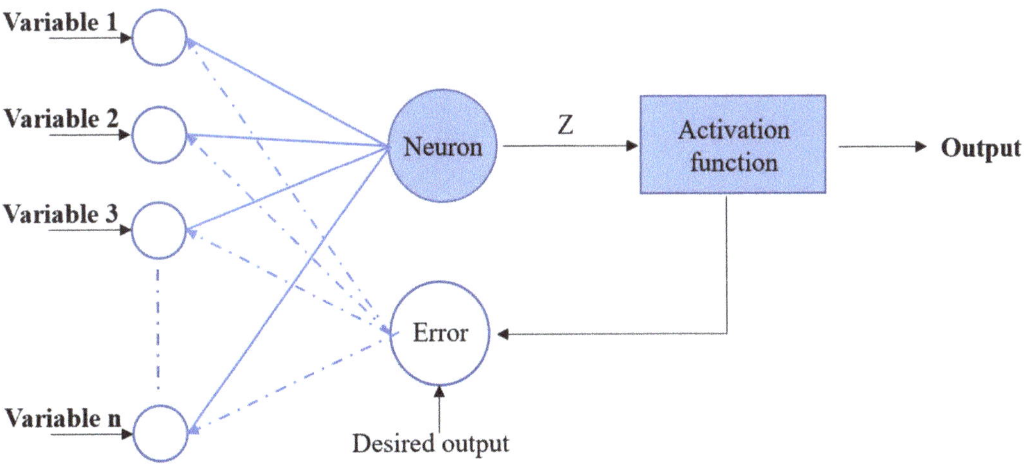

Figure 2. BP procedure.

2.4.3. Random Forest (RF)

As illustrated in Figure 3, RF generates multiple decision trees (RT) in which each RT is built based on a new training set oriented from the bagging and voting method [50]. The bagging method yields characteristics of independently training predictors through bootstrap and aggregation. Bootstrap indicates that RF models allow the duplicate value, which randomly resamples the original dataset by the number of predictors. Each split is built from the random subset selected from the input predictor variables, improving the diversity to achieve accurate estimation. Equation (6) shows the training set as R_n where X and Y, respectively, denote the input and output vector. The average result of RTs will be output as the ultimate prediction using the aggregation approach [50].

$$R_n = \{(X_1, Y_1), (X_2, Y_2), \ldots, (X_n, Y_n)\} \tag{6}$$

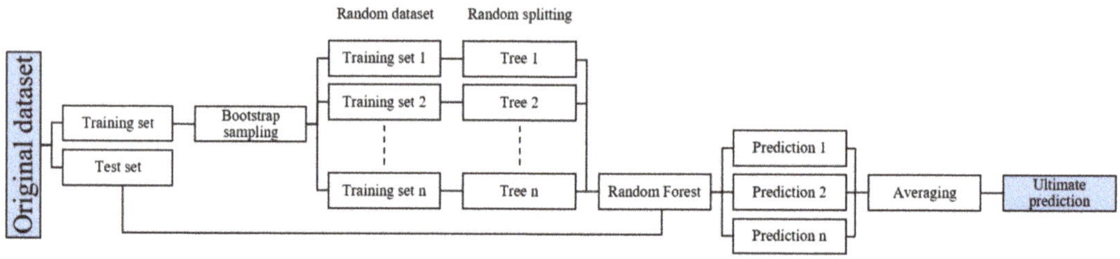

Figure 3. Structure of RF.

2.4.4. Beetle Antennae Search (BAS)

The BAS algorithm was proposed from the behavior of beetles, evolving the function to avoid the tedious effort of optimizing hyperparameters manually. The beetles cannot locate the accurate position while looking for food. As a result, beetles move towards the side which receives the greater intensity of odor. Inspired by the principle, the BAS algorithm simulates the goal hyperparameter as the food, rendering the ML models with the capability of automatically tuning [39]. As explained by Equation (7), the first step of BAS is to generate a random vector as the beetle antennae, where V indicates the direction and k represents the space dimensionality [40].

$$V = rand(k,1)/||rand(k,1)|| \tag{7}$$

Secondly, the algorithm determines the antennae coordinate based on the direction vector:

$$X_l = X_i + D \cdot V \tag{8}$$

$$X_r = X_i - D \cdot V \tag{9}$$

where X_l, X_r, and X_i, respectively, denote the coordinate of left, and right antennae and their centroid at the ith iteration; D is the distance between the left and right antennae. The concentration is then compared by the normalized function represented as the following:

$$X_{i+1} = X_i + S \cdot normal(X_l - X_r) \qquad f(X_l) < f(X_r) \tag{10}$$

$$X_{i+1} = X_i - S \cdot normal(X_l - X_r) \qquad f(X_l) > f(X_r) \tag{11}$$

where S represents the length of steps. The comparison iterates 50 times during the model training process to optimize the hyperparameters.

2.4.5. Cross-Validation

Figure 4 illustrates the 10-fold cross-validation which was applied in this research to mitigate the overfitting during the training and testing stages caused by the finite database. Firstly, the input variable is randomly resampled as training and test set, which, respectively, account for 70% and 30% of the original dataset. Then, the training set is separated into 10 equal folds. The 90% of folds yield the function of training the ML models, and the last fold validates the prediction performance by calculating the root means square error (*RMSE*) [51]. The 10 folds will take turns to be the validation fold. Specifically, in each cross-validation, the BAS algorithm is used to optimize the hyperparameters within

50 iterations. In each iteration, the *RMSE* is calculated for hyperparameter adjustment. Finally, the ML model with the minimum *RMSE* will be saved in each cross-validation (a total of 10 models). By comparing the *RMSE* values from each fold, the ML model with the lowest *RMSE* value and optimal hyperparameters will be chosen as the final ML model.

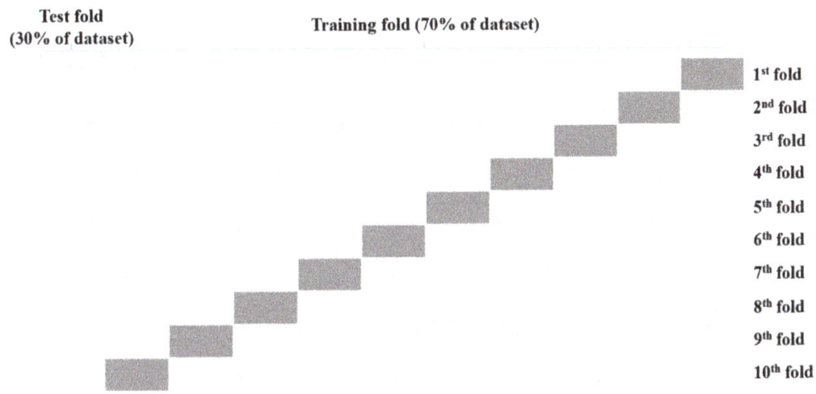

Figure 4. A 10-fold cross validation.

2.4.6. Performance Evaluation

Two evaluation indicators are applied in this study, aiming to estimate the precision of the baseline, BPNN, and RF models: correlation coefficient (*R*) and *RMSE*. The indexes are defined by the following equations:

$$R = \frac{\sum_{i=1}^{n}(y_i^* - \overline{y^*})(y_i - \overline{y})}{\sqrt{\sum_{i=1}^{n}(y_i^* - \overline{y^*})^2}\sqrt{\sum_{i=1}^{n}(y_i^* - \overline{y})^2}} \tag{12}$$

$$RMSE = \sqrt{\frac{1}{n}\sum_{i=1}^{n}(y_i^* - y_i)^2} \tag{13}$$

where n represents the quantity of data groups; y_i^* and y_i, respectively, denote the estimated and actual output; $\overline{y^*}$ and $\overline{y}$ are the mean values of y_i^* and y_i.

3. Results and Discussion

3.1. Effect of Portland Cement

The UCS and FS test results for control groups are illustrated in Figure 5. It is notable in Figure 5a that the CSS compressive strength increased with the curing age as evidenced by the observed increment up to 185.73% for 7- to 14-day curing time. This is mainly ascribed to cement hydration, which has been proven in Figure 6. Subfigures a and b illustrate the hardened sample photo taken by scanning electron microscope (SEM) and the energy-dispersive X-ray spectroscopy (EDX). The silicon content reached 41.14%, which was higher than that of conventional soil [8]. The results indicate the presence of the hydrate phase like calcium silicate hydrate (C-S-H). These products render strengthened bonding between soil particles and cement. The hydration slowed down to maintain a steady rate during the late curing age (after 14-day), as reflected by the declining increasement ranging from 7.45% to 36.18% in Figure 5. Meanwhile, the UCS test results increased with Portland cement content due to the same reason. The maximum increment during all curing ages is 450.34%, indicating that $CaSiO_3$ and C-S-H significantly promote the CSS compressive strength. Specifically, the colloidal hydrate products filled the porosity, contributing to mitigating entrapped air voids. Figure 5b represents a similar trend as FS results at 28-day increased 91.19% and 176.91%, respectively, when cement content rose 10%.

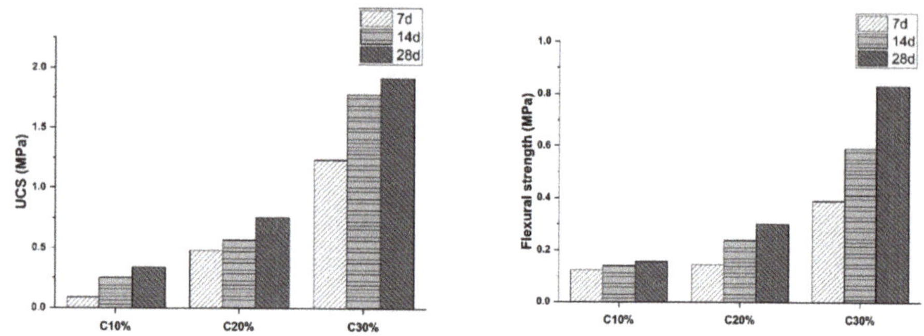

(a) (b)

Figure 5. Mechanical results of control groups: (**a**) UCS test, (**b**) FS test (Note: C represents Portland cement).

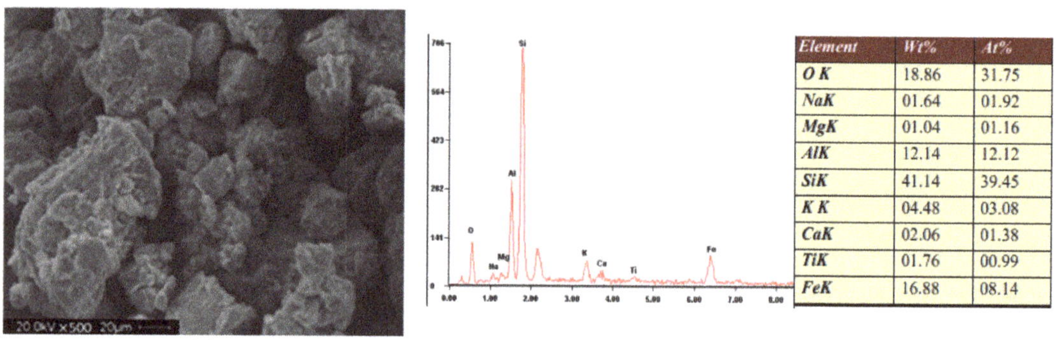

(a) (b)

Figure 6. Result of (**a**) SEM, (**b**) EDX.

On the other hand, increasing Portland cement dosing level causes an undesired alkali–silica reaction (ASR). It is shown in Figure 7 that brittle destruction occurred in control groups, leading to the evident fracture surface. Portland cement introduces brittleness along with compressive strength enhancement, impairing sample deformation to the external force. The conclusion can be demonstrated by the surrounding debris in Figure 7.

Figure 7. Fracture surface of conventional CSS sample.

3.2. Effect of C&D Waste

The 28-day UCS results are shown in Figure 8, with subfigures separated based on Portland cement content. The maximum strength for the three dosing levels as illustrated in the figures were 0.8048 MPa, 1.5008 Mpa, and 2.6572 Mpa, respectively. All the results were higher than those of the control groups, indicating mechanical performance increases with C&D waste incorporation. This is mainly ascribed to the old mortar attached to C&D waste particles which participate in cement hydration [52,53]. The consequent ITZ represents bonding strength that anchors the stiff C&D particles in the matrix to support and prevent soil collapse. Additionally, calcium hydroxide (CH) formed during the hydration process accelerates the hydration process of old mortar, which will generate more calcium silicate hydrates (C-S-H) to promote sample strength [5].

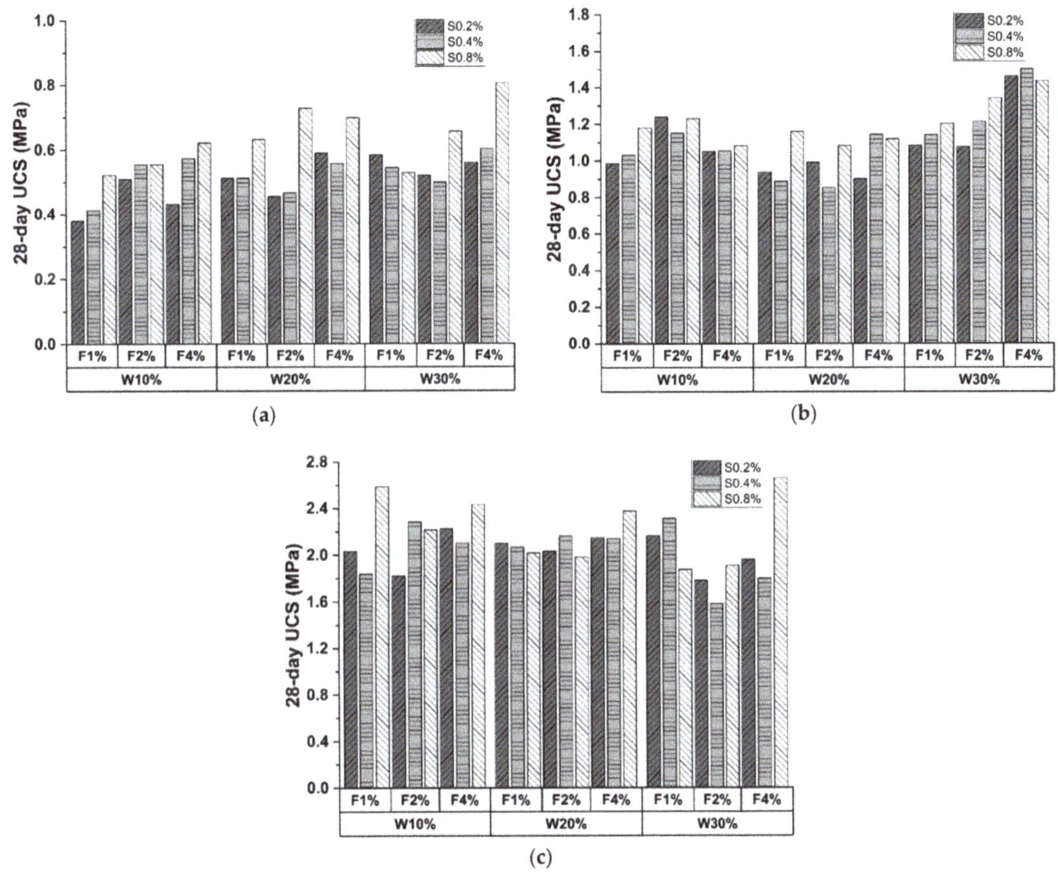

Figure 8. A 28-day UCS of CSS specimens with (**a**) 10% cement; (**b**) 20% cement; (**c**) 30% cement (Note: W represents C&D waste; S represents sodium sulfate; F represents polypropylene fiber).

Figure 9 shows the FS results at the 28-day curing age, indicating that C&D waste demonstrates a positive effect on FS performance. As shown in Figure 9b,c, C&D waste displayed better FS enhancement when cement content was high. The maximum improvements reached 56.83% and 57.2%, respectively, while the data in Figure 9a was 21.62%. This phenomenon can be explained as the increase in cement content enhances the degree of hydration of old mortar attached to the C&D waste surface. On the other hand, enhancement became insignificant when C&D waste content was high (30%). This is mainly due to superabundant large particles introduced into the matrix, resulting in porosity and entrapped air.

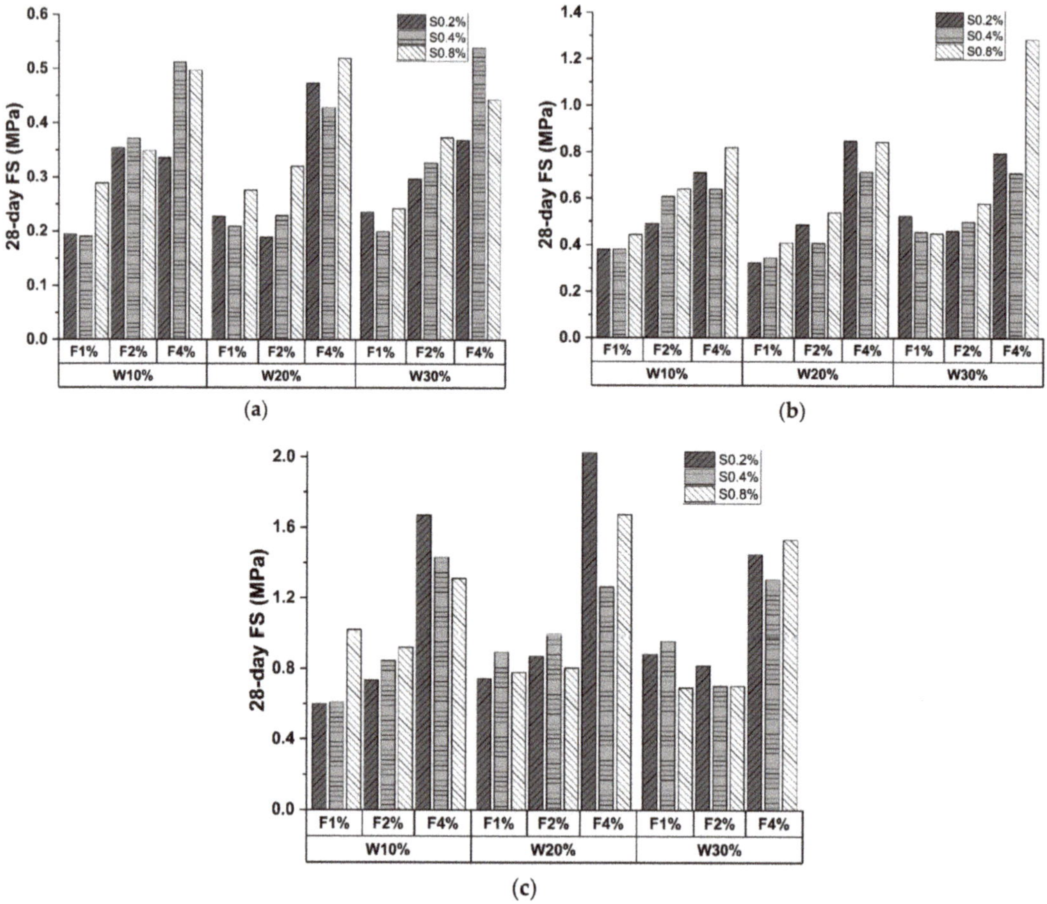

Figure 9. A 28-day FS of CSS specimens with (**a**) 10% cement; (**b**) 20% cement; (**c**) 30% cement (Note: W represents C&D waste; S represents sodium sulfate; F represents polypropylene fiber).

Direct shear tests were also conducted due to their cost-effectiveness and convenient operation. The relationship between normal stress (σ) and shear stress (τ) with various C&D waste proportions is illustrated in Figure 10. The inclusion of C&D waste significantly enhances the shear performance as proven by the increasing cohesion © and material angle of friction (φ). Moreover, the average increments were 33.52%, 43.28%, and 26.34% for each 10% increase in C&D waste dosage, indicating that 20% C&D waste content demonstrated the best-improving effect.

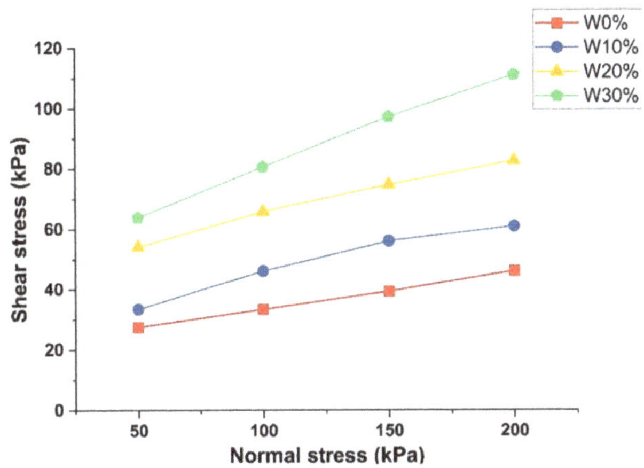

Figure 10. DS test result (Note: W represents C&D waste).

3.3. Effect of Polypropylene Fiber

In Figures 8 and 9, UCS and FS increased with polypropylene fiber proportion, demonstrating the positive effect. The enhancement function on UCS can be attributed to the higher particle friction provided by fiber than that inside the matrix [54]. Specifically, a rougher fiber surface prevents particle displacement, which impedes the generation of microscopic cracks. However, fiber incorporation introduces undesired porosity and compactness descending, leading to a strength reduction of 31.52% in Figure 8c. Figure 11 shows an electron microscope image of the sample fracture surface after the UCS test. Several porosities exist on the fiber periphery. These entrapped air voids caused by fiber agglomeration remarkably weaken the fiber enhancement on compressive strength [55].

Figure 11. Microstructure of UCS fracture surface.

Compared with UCS tests, fibrous material promoted FS results more evidently. As depicted in Figure 9a, the FS of CSS specimens increased up to 82.33% and 150.31% when the fiber content doubled. A similar trend can be observed in Figure 9b,c. The dominant enhancement of FS can be attributed to the bridging effect. The randomly distributed fibers demonstrate the outstanding inhibitory function on crack generation. In addition, polypropylene fiber will be pulled out of the fracture surface when the failure occurs, endowing the relict with certain flexural resistance. However, the bridging effective can be hindered by cement inclusion. The peak FS was recorded at 30% C&D waste dosed group when cement content was low (10% and 20%), whereas the 30% cement specimens reached the maximum at 20% C&D waste proportion. This is mainly ascribed to the excessive cement hydration that impeded the fibrous bridge formation.

3.4. Effect of Sodium Sulfate

The influence of sodium sulfate can be analyzed in Figures 8 and 9. UCS and FS results share a similar trend that mechanical performances strengthen with sodium sulfate proportion. The average increment of UCS and FS from 0.2% to 0.8% sodium sulfate were 16.94% and 16.29%, with the maximum increasement recorded as 59.61% and 69.96%, respectively. This is attributed to the reaction between sulfate ion and liquid phase (AlO^{2-}, Ca^{2+}, etc.). The main product is ettringite (AFt phase), revealing enhancing characteristics on early-age strength [56]. Moreover, metal ions (Na^{+-}, Ca^{2+}, etc.) demonstrate dominant effectiveness in improving alkalinity, which yields function of the reaction rate catalyzation along with SiO_2 and Al_2O_3 dissolution. However, samples incorporated with 0.4% sodium sulfate failed to follow the positive trend. The UCSs of the sample changed irregularly, as evidenced by the value fluctuating from −14% to 32.59% compared to 0.2% sodium sulfate. A similar phenomenon was also observed in Figure 9. The error source is probably ascribed to human error and material composition deviation.

Furthermore, based on Appendix A, the UCS enhancing rate varies from each curing age, as evidenced by the average increment of 15.41% and 30.49% for early (7-day to 14-day) and late (14-day to 28-day) curing times. The principle can be explained by Equations (14) and (15) [57]. Sulfate ions modified the conventional hydration reaction, resulting in the formation of ettringite which promotes the CSS mechanical performance. However, C-S-H exhibits the capability of absorbing sulfate in the early curing stage and releasing it during the later period, leading to rapid strength promotion from 14- to 28-day [58–60].

$$C_3A + 2CSH_2 + 26H \rightarrow C_6AS_3H_{32} \tag{14}$$

$$2C_3A + C_6AS_3H_{32} + 4H \rightarrow 3C_4ASH_{12} \tag{15}$$

4. Machine Learning Predicted Results

4.1. Prediction for UCS Performance

4.1.1. Hyperparameter Tuning

In total, 252 data (84 groups of experimental results) constituted the database, which reached the requirement of the database size for the traditional machine learning task. During the machine learning process, the contents of cement, C&D waste, fiber, and sulfate, and the curing age were set as features. The outputs were UCS and FS.

For BPNN models, hyperparameters that needed to be determined include the number of neurons and layers. BAS and 10-fold CV detected the optimal hyperparameters through iteration as illustrated in Figure 12. It is evident in Figure 12a,b that the third fold and BPNN network with three hidden layers obtained the lowest *RMSE* value. Figure 12c represents the BAS algorithm conducted in fold 3, indicating the *RMSE* value reduced with the iteration and the tuned hyperparameter was gained at the 36th iteration. The consequent BPNN hyperparameters were therefore determined as $numHiddenLayers = 3$, with $numNeuronsInEachLayers = 3, 11, 4$, respectively.

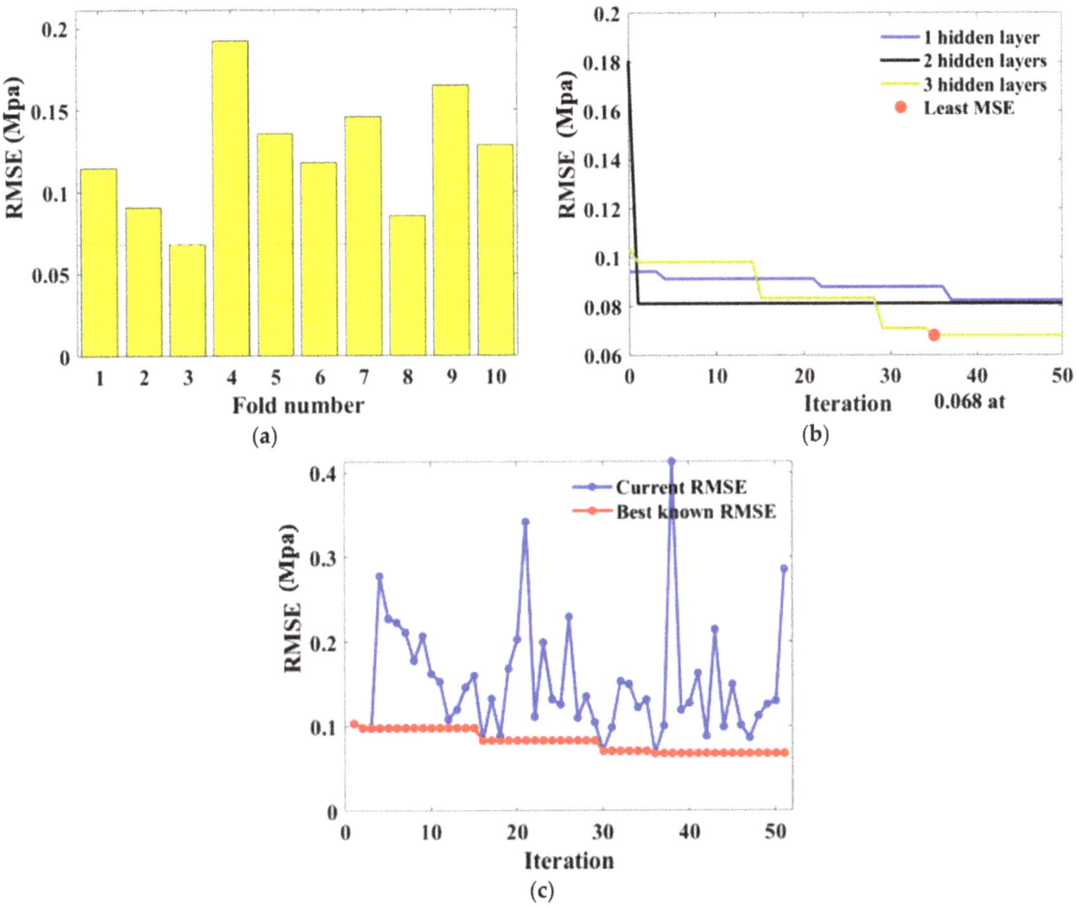

Figure 12. Hyperparameter tuning for BPNN: (**a**) RSME values obtained in 10 validation folds; (**b**) RSME convergency with various numbers of hidden layers; (**c**) Iteration conducted at the 3rd fold.

During the modeling setting, the amounts of trees (*ntree*) and the minimum number of leaves ($minNumleaf$) are fundamental parameters that needed to be adjusted for the RF algorithm. In this research, they were detected from the procedure as shown in Figure 13. It is noted that the *RMSE* value is basically convergent within 50 iterations for the traditional machine learning task. From Figure 12b,c and Figure 13b, the *RMSE* value's reduction can be clearly observed within the first 10–30 iterations and maintains the minimum value after 30 iterations, illustrating that the *RMSE* reaches the local minimum. Specifically, the minimum *RMSE* value was obtained at the 6th fold as 0.1015 which dropped significantly with iteration progress, demonstrating the obtainment of desired hyperparameters as $numTree = 88, minNumleaf = 1$.

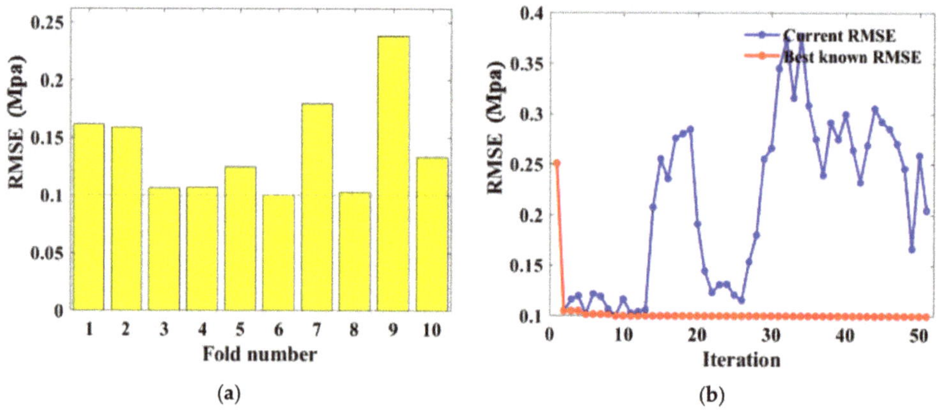

Figure 13. Hyperparameter tuning for RF: (**a**) RSME values obtained in 10 validation folds; (**b**) Iteration conducted at the 6th fold.

4.1.2. Performance of BAS-BPNN and BAS-RF for UCS

Figures 14 and 15 show the prediction performance of BAS-BPNN and BAS-RF, with subfigures (a) and (b) depicting the training and test sets, respectively. The yellow columns denoting the error of prediction were minor, so the consequent conclusion can be drawn that both BPNN and FR estimate the CSS strength accurately.

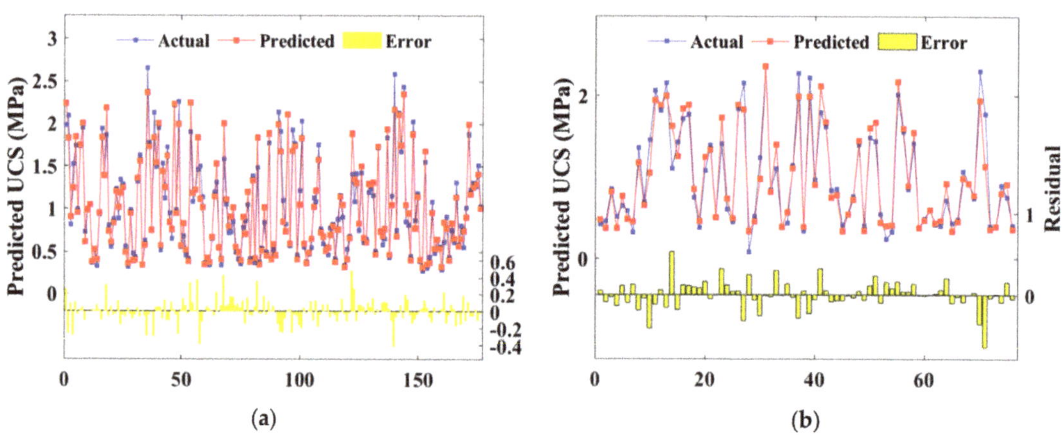

Figure 14. Scatter plot of predicted and actual UCS of BPNN model: (**a**) training set; (**b**) test set.

Additionally, prediction and actual results formed a great correlation as evidenced by the *R* value in Figure 16. As depicted in Figure 16a, the correlation coefficient (R) for the BPNN algorithm were 0.9717 and 0.9594 for the training and test set, respectively, which were both lower than that for the RF algorithm (0.9877 and 0.9685). Therefore, BPNN and RF simultaneously provided reliable predictions, whereas RF yielded enhanced accuracy. Moreover, a similar *R* value indicated that there was no overfitting problem in both algorithms.

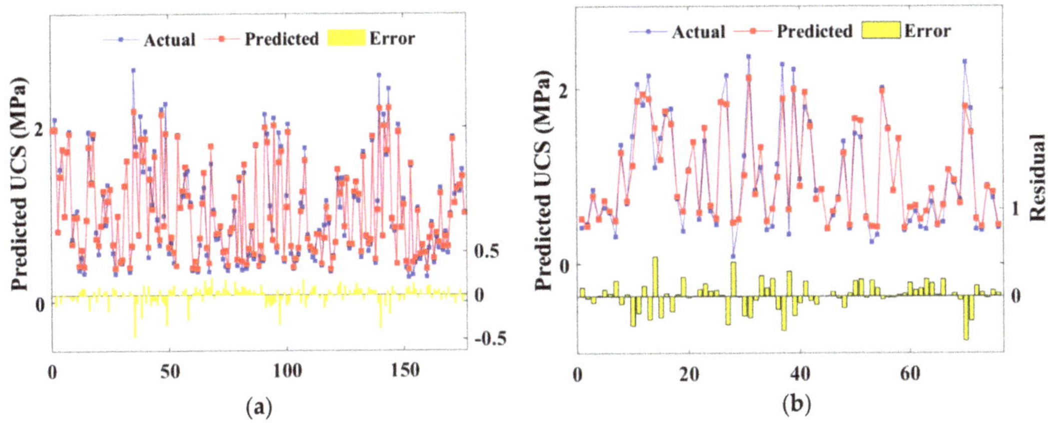

Figure 15. Scatter plot of predicted and actual UCS of RF model: (**a**) training set; (**b**) test set.

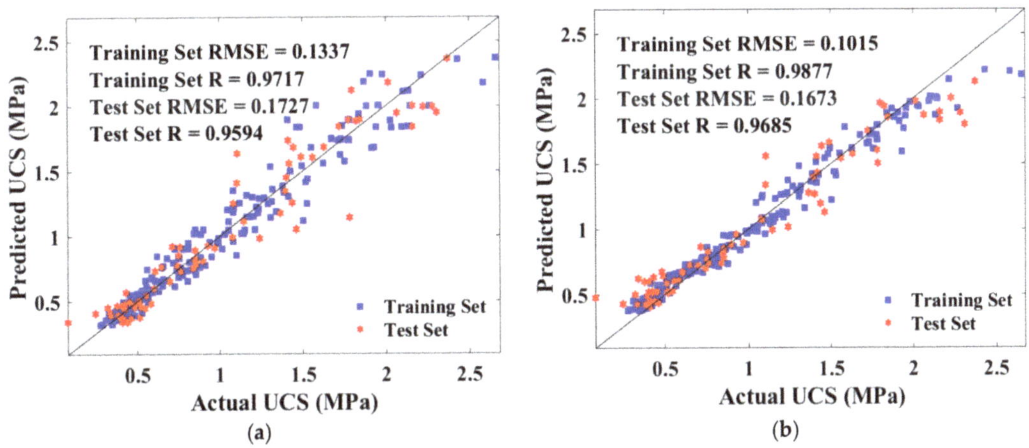

Figure 16. Scatter plot of predicted and actual UCS of training and test sets: (**a**) BAS-BPNN model; (**b**) BAS-RF model.

4.1.3. Comparison of BPNN. RF, LR, MLR, and KNN

Figure 17 is the box diagram representing the various between the actual strength and prediction. The boxes in the figure indicate the interquartile range for each model by the height between the upper and lower borders. It can be observed that BPNN yielded the best accuracy and shared the lowest median (the red line within the box) with the RF algorithm. Four outliers (read+) were defected in BPNN, which was more than that of other models. However, as the interquartile range and median affect the accuracy more significantly, BPNN and RF demonstrated relatively similar reliability among all five algorithms.

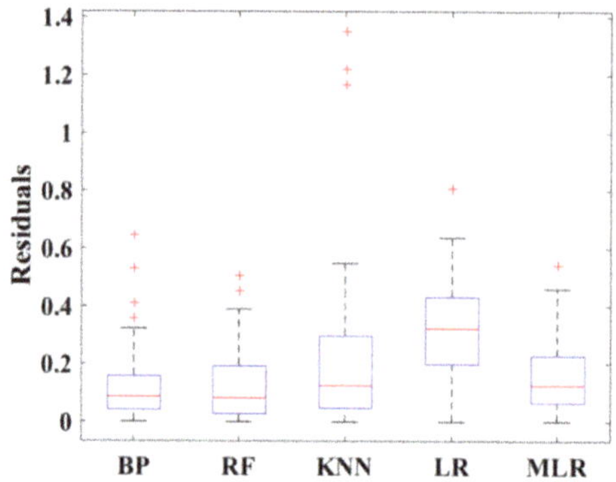

Figure 17. Box diagram of 5 models.

The Taylor diagram was also applied to evaluate the model performance through three assessment criteria including *R*, *RMSE*, and standard deviation, as shown in Figure 18. The dot denoted for RF was the nearest to the actual point with the minimum standard deviation, maximum *R*, and minor *RMSE*. Table 3 listed the specific value of *R* and *RMSE* for each algorithm, proving RF as the best-performed model in UCS prediction.

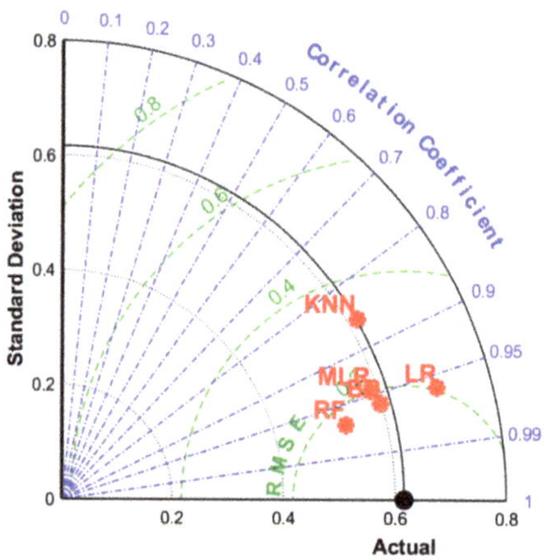

Figure 18. Taylor diagram of 5 models.

Table 3. Evaluation of 5 ML models on UCS test group.

Evaluation Index	Model				
	LR	MLR	KNN	BPNN	RF
RMSE (MPa)	0.3694	0.2014	0.3242	0.1727	0.0280
R	0.9598	0.9462	0.8599	0.9594	0.9685

4.2. Prediction for FS Performance

4.2.1. Hyperparameter Tuning

A similar procedure with UCS estimation was applied to optimize hyperparameters for FS prediction. The 3rd fold outputs the minimum *RMSE* value during the CV process as shown in Figure 19a. Moreover, the *numHiddenLayers* was examined as 1 because the *RMSE* reduced remarkably and reached the minimum when the iteration was processed for three times. The phenomenon can be ascribed to the effectiveness of BAS on hyperparameter tuning. The other desirable hyperparameter *numNeuronsInEachLayers* defected as 1.

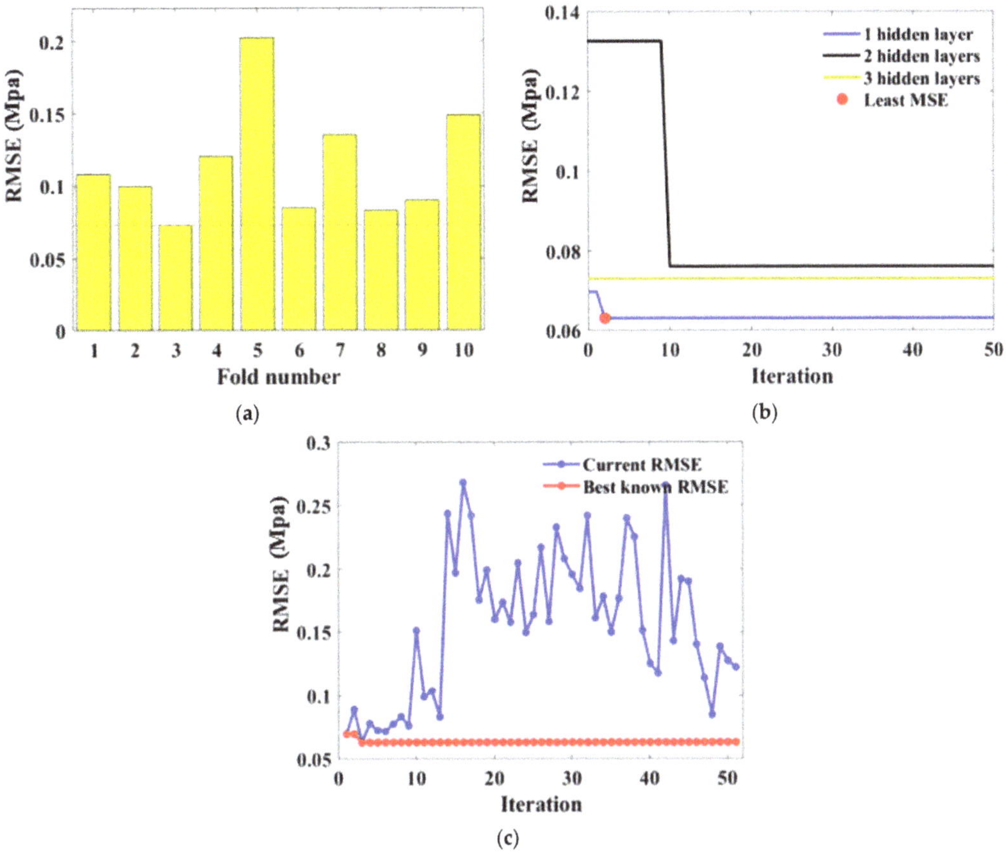

Figure 19. Hyperparameter tuning for BPNN: (**a**) RSME values obtained in 10 validation folds; (**b**) RSME convergency with various numbers of hidden layers; (**c**) Iteration conducted at the 3rd fold.

For the RF algorithm, the 9th fold had the minor *RMSE* as evidenced by Figure 20a. The hyperparameters optimized by this iteration were therefore applied to predict the FS performance. Figure 20b shows the *RMSE* scatter plot, indicating the decline of *RMSE* value until it maintained a minimum at the 41st repeat. The final tuned hyperparameters were $numTree = 29$, $minNumleaf = 1$.

4.2.2. Performance of BAS-BPNN and BAS-RF for FS

After being automatically tuned in the 70% training set, hyperparameters were applied in the 30% test set to predict the FS property of CSS. Figures 21 and 22 present the scatter plot of FS prediction from BPNN and RF models with the actual strength of the training and test set, respectively. It is noted from the figures that the prediction and the actual results fitted well, as the contract ratio of red and blue lines were relatively high. Furthermore, error bars located on the horizontal line proved that BPNN and RF algorithms demonstrated similar accuracy in FS prediction.

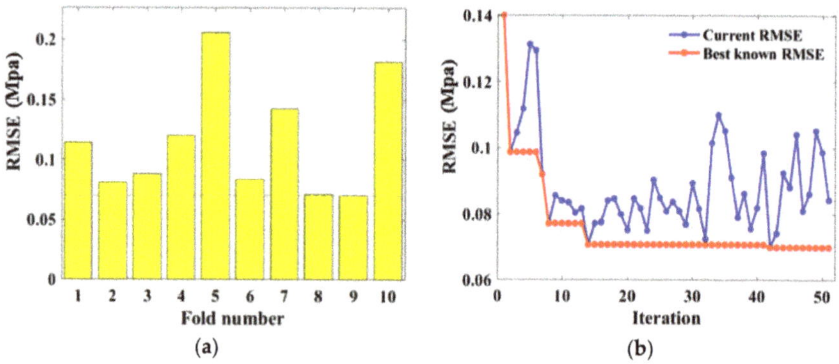

Figure 20. Hyperparameter tuning for RF: (**a**) RSME values obtained in 10 validation folds; (**b**) Iteration conducted at the 9th fold.

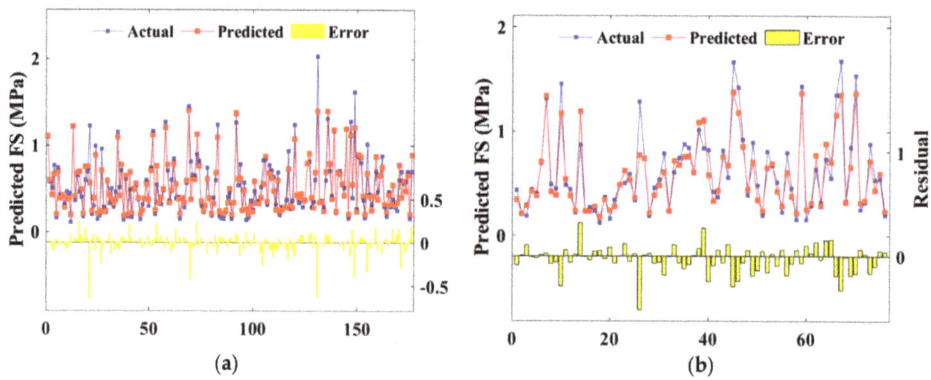

Figure 21. Scatter plot of predicted and actual FS of BPNN model: (**a**) training set; (**b**) test set.

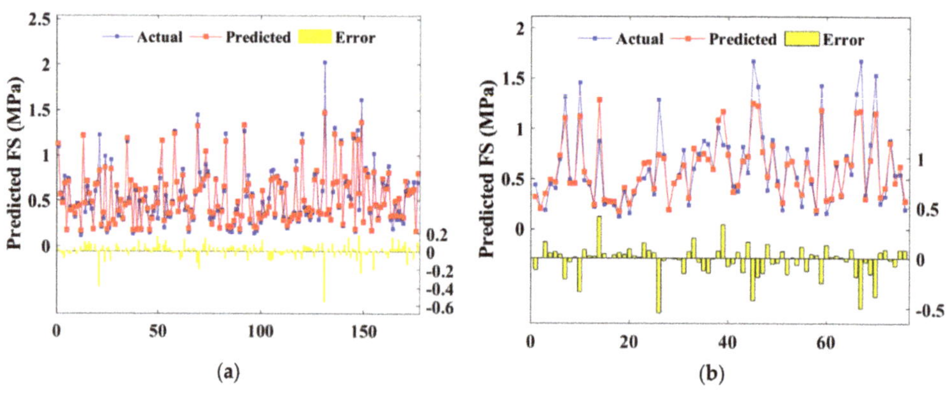

Figure 22. Scatter plot of predicted and actual FS of RF model: (**a**) training set; (**b**) test set.

Detailed *RMSE* and *R* value are illustrated in Figure 23, where subfigure (a) depicts the BPNN model, and (b) depicts the RF model. The *RMSE* values were ranging from 0.0841 to 0.1583, indicating that BPNN and RF models estimated the strength accurately. The training set of the RF algorithm defected the hyperparameters with the highest *R* and the lowest *RMSE*. However, the test set output the worst-performed value, which manifested that the RF model had a higher risk of overfitting compared with BPNN.

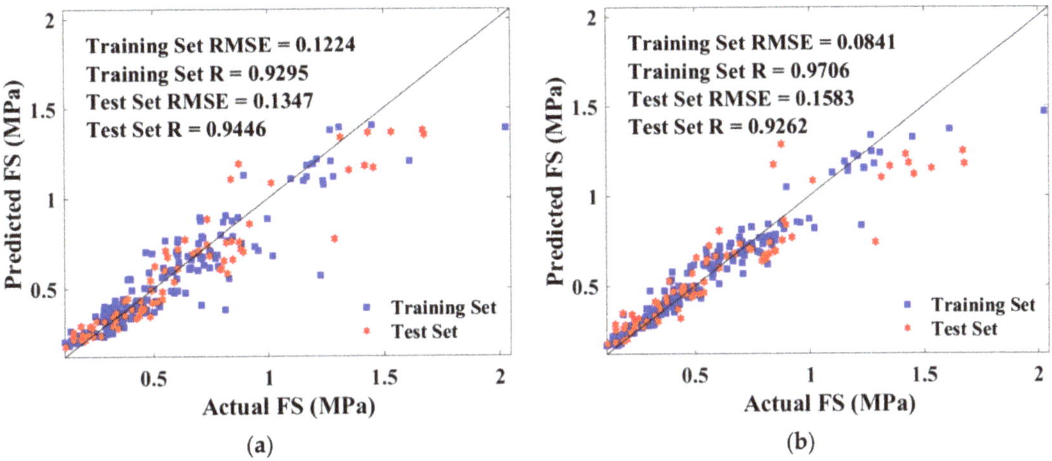

Figure 23. Scatter plot of predicted and actual FS of training and test sets: (**a**) BAS-BPNN model; (**b**) BAS-FR model.

4.2.3. Comparison of BPNN. RF, LR, MLR, and KNN

For FS prediction, MLR demonstrated high accuracy as evidenced by the condensed interquartile range in Figure 24a. However, BPNN exhibited better-integrated reliability because of fewer outliers and the lower median value. Figure 24b integrated *R*, *RMSE* and standard deviation into polar coordinates, obtaining the same conclusion owing to the closest distance between BPNN and the actual FS result. In addition, based on evaluation standards listed in Table 4, BAS-BPNN was also considered the most effective algorithm due to the least error and best degree of fitting.

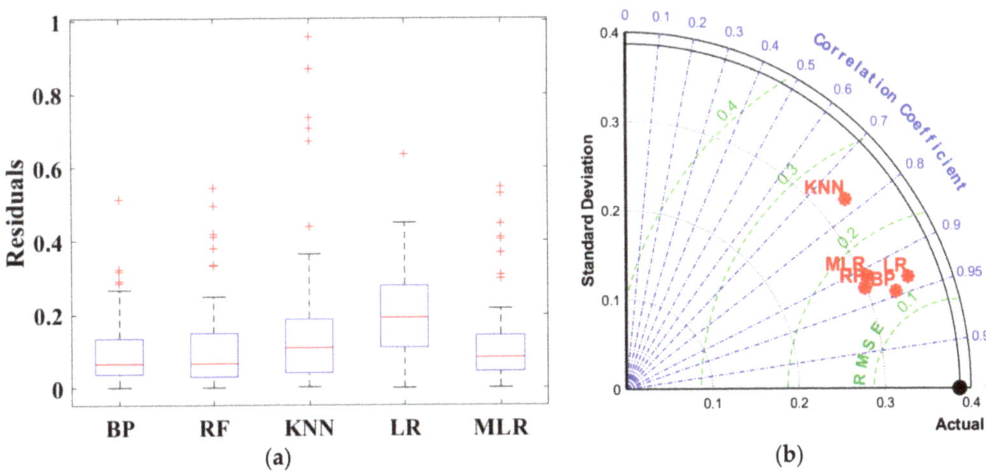

Figure 24. Prediction evaluation: (**a**) box diagram; (**b**) Taylor diagram.

Table 4. Evaluation of 5 ML models on FS test group.

Evaluation Index	Model				
	LR	MLR	KNN	BPNN	RF
RMSE (MPa)	0.2386	0.1677	0.2498	0.1347	0.1583
R	0.9341	0.9107	0.7678	0.9446	0.9262

5. Optimal Mixture Design

The RF algorithm defined the effect factor of each variable as depicted in Figure 25, which contributed to proposing the optimum mixture design. The consequent importance score for each content was similar to that obtained from laboratory tests. Water and soil proportion exhibited no influence on the mechanical property owing to the constant dosing level in all specimens. C&D waste and sodium sulfate had a similar introduction effect. It is noted that cement content, curing age, and fiber content yielded the best effectiveness on CSS mechanical strength. Combined with the UCS and FS results listed in Appendix A, specimens prepared with 30% cement, 20% C&D waste, 4% polypropylene fiber, and 0.8% sodium sulfate were considered the best performed. The conclusion can be attributed to the high ranking of the 28-day UCS and FS performance among all mixture designs.

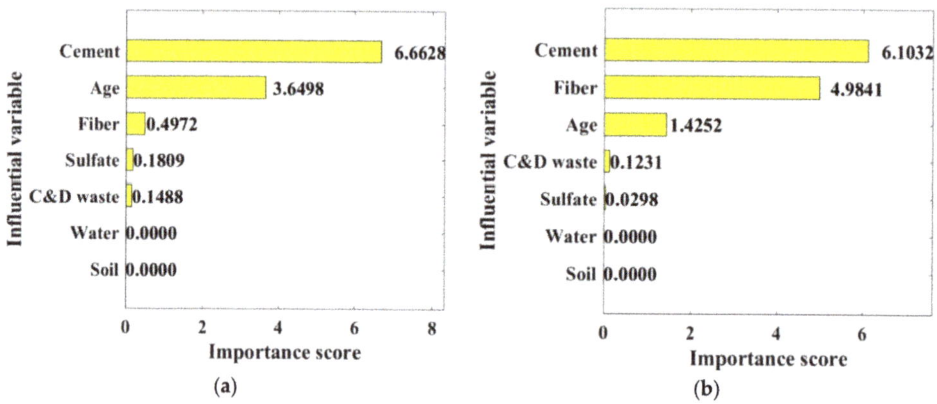

Figure 25. Importance score of influential variables: (**a**) UCS; (**b**) FS.

6. Conclusions

In this research, the inclusion effects of Portland cement, construction and demolition (C&D) waste, polypropylene fiber, and sodium sulfate on the mechanical properties were assessed through laboratory tests. Moreover, machine learning (ML) techniques were applied based on the 84 experimental results, predicting the unconfined compressive strength (UCS) and flexural strength (FS) of cement stabilized soil (CSS). Primary conclusions are drawn as follows:

(1) Portland cement demonstrates outstanding enhancement of mechanical strengths through cement hydration. The maximum increase in sample strength on 28-day when the curing time and admixture amounts were 450.34% and 176.91%.

(2) The C&D waste has a positive effect on both the compressive and flexural properties of the samples, with the largest increase in performance being 57.2%. A 20% C&D waste content demonstrates the best-improving effect.

(3) Polypropylene fiber brings a 150.31% increase in the flexural properties of the samples. However, the increase in compressive properties is not significant.

(4) Higher levels of sodium sulphate increase the mechanical properties of the cement soil by 59.61% and 69.96%, respectively. However, the 0.4% sodium sulphate fails to change the properties regularly, with a range of −14% to 32.59%.
(5) The influencing factors of each variable on CSS performance are ranked in descending order as: Portland cement, polypropylene fiber, C&D waste, sodium sulfate. The mixture design of 30% cement, 20% C&D waste, 4% fiber and 0.8% is considered as the best-performed combination.
(6) BPNN and RF acquired the most accurate prediction for UCS and FS, respectively. Baseline models generally are inferior to Machine Learning models with hyperparameters in mechanical strength prediction.

The research output from this article could lead to a wider application of CSS as an engineering material. Moreover, the concluded enhancement can be treated as a baseline model. Future research can extend the experiments to explore other properties such as slump, or to consider alternative aggregate ratios. Meanwhile, RF and BPNN can be employed to predict whether the designed proportion will achieve the mechanical strength requirements or to optimize the proportioning for a given strength.

Author Contributions: Conceptualization, X.W. and G.Z.; methodology, G.Z., Y.W. (Yufei Wang) and X.W.; writing—original draft, Z.D.; writing—review & editing, G.F., Y.W. (Yufei Wang), X.Z. and C.X.; formal analysis, Z.D. and X.L.; investigation, G.F., X.L. and Y.W. (Yan Wang); software, Y.W. (Yan Wang) and Y.Z.; resources, Y.Z., X.Z. and C.X. All authors have read and agreed to the published version of the manuscript.

Funding: The research was supported by Key Project of Hunan Education Department (grant number 21A0511), National Natural Science Foundation of China (grant numbers 51908201 and 51978254), Natural Science Foundation of Hunan Province (grant numbers 2020JJ5024; 2021JJ50142), and Hunan Provincial Science and Technology Plan (grant number 2021NK4273). This research is also supported by Academic Research Council of Australia Linkage Projects for Asset Intelligence: Maximising Operational Effectiveness for Digital Era, (Grant No. LP180100222). The study was also supported by State Key Laboratory for GeoMechanics and Deep Underground Engineering, China University of Mining & Technology/China University of Mining & Technology, Beijing (SKLGDUEK2105).

Institutional Review Board Statement: Not applicable.

Informed Consent Statement: Not applicable.

Data Availability Statement: The data presented in this study are openly available.

Conflicts of Interest: The authors declare no conflict of interest.

Appendix A

Table A1. Mixing proportion for all variable combinations.

ID	UCS (MPa)			FS (MPa)		
	7-Day	14-Day	28-Day	7-Day	14-Day	28-Day
Control 1	0.0876	0.2503	0.3408	0.1242	0.1403	0.1571
Control 2	0.4820	0.5676	0.7560	0.1451	0.2375	0.3003
Control 3	1.2340	1.7832	1.9160	0.3891	0.5912	0.8316
CWFS-1112	0.4612	0.3225	0.3783	0.1645	0.1537	0.1943
CWFS-1114	0.4476	0.3439	0.4128	0.1544	0.1481	0.1904
CWFS-1118	0.5304	0.4448	0.5220	0.1684	0.1918	0.2886
CWFS-1122	0.4212	0.3862	0.5092	0.2006	0.2064	0.3543
CWFS-1124	0.4128	0.4212	0.5544	0.2584	0.3167	0.3715

Table A1. *Cont.*

ID	UCS (MPa)			FS (MPa)		
	7-Day	14-Day	28-Day	7-Day	14-Day	28-Day
CWFS-1128	0.4024	0.4236	0.5544	0.1904	0.2996	0.3488
CWFS-1142	0.4212	0.3225	0.4320	0.2519	0.2890	0.3356
CWFS-1144	0.4448	0.4236	0.5728	0.3581	0.3670	0.5117
CWFS-1148	0.4984	0.4876	0.6208	0.3692	0.3423	0.4967
CWFS-1212	0.3676	0.4104	0.5116	0.1643	0.1817	0.2264
CWFS-1214	0.4076	0.3811	0.5116	0.1700	0.1581	0.2094
CWFS-1218	0.4448	0.4636	0.6316	0.1832	0.2064	0.2761
CWFS-1222	0.3648	0.3462	0.4556	0.1589	0.1567	0.1893
CWFS-1224	0.3488	0.3597	0.4664	0.2232	0.2314	0.2285
CWFS-1228	0.5144	0.5464	0.7272	0.2398	0.3235	0.3210
CWFS-1242	0.4848	0.4664	0.5892	0.3698	0.4548	0.4739
CWFS-1244	0.3488	0.4716	0.5568	0.3774	0.3058	0.4289
CWFS-1248	0.4048	0.5676	0.6980	0.3274	0.4652	0.5195
CWFS-1312	0.3860	0.4392	0.5836	0.1674	0.1772	0.2363
CWFS-1314	0.3352	0.4156	0.5436	0.1517	0.1904	0.1998
CWFS-1318	0.3116	0.3676	0.5276	0.1195	0.1615	0.2421
CWFS-1322	0.2900	0.3890	0.5196	0.2296	0.3000	0.2977
CWFS-1324	0.2796	0.3304	0.4984	0.1799	0.3218	0.3270
CWFS-1328	0.3940	0.5220	0.6556	0.2710	0.2516	0.3739
CWFS-1342	0.3888	0.4368	0.5596	0.3164	0.4713	0.3691
CWFS-1344	0.3752	0.4528	0.6020	0.3032	0.4713	0.5394
CWFS-1348	0.4904	0.5436	0.8048	0.4861	0.5032	0.4442
CWFS-2112	0.6100	0.6848	0.9860	0.2753	0.3355	0.3807
CWFS-2114	0.6476	0.8236	1.0316	0.2911	0.3504	0.3821
CWFS-2118	0.7728	0.8500	1.1804	0.3175	0.4010	0.4440
CWFS-2122	0.7700	0.8796	1.2392	0.3898	0.4295	0.4914
CWFS-2124	0.8024	0.8528	1.1512	0.4256	0.4452	0.6076
CWFS-2128	0.8472	0.9888	1.2312	0.8139	0.7081	0.6401
CWFS-2142	0.6500	0.6208	1.0524	0.5446	0.7073	0.7116
CWFS-2144	0.7196	0.7808	1.0552	0.6068	0.6676	0.6396
CWFS-2148	0.7596	0.8528	1.0820	0.5692	0.6123	0.8199
CWFS-2212	0.5916	0.5856	0.9380	0.2905	0.2862	0.3238
CWFS-2214	0.5464	0.6528	0.8872	0.2567	0.3296	0.3459
CWFS-2218	0.7648	0.7300	1.1596	0.3668	0.4259	0.4089
CWFS-2222	0.6100	0.7436	0.9916	0.4029	0.5192	0.4870
CWFS-2224	0.5596	0.6820	0.8528	0.3065	0.3741	0.4079
CWFS-2228	0.7516	0.8100	1.0820	0.4071	0.3687	0.5402
CWFS-2242	0.6584	0.7516	0.9008	0.5067	0.5630	0.8492

Table A1. *Cont.*

ID	UCS (MPa)			FS (MPa)		
	7-Day	14-Day	28-Day	7-Day	14-Day	28-Day
CWFS-2244	0.8180	0.8100	1.1432	0.6622	0.7713	0.7164
CWFS-2248	0.7156	0.8436	1.1196	0.5821	0.8092	0.8442
CWFS-2312	0.7516	0.8552	1.0820	0.3391	0.4377	0.5259
CWFS-2314	0.6580	0.9140	1.1404	0.2974	0.4101	0.4568
CWFS-2318	0.7640	0.9702	1.2020	0.2985	0.3597	0.4502
CWFS-2322	0.7408	0.8553	1.0768	0.2919	0.3759	0.4609
CWFS-2324	0.8368	0.8846	1.2128	0.4333	0.4249	0.5002
CWFS-2328	0.9516	0.9931	1.3404	0.4617	0.5216	0.5790
CWFS-2342	0.8648	1.0981	1.4608	0.7216	0.7460	0.7958
CWFS-2344	0.8980	1.0153	1.5008	0.7840	0.7140	0.7119
CWFS-2348	0.9272	1.1304	1.4364	0.7981	0.8484	1.2859
CWFS-3112	1.1968	1.4952	2.0340	0.3985	0.5504	0.6006
CWFS-3114	1.0848	1.4444	1.8388	0.3893	0.5857	0.6095
CWFS-3118	1.5244	2.1324	2.5880	0.5919	0.8247	1.0207
CWFS-3122	1.1032	1.4900	1.8232	0.4669	0.6060	0.7354
CWFS-3124	1.4152	1.7780	2.2816	0.6744	0.8850	0.8449
CWFS-3128	1.3856	1.8444	2.2148	0.6531	0.9415	0.9202
CWFS-3142	1.4180	1.7752	2.2256	1.0989	1.1942	1.6690
CWFS-3144	1.5164	1.7404	2.1032	1.1547	1.4189	1.4320
CWFS-3148	1.3085	1.9538	2.4336	1.2395	1.3497	1.3127
CWFS-3212	1.3140	1.6258	2.0976	0.5497	0.6575	0.7450
CWFS-3214	1.3647	1.5537	2.0656	0.6025	0.6948	0.8935
CWFS-3218	1.2685	1.4820	2.0148	0.4875	0.4820	0.7792
CWFS-3222	1.3192	1.7296	2.0284	0.6980	0.8248	0.8717
CWFS-3224	1.2926	1.7164	2.1592	0.7482	0.8778	0.9954
CWFS-3228	1.3778	1.7780	1.9804	0.6061	0.7348	0.8046
CWFS-3242	1.4253	1.7324	2.1456	1.1684	1.6124	2.0286
CWFS-3244	1.3940	1.7216	2.1376	0.8412	0.8754	1.2671
CWFS-3248	1.5592	1.8868	2.3720	1.0135	1.4556	1.6752
CWFS-3312	1.2182	1.7564	2.1592	1.2259	0.7466	0.8835
CWFS-3314	1.3085	1.6284	2.3084	0.5386	0.7910	0.9581
CWFS-3318	1.0819	1.4632	1.8712	0.4022	0.5725	0.6951
CWFS-3322	1.1595	1.1056	1.7780	0.6079	0.6103	0.8183
CWFS-3324	1.0448	1.4128	1.5808	0.5543	0.5529	0.7040
CWFS-3328	1.4020	1.4100	1.9072	0.4953	0.6974	0.7057
CWFS-3342	1.5276	1.6736	1.9560	0.8984	1.2113	1.4504
CWFS-3344	1.2368	1.9432	1.7964	1.2807	1.2696	1.3083
CWFS-3348	1.9296	2.2656	2.6572	1.2370	1.1690	1.5334

References

1. Anagnostopoulos, C.A. Strength properties of an epoxy resin and cement-stabilized silty clay soil. *Appl. Clay Sci.* **2015**, *114*, 517–529. [CrossRef]
2. Zhu, W.; Chen, X.; Struble, L.J.; Yang, E.-H. Characterization of calcium-containing phases in alkali-activated municipal solid waste incineration bottom ash binder through chemical extraction and deconvoluted Fourier transform infrared spectra. *J. Clean. Prod.* **2018**, *192*, 782–789. [CrossRef]
3. Liu, W.; Guo, Z.; Wang, C.; Niu, S. Physico-mechanical and microstructure properties of cemented coal Gangue-Fly ash backfill: Effects of curing temperature. *Constr. Build. Mater.* **2021**, *299*, 124011. [CrossRef]
4. Donatello, S.; Fernández-Jimenez, A.; Palomo, A. Very high volume fly ash cements. Early age hydration study using Na_2SO_4 as an activator. *J. Am. Ceram. Soc.* **2013**, *96*, 900–906. [CrossRef]
5. Sun, J.; Wang, Y.; Liu, S.; Dehghani, A.; Xiang, X.; Wei, J.; Wang, X. Mechanical, chemical and hydrothermal activation for waste glass reinforced cement. *Constr. Build. Mater.* **2021**, *301*, 124361. [CrossRef]
6. Tang, Z.; Li, W.; Ke, G.; Zhou, J.L.; Tam, V.W.Y. Sulfate attack resistance of sustainable concrete incorporating various industrial solid wastes. *J. Clean. Prod.* **2019**, *218*, 810–822. [CrossRef]
7. Feng, J.; Chen, B.; Sun, W.; Wang, Y. Microbial induced calcium carbonate precipitation study using Bacillus subtilis with application to self-healing concrete preparation and characterization. *Constr. Build. Mater.* **2021**, *280*, 122460. [CrossRef]
8. Tang, Y.; Feng, W.; Chen, Z.; Nong, Y.; Guan, S.; Sun, J. Fracture behavior of a sustainable material: Recycled concrete with waste crumb rubber subjected to elevated temperatures. *J. Clean. Prod.* **2021**, *318*, 128553. [CrossRef]
9. Sun, J.; Huang, Y.; Aslani, F.; Wang, X.; Ma, G. Mechanical enhancement for EMW-absorbing cementitious material using 3D concrete printing. *J. Build. Eng.* **2021**, *41*, 102763. [CrossRef]
10. Zhang, C.; Abedini, M. Development of PI model for FRP composite retrofitted RC columns subjected to high strain rate loads using LBE function. *Eng. Struct.* **2022**, *252*, 113580. [CrossRef]
11. Huang, H.; Guo, M.; Zhang, W.; Huang, M. Seismic Behavior of Strengthened RC Columns under Combined Loadings. *J. Bridge Eng.* **2022**, *27*, 05022005. [CrossRef]
12. Narani, S.; Abbaspour, M.; Hosseini, S.M.M.; Aflaki, E.; Nejad, F.M. Sustainable reuse of Waste Tire Textile Fibers (WTTFs) as reinforcement materials for expansive soils: With a special focus on landfill liners/covers. *J. Clean. Prod.* **2020**, *247*, 119151. [CrossRef]
13. Sun, J.; Aslani, F.; Wei, J.; Wang, X. Electromagnetic absorption of copper fiber oriented composite using 3D printing. *Constr. Build. Mater.* **2021**, *300*, 124026. [CrossRef]
14. Aslani, F.; Hou, L.; Nejadi, S.; Sun, J.; Abbasi, S. Experimental analysis of fiber-reinforced recycled aggregate self-compacting concrete using waste recycled concrete aggregates, polypropylene, and steel fibers. *Struct. Concr.* **2019**, *20*, 1670–1683. [CrossRef]
15. Aslani, F.; Sun, J.; Huang, G. Mechanical behavior of fiber-reinforced self-compacting rubberized concrete exposed to elevated temperatures. *J. Mater. Civ. Eng.* **2019**, *31*, 04019302. [CrossRef]
16. Sun, J.; Huang, Y.; Aslani, F.; Ma, G. Electromagnetic wave absorbing performance of 3D printed wave-shape copper solid cementitious element. *Cem. Concr. Compos.* **2020**, *114*, 103789. [CrossRef]
17. Sun, J.; Aslani, F.; Lu, J.; Wang, L.; Huang, Y.; Ma, G. Fibre-reinforced lightweight engineered cementitious composites for 3D concrete printing. *Ceram. Int.* **2021**, *47*, 27107–27121. [CrossRef]
18. Sun, J.; Huang, Y.; Aslani, F.; Ma, G. Properties of a double-layer EMW-absorbing structure containing a graded nano-sized absorbent combing extruded and sprayed 3D printing. *Constr. Build. Mater.* **2020**, *261*, 120031. [CrossRef]
19. Dobrovolski, M.E.G.; Munhoz, G.S.; Pereira, E.; Medeiros-Junior, R.A. Effect of crystalline admixture and polypropylene microfiber on the internal sulfate attack in Portland cement composites due to pyrite oxidation. *Constr. Build. Mater.* **2021**, *308*, 125018. [CrossRef]
20. Sun, J.; Ma, Y.; Li, J.; Zhang, J.; Ren, Z.; Wang, X. Machine learning-aided design and prediction of cementitious composites containing graphite and slag powder. *J. Build. Eng.* **2021**, *43*, 102544. [CrossRef]
21. Feng, W.; Wang, Y.; Sun, J.; Tang, Y.; Wu, D.; Jiang, Z.; Wang, J.; Wang, X. Prediction of thermo-mechanical properties of rubber-modified recycled aggregate concrete. *Constr. Build. Mater.* **2022**, *318*, 125970. [CrossRef]
22. Zhang, W.; Li, H.; Li, Y.; Liu, H.; Ding, X. Application of deep learning algorithms in geotechnical engineering: A short critical review. *Artif. Intell. Rev.* **2021**, *54*, 5633–5673. [CrossRef]
23. Zhang, W.; Wu, C.; Zhong, H.; Li, Y.; Wang, L. Prediction of undrained shear strength using extreme gradient boosting and random forest based on Bayesian optimization. *Geosci. Front.* **2021**, *12*, 469–477. [CrossRef]
24. Zhang, W.; Li, H.; Wu, C.; Li, Y.; Liu, Z.; Liu, H. Soft computing approach for prediction of surface settlement induced by earth pressure balance shield tunneling. *Undergr. Space* **2021**, *6*, 353–363. [CrossRef]
25. Cook, R.; Lapeyre, J.; Ma, H.; Kumar, A. Prediction of compressive strength of concrete: Critical comparison of performance of a hybrid machine learning model with standalone models. *J. Mater. Civ. Eng.* **2019**, *31*, 04019255. [CrossRef]
26. Zhang, J.; Sun, Y.; Li, G.; Wang, Y.; Sun, J.; Li, J. Machine-learning-assisted shear strength prediction of reinforced concrete beams with and without stirrups. *Eng. Comput.* **2020**, *38*, 1–15. [CrossRef]
27. Huang, H.; Huang, M.; Zhang, W.; Yang, S. Experimental study of predamaged columns strengthened by HPFL and BSP under combined load cases. *Struct. Infrastruct. Eng.* **2021**, *17*, 1210–1227. [CrossRef]
28. Sun, J.; Wang, Y.; Yao, X.; Ren, Z.; Zhang, G.; Zhang, C.; Chen, X.; Ma, W.; Wang, X. Machine-learning-aided prediction of flexural strength and ASR expansion for waste glass cementitious composite. *Appl. Sci.* **2021**, *11*, 6686. [CrossRef]

29. Kang, M.-C.; Yoo, D.-Y.; Gupta, R. Machine learning-based prediction for compressive and flexural strengths of steel fiber-reinforced concrete. *Constr. Build. Mater.* **2021**, *266*, 121117. [CrossRef]
30. Xu, H.; Wang, X.-Y.; Liu, C.-N.; Chen, J.-N.; Zhang, C. A 3D root system morphological and mechanical model based on L-Systems and its application to estimate the shear strength of root-soil composites. *Soil Tillage Res.* **2021**, *212*, 105074. [CrossRef]
31. Ma, G.; Sun, J.; Aslani, F.; Huang, Y.; Jiao, F. Review on electromagnetic wave absorbing capacity improvement of cementitious material. *Constr. Build. Mater.* **2020**, *262*, 120907. [CrossRef]
32. Wang, X.; Yang, Y.; Yang, R.; Liu, P. Experimental Analysis of Bearing Capacity of Basalt Fiber Reinforced Concrete Short Columns under Axial Compression. *Coatings* **2022**, *12*, 654. [CrossRef]
33. Huang, Y.; Zhang, J.; Ann, F.T.; Ma, G. Intelligent mixture design of steel fibre reinforced concrete using a support vector regression and firefly algorithm based multi-objective optimization model. *Constr. Build. Mater.* **2020**, *260*, 120457. [CrossRef]
34. Wei, J.; Xie, Z.; Zhang, W.; Luo, X.; Yang, Y.; Chen, B. Experimental study on circular steel tube-confined reinforced UHPC columns under axial loading. *Eng. Struct.* **2021**, *230*, 111599. [CrossRef]
35. Yu, Y.; Zhang, C.; Gu, X.; Cui, Y. Expansion prediction of alkali aggregate reactivity-affected concrete structures using a hybrid soft computing method. *Neural Comput. Appl.* **2019**, *31*, 8641–8660. [CrossRef]
36. Shariati, M.; Mafipour, M.S.; Mehrabi, P.; Bahadori, A.; Zandi, Y.; Salih, M.N.; Nguyen, H.; Dou, J.; Song, X.; Poi-Ngian, S. Application of a hybrid artificial neural network-particle swarm optimization (ANN-PSO) model in behavior prediction of channel shear connectors embedded in normal and high-strength concrete. *Appl. Sci.* **2019**, *9*, 5534. [CrossRef]
37. Kaveh, A.; Izadifard, R.; Mottaghi, L. Optimal design of planar RC frames considering CO_2 emissions using ECBO, EVPS and PSO metaheuristic algorithms. *J. Build. Eng.* **2020**, *28*, 101014. [CrossRef]
38. Chahnasir, E.S.; Zandi, Y.; Shariati, M.; Dehghani, E.; Toghroli, A.; Mohamad, E.T.; Shariati, A.; Safa, M.; Wakil, K.; Khorami, M. Application of support vector machine with firefly algorithm for investigation of the factors affecting the shear strength of angle shear connectors. *Smart Struct. Syst.* **2018**, *22*, 413–424.
39. Wang, J.; Chen, H. BSAS: Beetle swarm antennae search algorithm for optimization problems. *arXiv* **2018**, arXiv:1807.10470.
40. Liu, K.; Alam, M.S.; Zhu, J.; Zheng, J.; Chi, L. Prediction of carbonation depth for recycled aggregate concrete using ANN hybridized with swarm intelligence algorithms. *Constr. Build. Mater.* **2021**, *301*, 124382. [CrossRef]
41. Shi, T.; Lan, Y.; Hu, Z.; Wang, H.; Xu, J.; Zheng, B. Tensile and Fracture Properties of Silicon Carbide Whisker-Modified Cement-Based Materials. *Int. J. Concr. Struct. Mater.* **2022**, *16*, 1–13. [CrossRef]
42. *GB/T 50123-1999*; Standard for Soil Test Method. China Planning Press: Beijing, China, 1999.
43. Cunningham, P.; Delany, S.J. k-Nearest neighbour classifiers—A Tutorial. *ACM Comput. Surv. (CSUR)* **2021**, *54*, 1–25. [CrossRef]
44. Salami, B.A.; Rahman, S.M.; Oyehan, T.A.; Maslehuddin, M.; Al Dulaijan, S.U. Ensemble machine learning model for corrosion initiation time estimation of embedded steel reinforced self-compacting concrete. *Measurement* **2020**, *165*, 108141. [CrossRef]
45. Zhang, G.; Chen, C.; Sun, J.; Li, K.; Xiao, F.; Wang, Y.; Chen, M.; Huang, J.; Wang, X. Mixture optimisation for cement-soil mixtures with embedded GFRP tendons. *J. Mater. Res. Technol.* **2022**, *18*, 611–628. [CrossRef]
46. Sun, J.; Wang, J.; Zhu, Z.; He, R.; Peng, C.; Zhang, C.; Huang, J.; Wang, Y.; Wang, X. Mechanical Performance Prediction for Sustainable High-Strength Concrete Using Bio-Inspired Neural Network. *Buildings* **2022**, *12*, 65. [CrossRef]
47. Zhang, Y.; Aslani, F.; Lehane, B. Compressive strength of rubberized concrete: Regression and GA-BPNN approaches using ultrasonic pulse velocity. *Constr. Build. Mater.* **2021**, *307*, 124951. [CrossRef]
48. Xu, J.; Chen, Y.; Xie, T.; Zhao, X.; Xiong, B.; Chen, Z. Prediction of triaxial behavior of recycled aggregate concrete using multivariable regression and artificial neural network techniques. *Constr. Build. Mater.* **2019**, *226*, 534–554. [CrossRef]
49. Chaabene, W.B.; Flah, M.; Nehdi, M.L. Machine learning prediction of mechanical properties of concrete: Critical review. *Constr. Build. Mater.* **2020**, *260*, 119889. [CrossRef]
50. Han, Q.; Gui, C.; Xu, J.; Lacidogna, G. A generalized method to predict the compressive strength of high-performance concrete by improved random forest algorithm. *Constr. Build. Mater.* **2019**, *226*, 734–742. [CrossRef]
51. Chou, J.-S.; Tsai, C.-F. Concrete compressive strength analysis using a combined classification and regression technique. *Autom. Constr.* **2012**, *24*, 52–60. [CrossRef]
52. Dimitriou, G.; Savva, P.; Petrou, M.F. Enhancing mechanical and durability properties of recycled aggregate concrete. *Constr. Build. Mater.* **2018**, *158*, 228–235. [CrossRef]
53. Sun, J.; Lin, S.; Zhang, G.; Sun, Y.; Zhang, J.; Chen, C.; Morsy, A.M.; Wang, X. The effect of graphite and slag on electrical and mechanical properties of electrically conductive cementitious composites. *Constr. Build. Mater.* **2021**, *281*, 122606. [CrossRef]
54. Xu, D.; Liu, Q.; Qin, Y.; Chen, B. Analytical approach for crack identification of glass fiber reinforced polymer–sea sand concrete composite structures based on strain dissipations. *Struct. Health Monit.* **2020**, *13*, 1475921720974290. [CrossRef]
55. Xu, J.; Wu, Z.; Chen, H.; Shao, L.; Zhou, X.; Wang, S. Study on strength behavior of basalt fiber-reinforced loess by digital image technology (DIT) and scanning electron microscope (SEM). *Arab. J. Sci. Eng.* **2021**, *46*, 11319–11338. [CrossRef]
56. Marchon, D.; Flatt, R.J. Mechanisms of cement hydration. In *Science and Technology of Concrete Admixtures*; Elsevier: Amsterdam, The Netherlands, 2016; pp. 129–145.
57. Joseph, S.; Skibsted, J.; Cizer, Ö. A quantitative study of the C3A hydration. *Cem. Concr. Res.* **2019**, *115*, 145–159. [CrossRef]

58. Neto, J.d.S.A.; Angeles, G.; Kirchheim, A.P. Effects of sulfates on the hydration of Portland cement—A review. *Constr. Build. Mater.* **2021**, *279*, 122428. [CrossRef]
59. Zhang, G.; Wu, C.; Hou, D.; Yang, J.; Sun, D.; Zhang, X. Effect of environmental pH values on phase composition and microstructure of Portland cement paste under sulfate attack. *Compos. Part B Eng.* **2021**, *216*, 108862. [CrossRef]
60. Fu, J.; Jones, A.M.; Bligh, M.W.; Holt, C.; Keyte, L.M.; Moghaddam, F.; Foster, S.J.; Waite, T.D. Mechanisms of enhancement in early hydration by sodium sulfate in a slag-cement blend–Insights from pore solution chemistry. *Cem. Concr. Res.* **2020**, *135*, 106110. [CrossRef]

materials MDPI

Article

Experimental Study on Carbonation of Cement-Based Materials in Underground Engineering

Jun Zheng [1,2], Gang Zeng [3,4,*], Hui Zhou [1,2,*] and Guanghua Cai [5]

1 China Railway 11th Bureau Group Co., Ltd., Wuhan 430061, China; jzheng_crcc@126.com
2 State Key Laboratory of Geomechanics and Geotechnical Engineering, Institute of Rock and Soil Mechanics, Chinese Academy of Sciences, Wuhan 430071, China
3 School of Civil Engineering and Architecture, Hubei University of Arts and Science, Xiangyang 441053, China
4 Hubei Provincial Engineering Research Center of Slope Habitat Construction Technique Using Cement-Based Materials, China Three Gorges University, Yichang 443002, China
5 School of Civil Engineering, Nanjing Forestry University, Nanjing 210037, China; ghcai@njfu.edu.cn
* Correspondence: gangz@hbuas.edu.cn (G.Z.); hzhou@whrsm.ac.cn (H.Z.)

Abstract: The corrosive water environment has a decisive influence on the durability of a diversion tunnel lining. In this paper, the effects of carbonation on cement-based materials in water-immersion and saturated-humidity environments were studied by increasing the CO_2 concentration. The results show that under conditions of water-immersion and saturated humidity, the color of the non-carbonation region is dark, while the carbonation region is gray, and the color boundary is obvious. However, in an atmospheric environment, there is no zone with a dark color and the color boundary is not obvious. In a saturated-humidity environment, the carbonation depth increases over time and changes greatly, and its value is about 16.71 mm at 200 days. While in a water-immersion environment, the carbonation depth varies little with time and the value is only 2.31 mm. The carbonation depths of cement mortar samples in different environments generally follow a linear relationship with the square root of time. The carbonation coefficient in a saturated-humidity environment is more than nine times that in the water-immersion environment. In a water-immersion environment, the carbonation causes a large loss of calcium in cement-based materials, and their Ca/Si ratio obviously decreases. The calcium silicon ratio (Ca/Si) of cement-based materials in a water-immersion environment is 0.11, which is much less than 1.51 in a water-saturated environment and 1.49 in an atmospheric environment. In a saturated-humidity environment, the carbonation only reduces the pH of the pore solution in the carbonation region, and the structural stability of cement-based materials is not degraded. The number of pores of all radii after carbonation in a water-immersion environment exceeds that in a saturated-humidity environment, and the total pore volume and average pore radius in a water-immersion environment are also larger than in a saturated-humidity environment, so the water-immersion environment accelerates the development and expansion of pores. The research results can provide some theoretical and technical support for the design, construction, and safe operation of diversion tunnel linings.

Keywords: diversion tunnel; corrosion; cement-based materials; carbonation; experimental study

Citation: Zheng, J.; Zeng, G.; Zhou, H.; Cai, G. Experimental Study on Carbonation of Cement-Based Materials in Underground Engineering. *Materials* **2022**, *15*, 5238. https://doi.org/10.3390/ma15155238

Academic Editor: F. Pacheco Torgal

Received: 28 June 2022
Accepted: 25 July 2022
Published: 29 July 2022

Publisher's Note: MDPI stays neutral with regard to jurisdictional claims in published maps and institutional affiliations.

Copyright: © 2022 by the authors. Licensee MDPI, Basel, Switzerland. This article is an open access article distributed under the terms and conditions of the Creative Commons Attribution (CC BY) license (https://creativecommons.org/licenses/by/4.0/).

1. Introduction

In recent years, the erosion effect of CO_2 on concrete structures has received increasing attention [1–3]. The increased CO_2 concentration will intensify the carbonation of concrete structures [4,5]. Carbonation is considered one of the most disruptive factors that can affect concrete durability, potentially causing a significant reduction in the service life [6,7]. The degradation of cement-based materials due to CO_2 attack mainly includes the following chemical reactions [8].

$$CO_{2(g)} \leftrightarrow CO_{2(aq)} \tag{1}$$

$$CO_{2(g)} + H_2O \leftrightarrow H_2CO_{3(aq)} \leftrightarrow H^+(aq) + HCO_3{}^-{}_{(aq)} \leftrightarrow CO_3{}^{2-}{}_{(aq)} + 2H^+{}_{(aq)} \quad (2)$$

$$Ca(OH)_2(s) \leftrightarrow Ca^{2+}(aq) + 2OH^-(aq) \quad (3)$$

$$Ca^{2+}{}_{(aq)} + HCO_3{}^-{}_{(aq)} + OH^-{}_{(aq)} \rightarrow CaCO_{3(s)} + H_2O \quad (4)$$

$$Ca^{2+}{}_{(aq)} + CO_3{}^{2-}{}_{(aq)} \rightarrow CaCO_{3(s)} \quad (5)$$

$$C\text{-}S\text{-}H_{(s)} + H^+{}_{(aq)} + HCO_3{}^-{}_{(aq)} \rightarrow CaCO_{3(s)} + SiO_xOH_{x(amorphous)} + H_2O \quad (6)$$

$$C\text{-}S\text{-}H_{(s)} + 2H^+{}_{(aq)} + CO_3{}^{2-}{}_{(aq)} \rightarrow CaCO_{3(s)} + SiO_xOH_{x(amorphous)} + H_2O \quad (7)$$

$$CaCO_{3(s)} + CO_{2(aq)} + H_2O \leftrightarrow Ca^{2+}{}_{(aq)} + HCO_3{}^-{}_{(aq)} \quad (8)$$

$$H^+{}_{(aq)} + CaCO_{3(s)} \leftrightarrow Ca^{2+}{}_{(aq)} + HCO_3{}^-{}_{(aq)} \quad (9)$$

There are four stages in the carbonation process of cement-based materials. In the first stage, some of the CO_2 dissolves into water (Equation (1)) and some reacts with H_2O to form carbonic acid (Equation (2)) in the presence of water or humidity. In the second stage, the carbonic acid reacts with hydration products such as $Ca(OH)_2$ and C-S-H (Equations (3)–(7)). Carbonation in cement-based materials does not occur at the same time in each phase. When $Ca(OH)_2$ is consumed, C-S-H is decomposed into $CaCO_3$ and silica gel, reducing the strength of cement-based materials [9]. As the $Ca(OH)_2$ is reduced, the pH value of the pore solution inside the cement-based materials falls below about 9.5, which will cause the corrosion of embedded steel in the cement-based materials. In Equations (6) and (7), SiO_xOH_x is an amorphous silica gel [10,11], and the C-S-H gel is converted to a low-C/S gel because of the decalcification, causing the conversion of a portion of the high-density (HD) C-S-H gel to a low-density (LD) C-S-H gel [12,13]. In the third stage, the resulting insoluble $CaCO_3$ is transformed into a soluble phase, and this process eventually causes the calcium in the major cement phases to dissolve out of the cement, leaving only a porous silica gel [10]. The fourth stage is referred to as full carbonation.

Carbonation is a diffusion process involving the transportation of CO_2 from one area to another by way of random molecular motion. Under steady-state conditions, the diffusion process follows Fick's first law, set out in Equation (10) [14]:

$$X = K_{acc.} t^n, \quad (10)$$

where X is the carbonation depth (mm), $K_{acc.}$ is the carbonation coefficient (mm/$\sqrt{}$year), t is the time duration of exposure (years), and n is an exponent that is smaller than 1.0 (usually taken as 0.5).

At present, various laboratory experiments and models have been developed to understand the erosion of CO_2 on cement-based materials. Mohammed et al. [15] used 100% CO_2 concentration to achieve an accelerated carbonation test and studied the carbonation depth and carbonation products. In the research above, the carbonation test had been achieved with a certain humidity of CO_2, called "dry carbonation" [10]. In this case, the carbonation process generally stops after the creation of $CaCO_3$ and acts to increase the strength of the cement and decrease its permeability [16–18]. But in some cases, cement-based materials are in a water-immersion environment such as a diversion tunnel. In contrast, "wet carbonation" occurs on cement-based materials that are submerged in liquid (water or brine), and it is similar to the in situ condition of cement-based materials in diversion tunnels.

2. Materials and Methods

2.1. Materials

The coarse aggregate in the concrete of a diversion tunnel lining is large compared to the size of samples that are used in carbonation tests. In addition, CO_2 only has an erosion effect on the cement composition of the concrete. Therefore, cement mortar was used as the cement-based material to study the erosion effect of CO_2 on concrete.

Yellow sand, passed through a 3 mm sieve, was used as the aggregate of cement mortar. The mineral composition of yellow sand was analyzed by a D8-ADVANCE X-ray diffractometer; the main mineral composition is shown in Table 1. Composite Portland cement P•C 32.5R from Huaxin Cement (Wuhan, China) Co., Ltd. was used to produce the cement mortar.

Table 1. Main mineral composition of sand.

Mineral Composition	SiO_2	Albite ($Na_{0.98}Ca_{0.02}$) ($Al_{1.02}Si_{2.98}O_8$)	Microcline Maximum ($KAlSi_3O_8$)
Mass Fraction/%	74.66	7.27	18.07

2.2. *Mixture Design and Sample Preparation*

The cement mortar mixture was prepared with an effective water/cement ratio (W/C) of 0.4, and the mass ratio of the cement to yellow sand was 1:1. For the cement mortar mix above, a series of cement mortar samples were cast in plastic molds with dimensions of $150 \times 150 \times 150$ mm^3 and were then vibrated on a vibration table. After being cured in the laboratory for 24 h, all the cement mortar samples were demolded and further cured in a saturated solution of $Ca(OH)_2$ at a temperature of 20 ± 2 °C. After curing for 28 days, the cylindrical samples with dimensions of $\Phi 50 \times 100$ mm were prepared by core drilling and cutting machines.

2.3. *Accelerated Carbonation Test Method*

It is well known that carbonation is a very slow process in the natural environment, and the service environment of a diversion tunnel is complex. An accelerated carbonation test is crucial to simulate the erosion of the concrete lining in a diversion tunnel. There were two types of accelerated carbonation environments that were utilized in the present research according to the service environment of a diversion tunnel [19]. The schematic diagram of the accelerated carbonation test is shown in Figure 1; a CO_2 tank was used to provide the vessel with a 99.99% CO_2 concentration at a pressure of 2 bars. The samples of group HA were placed on a stainless-steel shelf above the deionized water, and the samples of group WB were in the water. Groups HA and WB simulated a saturated-humidity environment and water-immersion environment, respectively. A third group, AI was the control group, and its samples were in an atmospheric environment. The apparatus of the accelerated carbonation test is shown in Figure 2. The test procedure was as follows:

(1) The tops and bottoms of the standard cylindrical samples were sealed with dissolved paraffin to ensure the radial movement of the carbonation.

(2) The samples of group WB were placed in the bottom of the vessel and deionized water was injected into the vessel. The water surface was 50 mm from the stainless-steel shelf, and the samples of group HA were placed on the stainless-steel shelf as shown in Figure 1. The samples of group AI were placed in an atmospheric environment.

(3) Before CO_2 gas injection, the vessel was vacuumed to the -1 bar by a vacuum pump. The CO_2 pressure in the vessel was then controlled by a gas regulator, and kept at a required constant level. Keeping the room temperature constant at 20 ± 5 °C, the samples of group AI were in the same laboratory environment.

(4) We checked the value of the pressure gauge and the temperature of the laboratory regularly, and changed the water every five days.

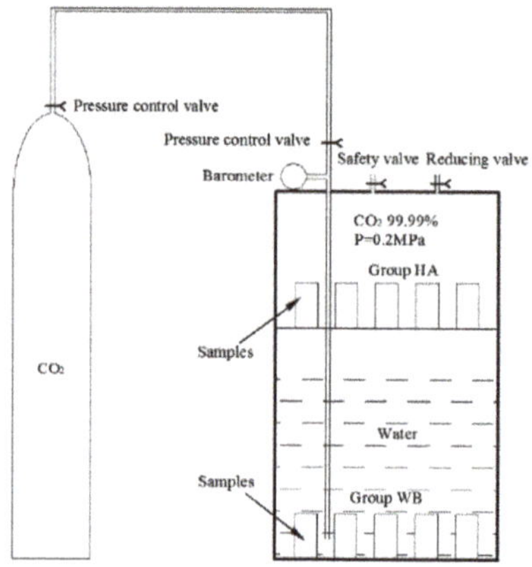

Figure 1. Schematic diagram of the pressurized accelerated carbonation test equipment.

Figure 2. Apparatus of the accelerated carbonation test.

2.4. Measurement of Carbonation Depth

Cylindrical cement mortar samples were prepared for carbonation depth measurement. At ages of 6, 36, 72, 100, and 200 days, the samples were removed from the vessel and dried in an oven at 60 °C for 48 h. The dried samples were then cut through the middle with a cutting machine. The section surface was cleaned with a brush and the carbonation depth was detected by spraying 1% phenolphthalein-alcohol solution. After 30 s, phenolphthalein pink disappeared in the full-carbonation region that indicated a drop of the pH value below 9.5, as recommended by the International Union of Laboratories and Experts in Construction Materials, Systems, and Structures (RILEM). The carbonation depth was measured with a vernier caliper, and the average carbonation depth was taken as the carbonation depth of the samples.

2.5. Phase Changing Monitoring

The phase changing of the carbonation layers of cement mortar samples in different groups was detected by X-ray diffraction (XRD) and thermo gravimetric-differential thermal analysis (TG-DTA). The powder samples were obtained from the carbonation layers by a grinding machine, and were passed through a 75 μm sieve, which was used in XRD and TG-DTA to detect the carbonation products in different groups. The powder samples were first subjected to XRD analysis. They were analyzed by means of a D8-ADVANCE (German)

X-ray diffracto meter. The same powder samples that were used in XRD analysis were used in the TG-DTA analysis using a DTG-60 thermo gravimetric-differential thermal analyzer. The DTG-60 series from Shimadzu performed in the range from ambient to 1100 °C with an increase rate of 10 °C/min using a parallel guide differential top pan balance mechanism that simultaneously measured the temperature changes and mass changes between an inert reference and the sample.

2.6. Energy Dispersive Spectroscopy (EDS)

The carbonation layers of the samples were cut into small blocks with a particle size of 3–5 mm; a relatively flat surface in the small blocks was required for the test. Before the experiment, the flat surface had to be sprayed with a 10 nm layer of gold-palladium to improve the conductivity. The microstructure of the samples was observed with an FEI QUANTA-250 environmental scanning electron microscope (SEM) that was operated at an accelerating voltage of 30 kV. Energy-dispersive spectroscopy (EDS) was applied to detect the distribution of specific elements such as Ca in the samples.

2.7. Quantitative Analysis of the Pore Structure

Low-pressure N_2 adsorption/desorption experiments were completed with a Quantachrome NOVA1000e series surface area analyzer for surface area and pore analysis. The carbonation layers of the samples were crushed to pass through a 3 mm sieve, and then were dried at 105 °C for 24 h in a vacuum oven. The surface area was calculated from the N_2 adsorption data in the relative pressure (P/P_0) range of 0.05–0.35 using the Brunauer-Emmete-Teller (BET) method. The pore volume and pore radius distribution were obtained from the N_2 adsorption data in the relative pressure (P/P_0) range of 0.06–0.99 by applying the Barrette–Joynere–Halenda (BJH) method.

3. Results

3.1. Visual Inspection and Carbonation Depth

Figure 3 presents an exposed sample in the group WB; there was an unknown material precipitation on the surface of the sample. The aggregate of the sample surface was exposed. The unknown precipitation of the sample surface was rinsed with deionized water, and the rinsed liquid was precipitated as shown in Figure 3b. According to a visual inspection, the precipitation contained yellow sand from the cement mortar. Due to the erosion of the saturated carbon dioxide aqueous solution, the cement composition on the surface of the sample was damaged, and the sand fell from its surface. The precipitated solution was oven-dried as shown in Figure 3c.

(**a**) (**b**) (**c**)

Figure 3. Photos of the exposed sample in group WB (water-immersion environment); (**a**) sample, (**b**) precipitate, and (**c**) oven dry.

Under the influence of the infiltration of water pressure and current scour, the effect of CO_2 on the degradation of concrete was exacerbated. Photos of the degraded diversion tunnel are shown in Figure 4. Figure 4a shows the diversion tunnel of the Lanzhou water source project under construction; the presence of CO_2 in the groundwater caused the surface of the diversion tunnel lining to precipitate white matter under the action of groundwater seepage pressure. Figure 4b shows the diversion tunnel of the Guangzhou water storage power plant in operation; the aggregate on the concrete surface was exposed under the action of degradation and current scour. The degradation phenomena that are shown in Figures 3 and 4 are very similar.

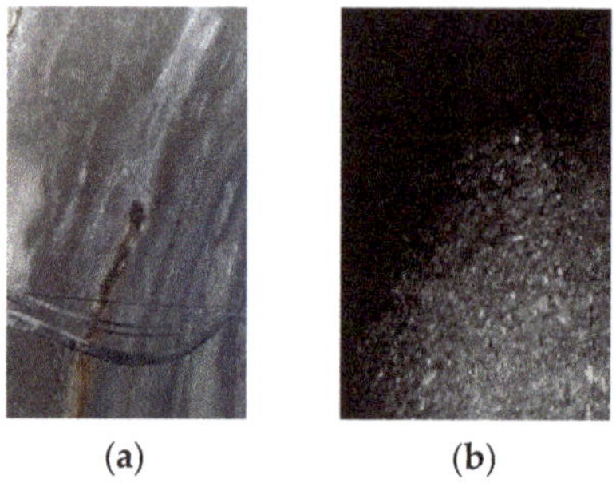

Figure 4. Photos of the degraded diversion tunnels: (**a**) the diversion tunnel of Lanzhou water source project, and (**b**) the diversion tunnel of Guangzhou energy storage power plant.

Figure 5a shows cross-sectional drawings of the samples in different exposed environments (different groups) after cutting. The carbonation layers appear in the marginal region with a white color, which is a typical phenomenon for carbonation. The colors of the internal regions of the exposed samples obviously were quite different. The internal regions of the samples in the atmospheric environment (group AI) were light grey, while those of the samples in the other two environments (groups WB and HA) were dark grey. The color boundary was not obvious in the atmospheric environment but was obvious in the other two environments (groups WB and HA). Figure 5a shows that the color boundary was homogeneous in the water-immersion environment (group WB) and inhomogeneous in the saturated-humidity environment (group HA). The region with a white color in the water-immersion environment (group WB) was smaller than that in the saturated-humidity environment. Figure 5b shows photos and dye testing images of fracture surfaces of the exposed samples after the Brazil splitting test. Half of the fracture surfaces were dropped in the phenolphthalein reagent, while the other half were not dropped. It can be seen from Figure 5b that the white-colored area of a sample is the carbonation area. The sample of group HA shows more uneven visual degradation than the sample of group WB, suggesting a deeper carbonation depth in the former.

Figure 5. Sectional drawings of the eroded samples: (**a**) cross-section and (**b**) vertical section.

Figure 6 shows the carbonation depth with exposure time in different environmental conditions (groups HA and WB). As shown in Figure 6a, the carbonation depth changes little with exposure time in the water-immersion environment (group WB) and is relatively uniform. It can be seen from Figure 6b that the carbonation depth increases with the exposure time in the saturated-humidity environment (group HA), and the sample of the group HA shows uneven carbonation. Comparing Figure 6a,b, we see that carbon dioxide on the carbonation of the cement mortar is more obvious in the saturated-humidity environment (group HA) than in the water-immersion environment (group WB).

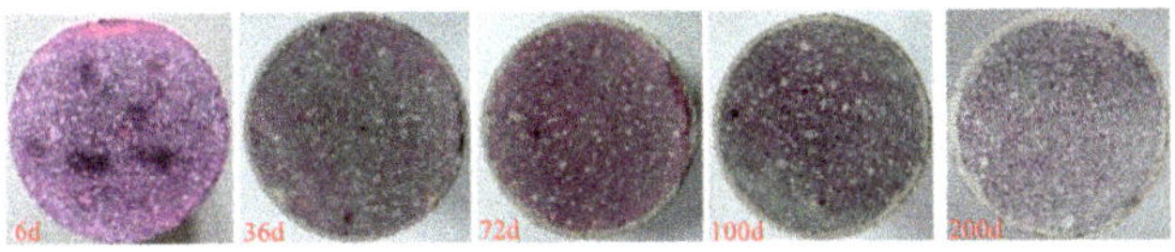

(**a**)

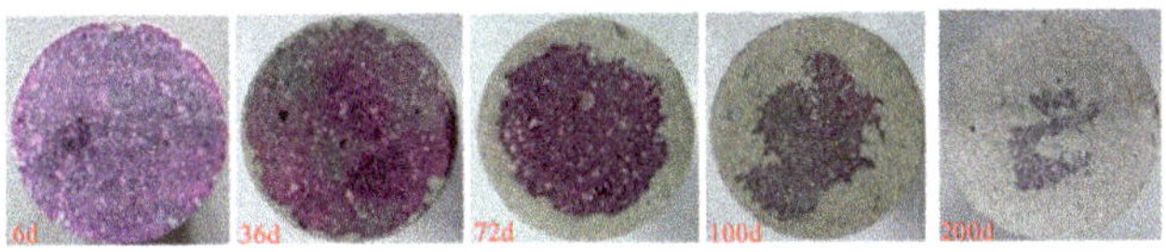

(**b**)

Figure 6. Carbonation depth measurements: (**a**) Group WB (water-immersion environment) and (**b**) Group HA (saturated-humidity environment).

Figure 7 shows the change of the carbonation depth of cement mortar with exposure time due to an accelerated carbonation test with 100% CO_2 for up to 200 days. The samples in the saturated-humidity environment (group HA) show a higher carbonation depth over the entire time than the samples in the water-immersion environment (group WB). The carbonation depth of the samples in the saturated-humidity environment is about 16.71 mm, whereas the carbonation depth of the samples in the water-immersion environment is only 2.31 mm. In Figure 7, the carbonation depths of the cement mortar samples in different environments generally follow a linear relationship with the square root of time. The experimental results show an excellent correlation coefficient with linear regression. The accelerated coefficient of diffusion for the cement mortar samples in different environments was calculated as the slope of the relationship between the carbonation depth and the square root of time, as shown in Figure 7, according to Fick's first law (Equation (10)). The carbonation coefficient in a saturated-humidity environment is more than nine times that in the water-immersion environment from the linear fitting equation in Figure 7.

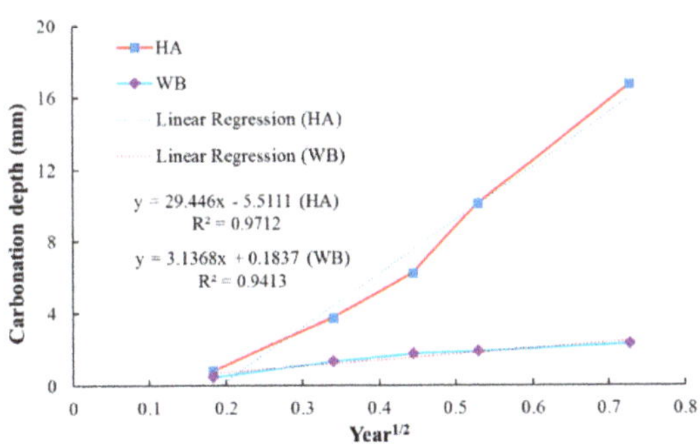

Figure 7. Average carbonation depth versus the square root of exposure time (year) relationships.

3.2. X-ray Diffraction Pattern Analysis

X-ray diffraction was used to identify the crystalline phases that were present in the precipitate (Figure 3c). The X-ray diffraction pattern is shown in Figure 8. The most intense peak on the pattern is located at 26.6° (2θ) and corresponds to the crystallographic plane in quartz (SiO_2). Several other quartz (SiO_2) peaks are also shown. Except for quartz (SiO_2) peaks, the most intense peak on the pattern is located at 27.9° (2θ) and corresponds to the crystallographic plane in albite. Less intense peaks corresponding to microcline maximum ($KAlSi_3O_8$), calcium aluminum oxide, calcite ($CaCO_3$), and dolomite [$CaMg(CO_3)_2$] that are are also present in this pattern, and are labeled. The strongest calcite ($CaCO_3$) peak is located at 29.4° (2θ).

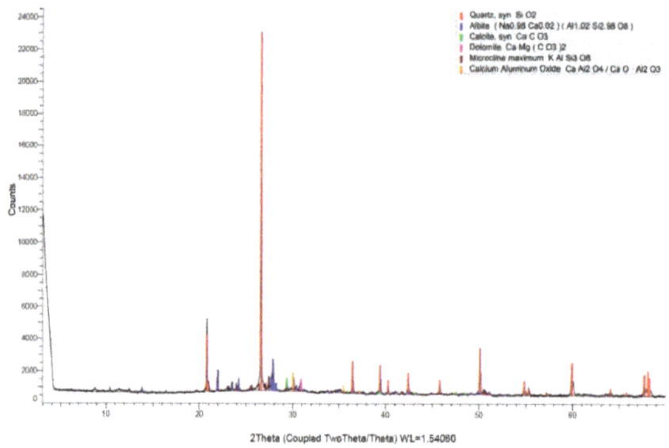

Figure 8. X-ray diffraction pattern of the precipitate.

The principle of obtaining the mass percentages of chemical phases is that the intensity of X-ray diffraction is proportional to the mass percentages of the detected phases in the samples. The contribution of chemical phases of the precipitate (Figure 3) is shown in Figure 9. More than half of the precipitate is composed of quartz (SiO_2). The precipitate also contains albite and microcline maximum, with mass percentages of 22.69% and 11.67%, respectively. It can be seen from Table 1 that the sand contains albite and microcline maximum, while there is no albite or microcline maximum in the cement, so the precipitate contains a certain amount of yellow sand. As shown in Figure 3, the liquid after rinsing the surface of the sample that was exposed to a water-immersion environment contains sand, and the apparent result is consistent with the XRD result. The precipitate contains small amounts of calcium aluminum oxide (9.14%) and calcite ($CaCO_3$) (3.25%).

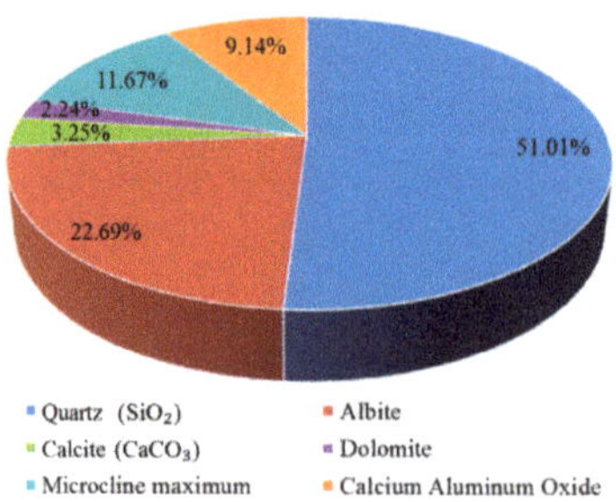

Figure 9. Contribution of chemical phases of the precipitate.

Figure 10 presents the XRD patterns of the carbonation layers of the samples that were exposed to different environments. Peaks are identified in the diffraction patterns relating to quartz (SiO_2), calcite ($CaCO_3$), albite, microcline maximum ($KAlSi_3O_8$), and dolomite [$CaMg(CO_3)_2$] in different groups. Due to carbonation, the calcium hydroxide

($Ca(OH)_2$) peaks are of lower intensity or are not visible in the patterns that are shown in Figure 10. The patterns corresponding to the powder of carbonation layers in different environments all show the most intense peak at 26.6° (2θ), and the peak corresponds to the crystallographic plane in quartz (SiO_2). Except for the quartz (SiO_2) peaks, the most intense peaks on the patterns are all located at 29.4° (2θ) and correspond to the crystallographic plane in $CaCO_3$. It can be observed from Figure 10 that the peak intensity of $CaCO_3$ in the atmospheric environment is smaller than in the other two environments.

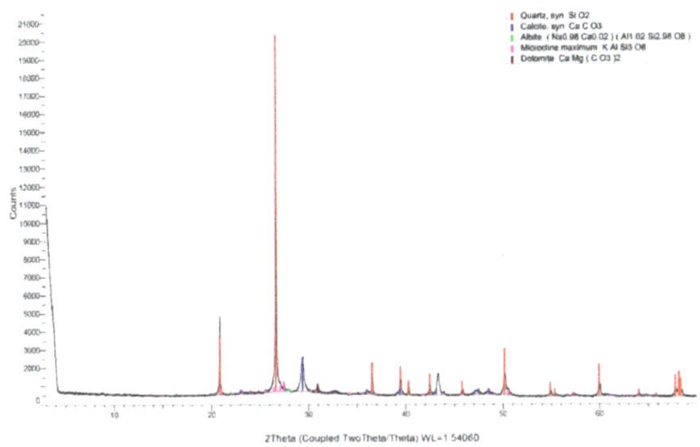

(a)

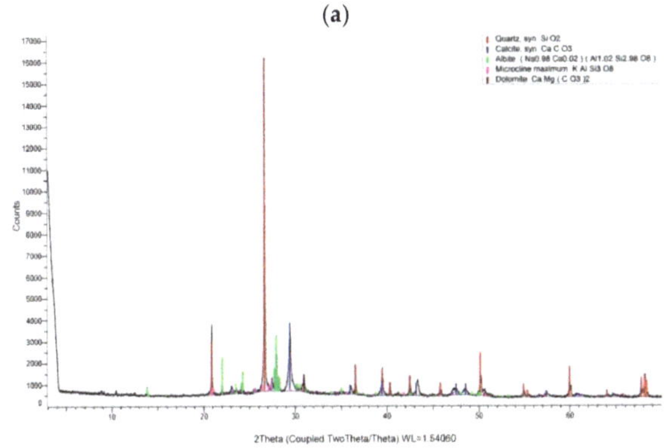

(b)

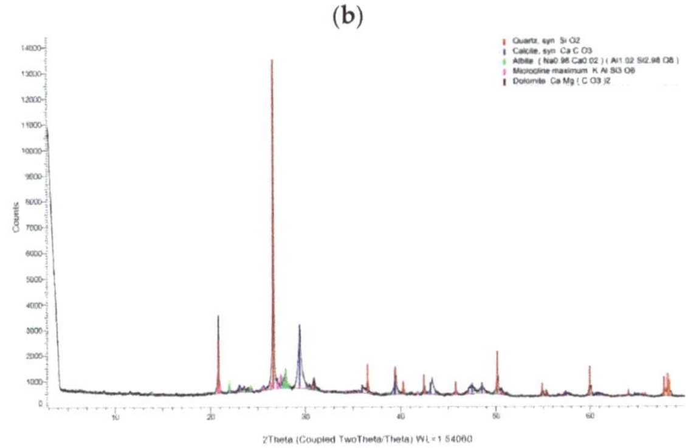

(c)

Figure 10. X-ray diffraction patterns of carbonation layers in different environments: (**a**) group AI (atmospheric environment), (**b**) group WB (water-immersion environment), and (**c**) group HA (saturated-humidity environment).

The contribution of chemical phases of carbonation layers in different environments is shown in Figure 11. As both the cement and sand contain quartz (SiO_2), the mass percentage of quartz (SiO_2) is higher than other chemical phases in different exposed environments. As shown in Figure 11, there are some differences in the mass percentages of carbonation product $CaCO_3$ in different environments. The mass percentage of $CaCO_3$ of the carbonation layer in the saturated-humidity environment is the highest, about 17.34%, followed by the water-immersion environment, about 13.91%, while the mass percentage of $CaCO_3$ in an atmospheric environment is the least, about 11.59%.

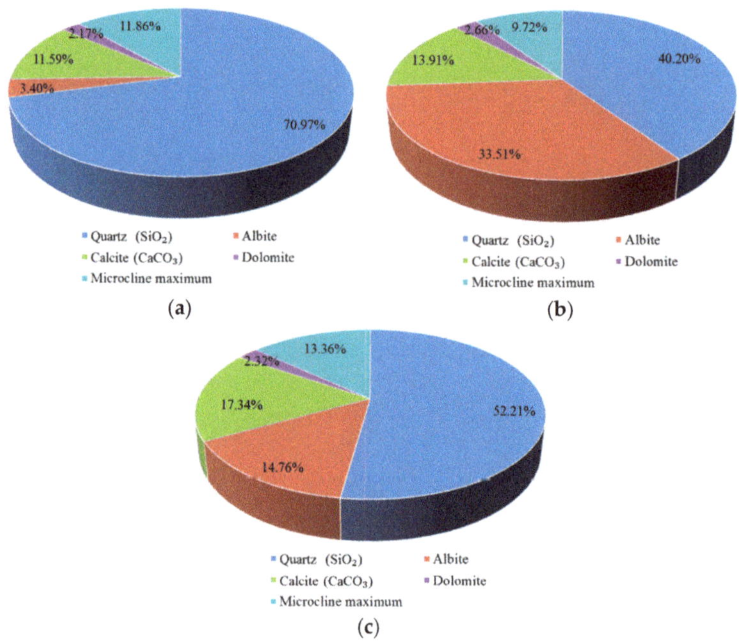

Figure 11. Contribution of chemical phases of carbonation layers in different exposed environments: (**a**) group AI (atmospheric environment), (**b**) group WB (water-immersion environment), and (**c**) group HA (saturated-humidity environment).

3.3. *Thermogravimetric-Differential Thermal Analysis (TG-DTA)*

Thermo gravimetric-differential thermal analysis (TG-DTA) is a method for analyzing the chemical composition and content of chemical substances by measuring the change of the mass at different temperatures. As the temperature rises, the bound water in calcium hydroxide ($Ca(OH)_2$) is generally decomposed between 300 °C and 500 °C, while CO_2 in calcium carbonate ($CaCO_3$) is decomposed between 600 °C and 700 °C. Due to the above decomposition, calcium hydroxide and calcium carbonate are decomposed into calcium oxide.

Many factors affect the thermal decomposition temperature of $Ca(OH)_2$ and $CaCO_3$. Since the range of thermal decomposition temperature is not fixed, it should be determined by the endothermic peak of the differential thermal analysis (DTA) curve. The difference in the thermo gravimetric analysis (TGA) curves is calculated from the range of the thermal decomposition temperature, which is the mass percentage of the bound water that is released by the $Ca(OH)_2$ decomposition process and the mass percentage of CO_2 that is released during the CaCO3 decomposition process. Depending on the mass percentage of H_2O and CO_2, the mass percentage of $Ca(OH)_2$ and $CaCO_3$ can be determined. The TG-DTA curves of the carbonation layers in different exposed environments are shown in Figure 12.

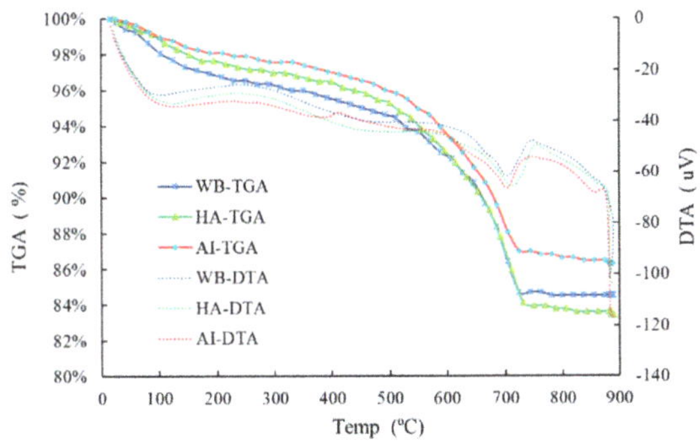

Figure 12. TG-DTA curves of the carbonation layers in different exposed environments.

In Figure 12, the endothermic peak of thermal decomposition of $CaCO_3$ is visible, while the endothermic peak of thermal decomposition of $Ca(OH)_2$ is weak. The temperature range of the thermal decomposition of $CaCO_3$ was determined according to the endothermic peak, and the content of $CaCO_3$ was further determined as shown in Table 2. It can be seen that the percentage of $CaCO_3$ in the carbonation layer of the sample in the saturated-humidity environment is about 15.58%, and the content of $CaCO_3$ in the carbonation layer of the sample in the water-immersion environment comes second, about 13.91%; meanwhile, that in the atmospheric environment is the least, about 9.47%.

Table 2. The mass percentage content of calcium carbonate in the carbonation layer of the samples.

Group	T_1 (°C)	T_2 (°C)	$CaCO_3$ (%)
AI	658.91	736.58	9.47
WB	645.69	741.74	13.91
HA	640.96	747.23	15.58

3.4. Energy Dispersive Spectroscopy (EDS) Analysis

The energy dispersive spectroscopy (EDS) spectra were used to verify the elemental compositions of crystals on the surfaces of carbonation layers of the samples by an energy dispersive spectrometer using a scanning electron microscope (SEM). Energy dispersive spectroscopy (EDS) analysis can only test the type and content of elements of a tiny region on the surface of a sample. The results of a single point are very random and cannot represent the overall properties of the sample. Therefore, several spectral points were selected. Figure 13 presents the SEM images of the carbonation layers of the samples and the selected spectral points on the SEM images for EDS analysis.

Figure 14 presents the typical EDS spectra that were acquired at spectral points 5, 3, and 3 in Figure 13a–c, respectively. Figure 14a,b show the presence of carbon (C), oxygen (O), magnesium (Mg), aluminum (Al), silicon (Si), and calcium (Ca), which are the main elements of spectra 5 and 3 in Figure 13a,b, respectively. Figure 14c shows the presence of carbon (C), oxygen (O), silicon (Si), and calcium (Ca), which are the main elements of spectrum 3 in Figure 13c. In Figure 14, the EDS spectra clearly show that the intensity of the Ca peak in the water-immersion environment is weaker than in the other two environments.

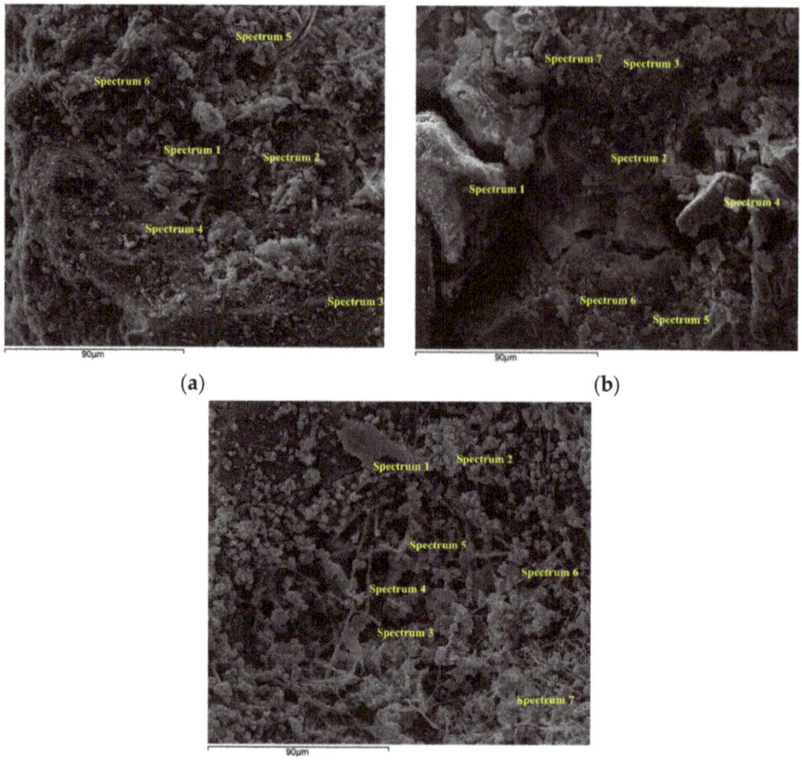

(a) (b) (c)

Figure 13. SEM of the carbonation layers of the samples with the selected spectral points for EDS analysis: (**a**) group AI (atmospheric environment), (**b**) group WB (water-immersion environment), and (**c**) group HA (saturated-humidity environment).

Tables 3–5 show the element composition that was obtained from the EDS spectra, expressed as the weight percentage (wt%) of the selected spectral points at the carbonation layers of samples in different environments. Since these selected spectral points are limited and the distribution of chemical compounds is rather heterogeneous in these materials, these results should be considered only relative and not absolute. According to these results, the average mass percentage content (wt%) of calcium in the saturated-humidity environment is about 18.55%, it is about 19.43% in the atmospheric environment, and each selected spectral point in the two environments contains the Ca element. Not all the selected spectral points contain Ca in the water-immersion environment, and the average mass percentage content (wt%) of Ca is only 3.46%, which is much lower than in the other two environments. The calcium-silicon ratio (Ca/Si) represents the mass ratio of calcium and silicon; the calcium-silicon ratios (Ca/Si) are about 1.49 and 1.51 in the atmospheric and saturated-humidity environments, respectively. Due to the calcium leaching process of the cement-based materials in the water-immersion environment, the calcium-silicon ratio (Ca/Si) is much less than in the other two environments, with a value of approximately 0.11.

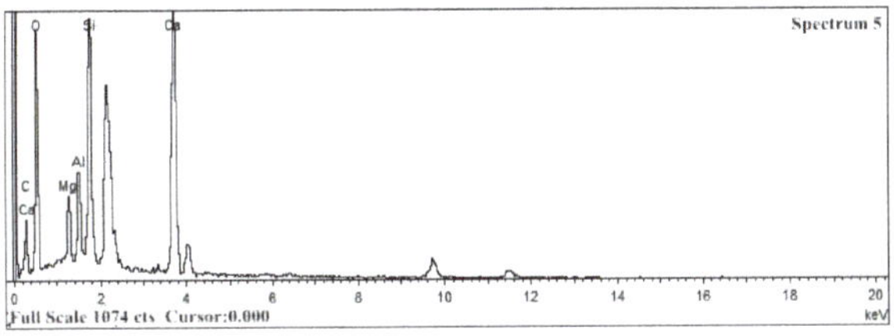

(**a**)

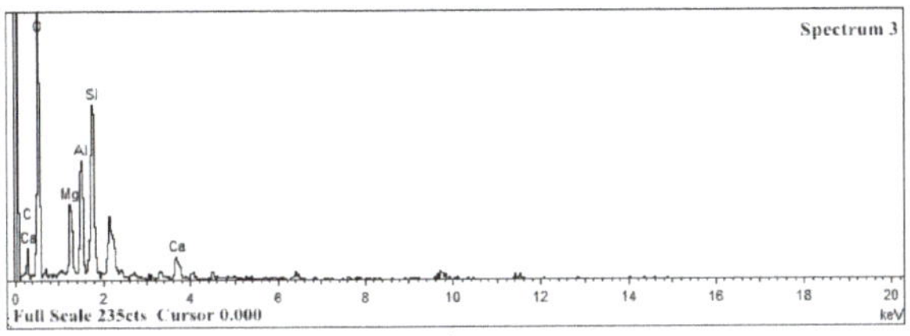

(**b**)

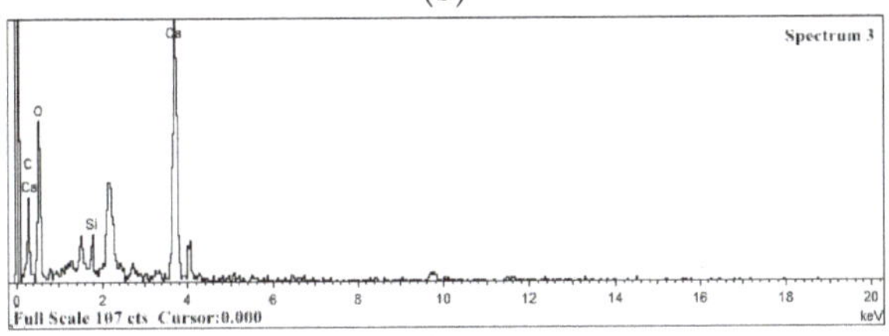

(**c**)

Figure 14. Typical EDS spectra for the marked points in Figure 13: (**a**) group AI (atmospheric environment), (**b**) group WB (water-immersion environment), and (**c**) group HA (saturated-humidity environment).

Table 3. Element mass percentage content of the carbonation layer in group AI (wt%).

Spectrum	C	O	Mg	Al	Si	Ca	Fe
1	8.58	54.54	1.25	3.90	6.57	25.16	
2	12.28	47.23	3.17	4.83	12.61	19.29	0.59
3	8.72	47.82	0.40	6.62	21.72	14.72	
4	15.19	50.13		5.81	18.30	10.57	
5	15.07	51.83	2.80	3.54	9.42	17.34	
6	8.81	45.30		6.98	9.41	29.50	
Max	15.19	54.54	3.17	6.98	21.72	29.50	0.59
Min	8.58	45.30	0.00	3.54	6.57	10.57	0.00
Average	11.44	49.48	1.27	5.28	13.01	19.43	0.10

Table 4. Element mass percentage content of the carbonation layer in group WB (wt%).

Spectrum	C	O	Mg	Al	Si	Ca	Fe
1		54.56			45.44		
2	6.83	49.94	4.56	10.95	20.24	4.16	3.32
3	17.73	56.98	4.77	7.5	11.19	1.83	
4	9.44	39.32		11.19	32.91	7.14	
5	10.01	40.06	3.48	11.42	27.41	7.62	
6	5.64	56.82		9.48	24.57	3.49	
7		45.4			54.6		
Max	17.73	56.98	4.77	11.42	54.6	7.62	3.32
Min	0	39.32	0	0	11.19	0	0
Average	7.09	49.01	1.83	7.22	30.91	3.46	0.47

Table 5. Element mass percentage content of the carbonation layer in group HA (wt%).

Spectrum	C	O	Mg	Al	Si	Ca	Fe
1	9.51	44.34	5.65	6.74	18.77	8.58	6.41
2	19.08	46.12			15.56	19.24	
3	21.96	57.42			9.75	10.87	
4	10.51	37.28		4.63	22.53	25.05	
5	15.08	62.5		3.5	9.71	9.21	
6	13.41	52.44			4.66	29.49	
7	10.13	53.38		3.85	5.24	27.4	
Max	21.96	62.50	5.65	6.74	22.53	29.49	6.41
Min	9.51	37.28	0	0	4.66	8.58	0
Average	14.24	50.50	0.81	2.67	12.32	18.55	0.92

3.5. Quantitative Analysis of the Pore Structure

In general, carbonation can improve the pore structure of the matrix for cement-based materials. However, the impact of carbonation on the pore structure varies in the different exposed environments, as shown in Figure 15, which represents the pore radius distribution from the adsorption branch of isotherms using the BJH method. It is observed that the pore radius in the pore structure of carbonation layers in the different groups was between 1.6 and 50.4 nm, and all presented a major peak at around 13 nm. The pore radius distribution curve of the carbonation layers of samples in the water-immersion environment (group WB) is above that in the saturated-humidity environment (group HA). The pore radius distribution curves of groups AI and HA intersect at a pore radius of 11 nm, while the pore radius distribution curves of Group AI and WB intersect at a pore radius of 30 nm. For a pore radius that is less than 11 nm, the lower values of pore volume are in group AI, while the group HA has higher values of pore volume. However, we see the opposite result for a pore radius that is greater than 11 nm. In most of the pore radius ranges (less than 30 nm), the pore volume of group WB is larger than that of group AI, and only in the range of pore radii that are greater than 30 nm is the pore volume of group AI larger.

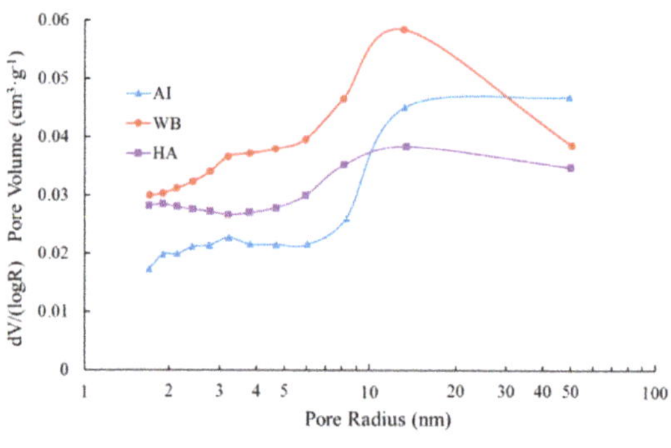

Figure 15. Pore radius distribution of the carbonation layers in different exposed environments.

The pore structure parameters of the carbonation layers in different environments that were determined by the nitrogen adsorption method are shown in Table 6. The specific surface area that was calculated from N_2 adsorption data using the multipoint BET model ranges from 23.6899 to 35.3504 m^2/g, and the specific surface area of group HA is the largest, at about 35.3504 m^2/g, while that of group AI is the smallest, at about 23.6899 m^2/g. The specific surface area of group HA is 1.5 times that of group AI, and it is only 3.058 m^2/g larger than that of group WB. The total pore volumes (cm^3/g) were found to be 0.0671 for group AI, 0.0787 for group WB, and 0.0681 for group HA (Table 6). The average pore radius of group AI, 5.6623 nm, is larger than those of groups WB and HA, and the average pore radius of group HA is the smallest of all groups.

Table 6. Pore structure parameters of the carbonation layers of the samples.

Group	Specific Surface Area ($m^2 \cdot g^{-1}$) [a]	Total Pore Volume ($cm^3 \cdot g^{-1}$) [b]	Relative Pressure (P/P_0) [c]	Average Pore Radius (nm) [d]
AI	23.6899	0.0671	0.9882	5.6623
WB	32.2924	0.0787	0.9884	4.8736
HA	35.3504	0.0681	0.9883	3.8546

[a] Specific surface area is calculated by the BET method. [b] Total pore volume is estimated to be the liquid volume of nitrogen at the relative pressure in Table 6. [c] Relative pressure is the pressure that is used to calculate the pore volume. [d] Average pore radius = 2 ∗ total pore volume/specific surface area.

4. Discussion

Carbonation is a long-term durability problem of reinforced concrete, predominantly for two reasons. It causes the pH value of a pore solution inside cement-based materials to fall, which leads to corrosion of the steel bar in the cement-based materials, and it also causes degradation of the cement-based materials. As shown in Figure 3, the cement composition on the surface of the sample is damaged, and hence continuous degradation of the surface results in exposure of the sand in the water-immersion environment. Initially, CO_2 causing the erosion of cement is mainly due to the carbonation of $Ca(OH)_2$, and $CaCO_3$ is formed, which fills the pores of the cement-based materials, reducing the porosity and permeability but increasing strength. As erosion continues, $Ca(OH)_2$ is consumed, and C-S-H gel is decomposed into $CaCO_3$ (Equation (7)). In the water-immersion environment, $CaCO_3$ is converted to water-soluble $Ca(HCO_3)_2$, which will increase the porosity and permeability but result in the loss of strength. As shown in Figure 3, the sand on the surface is easily rinsed off with deionized water. The presence of water causes CO_2 to have a severe erosion effect on the cement of cement mortar, destroying the chemical stability of cement hydration products, resulting in the decrease in the cementation capacity of cement. Much research has shown that a water environment greatly influences the corrosion process of cement-based materials [9]. However, the carbonation depth in a saturated-humidity environment is much deeper than in a water-immersion environment (Figures 5 and 6). As shown in Figure 7, the carbonation depth and square-root time follow a linear relationship in the water immersion and saturated-humidity environments, but the carbonation rate in the saturated-humidity environment is more than nine times that in the water-immersion environment. El-Turki et al. [20] proposed that the rate of carbonation rate increases with the relative humidity. High humidity provides the most favorable conditions for carbonation (Equation (2)), and it improves the carbonation rate. However, the carbonation rate dropped sharply in the water-immersion environment. The carbonation process is ongoing due to the diffusion of CO_2 in the pore network according to Fick's first law, and the diffusion is much faster in air than in water. The presence of water decreased the carbonation rate due to the slow diffusion of CO_2 within the saturated pore network, whereas the excessive water could cause deterioration to the cement-based materials in the longer term. Figure 5 presents color-coded images before dropping the phenolphthalein reagent, and color changes, through visual inspection, after dropping the phenolphthalein reagent. These pictures show the interface between the carbonation and non-carbonation zones. Kutchko et al. [8] pointed out that un-hydrated cement compounds are usually the brightest, while calcium hydroxide (CH) is less bright, and calcium silicate hydrate(C-S-H) is darker still. Water is the most important factor affecting the hydration degree of cement components. Since the moisture content is higher in the water-immersion and saturated-humidity environments, the degree of hydration is greater in these groups. As a result, the color of the middle zone of samples in the water-immersion and saturated-humidity environments is darker than in the atmospheric environment.

According to the XRD and TG-DTA analysis results, the $CaCO_3$ increased in all the groups. The content of $CaCO_3$ of the carbonation layers in the saturated-humidity and water-immersion environments is higher than in the atmospheric environment, and it is highest in the saturated-humidity environment. As shown in Equations (1)–(5), the content of the carbonation product $CaCO_3$ is controlled by several parameters, such as the water content and CO_2 concentration. The high concentration of CO_2 and large amounts of water provide favorable conditions for carbonation in the water-immersion and saturated-humidity environments, so the content of $CaCO_3$ is high. Water also affects the degree and speed of cement hydration, so the content of hydration products such as $Ca(OH)_2$ in the saturated-humidity and water-immersion environments is higher than in the atmospheric environment, which is also why the content of carbonation product $CaCO_3$ is lowest in the atmospheric environment. In the water-immersion environment, $CaCO_3$ is converted to water-soluble $Ca(HCO_3)_2$ (Equations (1)–(5)), and then the $Ca(HCO_3)_2$ is dissolved in water. Since the stability of $Ca(HCO_3)_2$ is poor, part of $Ca(HCO_3)_2$ is decomposed into $CaCO_3$ and covers the surface of the sample. As shown in Figure 9, the precipitate on the surface of the sample in the water-immersion environment contains 3.25% $CaCO_3$. However, in the saturated-humidity environment, $CaCO_3$ fills the pores of the cement-based materials and does not precipitate from the pores. Therefore, the content of $CaCO_3$ of the carbonation layer in the saturated-humidity environment is higher than in the water-immersion environment.

The carbonation layer of the sample had relatively less Ca content in the water-immersion environment, while the Ca content of the carbonation layers in the saturated-humidity and atmospheric environment was higher, and the values in these two environments were almost the same. In the saturated-humidity environment, the erosion effect of CO_2 on the cement-based materials is mainly the rapid increase of the carbonation depth, which leads to the corrosion of the embedded steel, while in the water-immersion environment, the leaching process of cement based materials is the main cause of degradation of cement-based materials. Both Ca and Si in cement-based materials are dissolved in the water-immersion environment, but compared with Ca, Si occupies only a small portion, and its dissolution rate is extremely slow [21]. Thus, the silicon (Si) leaching process is ignored, and the calcium-silicon ratio (Ca/Si) can be used to characterize the calcium leaching process of cement-based materials in the water-immersion environment. The average Ca/Si ratio in the water-immersion environment is only 0.14, which is much smaller than the average Ca/Si ratio, which is about 1.5 in the other two environments. Dauzères et al. [22] proposed that if the Ca/Si ratio decreases, the chemical stability of cement hydrated products is destroyed. As shown in Figure 6, the carbonation depth of the sample in the water-immersion environment is much shallower than in the saturated-humidity environment. The decalcification in the water-immersion environment leads to damage of the carbonation layer, while the carbonation layer in the saturated-humidity environment is not damaged, exhibiting only a decrease of the pH value.

Figure 15 shows that the pore volume of small pores in the water-immersion and saturated-humidity environments exceeds that in the atmospheric environment. As shown in Table 6, the specific surface area and total pore volume in the water-immersion and saturated-humidity environments exceed those in the atmospheric environment, while the average pore radius is less than in the atmospheric environment. This indicates that the number of small pores in the water-immersion and saturated-humidity environments is more than in the atmospheric environment. This is mainly due to the crystallization and precipitation of $CaCO_3$ in the pores, where large pores are divided into small pores. As carbonation continues to consume the $Ca(OH)_2$ of the pore solution, the $Ca(OH)_2$ crystal of the cement hydration products is dissolved to supplement the loss of Ca^{2+} in the pore solution, resulting in an increase in the number of small pores in the cement-based materials. Compared with the atmospheric environment, water is abundant in the environments of water-immersion and saturated-humidity, and water is the basic condition for cement hydration. Therefore, the content of $Ca(OH)_2$ and C-S-H gels in the later stage are improved.

At the same time, water accelerates the rate at which $Ca(OH)_2$ crystal dissolves into the pore solution and provides sufficient moisture that is required for the carbonation reaction, and thus the degree of carbonation is more adequate. With the continuous consumption of $Ca(OH)_2$ crystals, C-S-H gel will continue to be carbonized by CO_2, and the high-density C-S-H gel will be converted to low-density C-S-H gel after decalcification of the C-S-H gel. The pore volume of all the pore radii in the water-immersion environment is larger (Figure 15) than in the saturated-humidity environment, and the total pore volume and average pore size in the water-immersion environment exceeds those in the saturated-humidity environment (Table 6). This is mainly because CO_2 further reacts with the carbonation product $CaCO_3$ to form soluble $Ca(HCO_3)_2$, and then precipitates out in the water-immersion environment, leading to the loss of Ca in the cement-based materials. Thus, the water-immersion environment will aggravate the further development and expansion of the pores.

5. Conclusions

Diversion tunnels belong to underground structures, the service environments of which are more complicated than concrete structures on the ground. During the operation period, the concrete lining of the diversion tunnel is in a water-immersion environment, while during the maintenance period, it is in a saturated-humidity environment. In this paper, we studied the effect of carbonation on cement-based materials in water-immersion and saturated-humidity environments by using an accelerated CO_2 erosion test instrument that simulates the service environment of a diversion tunnel. Based on the experimental work in the present study, the following concluding remarks can be made:

(1) In the saturated-humidity environment, the carbonation depth increases with exposure time, which changes greatly, while in the water-immersion environment, the carbonation depth changes little with exposure time. However, the water-immersion environment will cause degradation of the cement-based materials.

(2) The color of the non-carbonation region is dark, and that of the carbonation region is gray in the water-immersion and saturated-humidity environments, while there is no dark-colored region in the atmospheric environment. This is mainly due to adequate moisture, which promotes the continued hydration of cement; thus, in the saturated-humidity and water-immersion environments, the cement has a greater degree of hydration.

(3) The content of the carbonation product $CaCO_3$ is higher in the water-immersion and saturated-humidity environments than in the atmospheric environment. In the water-immersion environment, CO_2 continues to react with the carbonation product $CaCO_3$ to produce soluble Ca $(HCO_3)_2$, which results in a lower content of $CaCO_3$ in the water immersion environment than in the saturated-humidity environment. Therefore, in the water-immersion environment, carbonation will cause decalcification of the cement-based materials, which leads to a significant decrease of the Ca/Si ratio in these materials, ultimately destroying the structural stability of cement-based materials.

(4) In the water-immersion and saturated-humidity environments, the number of small pores in the cement-based materials is greater than in the atmospheric environment. The number of pores of all the pore radii in the water-immersion environment exceeds that in the saturated-humidity environment, and the pore volume and average pore size are also greater than in the saturated-humidity environment. Thus, the water-immersion environment will aggravate the further development and expansion of the pores, resulting in the increased permeability and reduced strength of cement-based materials.

Author Contributions: Conceptualization, G.Z. and H.Z.; methodology, J.Z.; software, G.C.; validation, J.Z., G.Z. and H.Z.; formal analysis, J.Z.; investigation, J.Z.; resources, H.Z.; data curation, J.Z.; writing—original draft preparation, J.Z.; writing—review and editing, G.Z.; visualization, J.Z.;

supervision, H.Z.; project administration, G.Z.; funding acquisition, H.Z. All authors have read and agreed to the published version of the manuscript.

Funding: This research was funded by the Natural Science Foundation of Hubei Province, grant number 2021CFB491; the Open Fund for Hubei Provincial Engineering Research Center of Slope Habitat Construction Technique Using Cement-based Materials, grant number 2022SNJ08 and the Open Fund for State Key Laboratory of Geotechnical Mechanics and Engineering (grant number SKLGME021029, Z019026); the Research project of China Railway Construction Corporation, grant number 2022-B02; the National Natural Science Foundation of China (grant number 41941018, E111420101).

Institutional Review Board Statement: Not applicable.

Informed Consent Statement: Not applicable.

Data Availability Statement: The data that support the findings of this study are available on request from the corresponding authors.

Acknowledgments: The authors are sincerely thankful for the funding support.

Conflicts of Interest: The authors declare no conflict of interest.

References

1. Min, L.H.; Wu, J.B.; Zhu, X.D.; He, H.L.; Jia, W.X.; Xiang, W.N. Evolution and variation of atmospheric carbon dioxide concentration over terrestrial ecosystems as derived from eddy covariance measurements. *Atmos. Environ.* **2015**, *114*, 75–82.
2. Luo, X.; Keenan, T.F. Tropical extreme droughts drive long-term increase in atmospheric CO_2 growth rate variability. *Nat. Commun.* **2022**, *13*, 1193. [PubMed]
3. Wang, J.; Li, Q.; Lu, Y.; Luo, S. Effect of waste-oil regenerant on diffusion and fusion behaviors of asphalt recycling using molecular dynamics simulation. *Constr. Build. Mater.* **2022**, *343*, 128043. [CrossRef]
4. Lin, Y.H.; Zhu, D.J.; Zeng, D.Z.; Yang, Y.G.; Shi, T.H.; Deng, K.H.; Ren, C.Q.; Zhang, D.P.; Wang, F. Experimental studies on corrosion of cement in CO_2 injection wells under supercritical conditions. *Corros. Sci.* **2013**, *74*, 13–21.
5. Gu, T.; Guo, X.Y.; Li, Z.Y.; Cheng, X.W.; Fan, X.X.; Asghar, K.; Duan, W.H. Coupled effect of CO_2 attack and tensile stress on well cement under CO_2 storage conditions. *Constr. Build. Mater.* **2017**, *130*, 92–102.
6. Su, P.F. Investigation of the mechanical and shrinkage properties of plastic-rubber compound modified cement mortar with recycled tire steel fiber. *Constr. Build. Mater.* **2022**, *334*, 127391.
7. Zhao, X.; Ge, D.; Wang, J.; Wu, D.; Liu, J. The performance evaluation of asphalt mortar and asphalt mixture containing municipal solid waste incineration fly ash. *Materials* **2022**, *15*, 1387.
8. Kutchko, B.G.; Strazisar, B.R.; Dzombak, D.A. Degradation of well cement by CO_2 under geologic sequestration conditions. *Environ. Sci. Technol.* **2007**, *41*, 4787–4792. [CrossRef]
9. Abid, K.; Gholami, R.; Choate, P.; Nagaratnam, B.H. A review on cement degradation under CO_2-rich environment of sequestration projects. *J. Nat. Gas Sci. Eng.* **2015**, *27*, 1149–1157.
10. Li, Q.Y.; Lim, Y.M.; Jun, Y.S. Effects of sulfate during CO_2 attack on Portland cement and their impacts on mechanical properties under geologic CO_2 sequestration conditions. *Environ. Sci. Technol.* **2015**, *49*, 7032–7041.
11. Omosebi, O.; Maheshwari, H.; Ahmed, R.; Shah, S.; Osisanya, S.; Hassani, S.; Bruijn, G.D.; Cornell, W.; Simon, D. Degradation of well cement in HPHT acidic environment: Effects of CO_2 concentration and pressure. *Cem. Concr. Comp.* **2016**, *74*, 54–70. [CrossRef]
12. Guo, Y.Y.; Zhang, Y.X.; Soe, K. Effect of fly ash on mechanical properties of magnesium oxychloride cement under water attack. *Struct. Concr.* **2020**, *21*, 1181–1199. [CrossRef]
13. Tang, K.; Mao, X.S.; Wu, Q.; Liu, L.; Zhao, Y.; Huang, W.J.; Zhang, J.X. The concrete performance with iron tailings sand modified by polypropylene fibers under aggressive environment. *Adv. Mater. Sci. Eng.* **2020**, *2020*, 4368592. [CrossRef]
14. Tommy, Y.L.; Liao, W.Y.; Wong, C.K.; Tang, W.C. Evaluation of carbonation resistance of paint coated concrete for buildings. *Constr. Build. Mater.* **2016**, *107*, 299–306.
15. Mohammed, M.K.; Dawson, A.R.; Thom, N.K. Carbonation of filler typed self-compacting concrete and its impact on the microstructure by utilization of 100% CO_2 accelerating techniques. *Constr. Build. Mater.* **2014**, *50*, 508–516. [CrossRef]
16. Gervais, C.; Garrabrants, A.C.; Sanchez, F.; Barna, R.; Moszkowicz, P.; Kosson, D.S. The effects of carbonation and drying during intermittent leaching on the release of inorganic constituents from a cement based matrix. *Cem. Concr. Res.* **2004**, *34*, 119–131. [CrossRef]
17. Zhang, J.; Shi, C.; Zhang, Z. Carbonation induced phase evolution in alkali-activated slag/fly ash cements: The effect of silicate modulus of activators. *Constr. Build. Mater.* **2019**, *223*, 566–582. [CrossRef]
18. Mccaslin, E.R.; White, C.E. A parametric study of accelerated carbonation in alkali-activated slag. *Cem. Concr. Res.* **2021**, *145*, 106454. [CrossRef]
19. He, P.P.; Poon, C.S.; Tsang, D.C.W. Effect of pulverized fuel ash and CO_2 curing on the water resistance of magnesium oxychloride cement(MOC). *Cem. Concr. Res.* **2017**, *97*, 115–122. [CrossRef]

20. El-Turki, A.; Ball, R.J.; Allen, G.C. The influence of relative humidity on structural and chemical changes during carbonation of hydraulic lime. *Cem. Concr. Res.* **2007**, *37*, 1233–1240. [CrossRef]
21. Tang, Y.J.; Zuo, X.B.; He, S.L.; Ayinde, O.; Yin, G.J. Influence of slag content and water-binder ratio on leaching behavior of cement pastes. *Constr. Build. Mater.* **2016**, *129*, 61–69.
22. Dauzeres, A.; Le, B.P.; Cau, D.C.C.; Brunet, F.; Bourbon, X.; Timonen, J.; Voutilainen, M.; Voutilainen, M.; Sardini, P. On the physico-chemical evolution of low-pH and CEM I cement pastes interacting with Callovo-Oxfordian pore water under its in situ CO_2 partial pressure. *Cem. Concr. Res.* **2014**, *58*, 76–88. [CrossRef]

materials

MDPI

Article

Study on the Performance of Steel Slag and Its Asphalt Mixture with Oxalic Acid and Water Erosion

Xiaoming Huang [1], Feng Yan [1,*], Rongxin Guo [1] and Huan He [2]

1 Yunnan Key Laboratory of Disaster Reduction in Civil Engineering, Faculty of Civil Engineering and Mechanics, Kunming University of Science and Technology, Kunming 650500, China
2 Faculty of Environmental Science and Engineering, Kunming University of Science and Technology, Kunming 650500, China
* Correspondence: yanfeng@kust.edu.cn

Abstract: The reuse of steel slag, a large-scale solid waste from steel production, has good social and environmental benefits. The application of a steel slag asphalt mixture is mainly hindered by its volume expansion in water. The expansion of steel slag can be inhibited by oxalic acid. The expansion rate and adhesion of steel slag were investigated, and the immersion stability of steel slag and its asphalt mixture was evaluated by water erosion. By means of XRD, XRF, TG, SEM, etc., the influence mechanism of oxalic acid and water erosion on the properties of steel slag and its asphalt mixture was discussed. The results show that oxalic acid can not only inhibit the expansion of steel slag but also improve its crush resistance, with a reduction in the expansion rate of steel slag by 53%. Oxalic acid is able to leach alkaline metal elements, reducing its adhesion with asphalt. After 10 days of water erosion, the rutting stability and bending crack resistance of the treated steel slag mixture decreased by 37% and 43.2%, respectively. Calcium oxalate is generated on the surface of treated steel slag, which improves the surface compactness, effectively inhibits the expansion of steel slag caused by water erosion, and improves the performance of steel slag and its asphalt mixture. Water erosion can accelerate the hydration and shedding of calcium-containing substances on the surface of steel slag, reduce the adhesion of steel slag, and lead to degradation in the performance of steel slag and its asphalt mixture. Oxalic acid is able to effectively inhibit the expansion of steel slag, and the treated steel slag can be used as recycled aggregate in asphalt mixture, effectively solving the problems of road aggregate deficiency and environmental pollution caused by steel slag.

Keywords: steel slag; oxalic acid; expansion; water erosion; asphalt mixture; road performance; influence mechanism

Citation: Huang, X.; Yan, F.; Guo, R.; He, H. Study on the Performance of Steel Slag and Its Asphalt Mixture with Oxalic Acid and Water Erosion. *Materials* **2022**, *15*, 6642. https://doi.org/10.3390/ma15196642

Academic Editors: Qiang Li, Jiaqing Wang, Kaijian Huang, Dongdong Ge and Fangyuan Gong

Received: 2 August 2022
Accepted: 20 September 2022
Published: 25 September 2022

Publisher's Note: MDPI stays neutral with regard to jurisdictional claims in published maps and institutional affiliations.

Copyright: © 2022 by the authors. Licensee MDPI, Basel, Switzerland. This article is an open access article distributed under the terms and conditions of the Creative Commons Attribution (CC BY) license (https://creativecommons.org/licenses/by/4.0/).

1. Introduction

Steel slag accounts for a large amount of the solid waste produced by steel making. With the development of China's steel industry, the output of steel slag in China was as high as 100 million tons in 2021, and the total accumulated amount is as high as 1 billion tons at present. The multipurpose use rate of steel slag in China is less than 30%, of which less than 10% is used in building materials. Most steel slag is regarded as industrial waste, so it is filled or accumulated, resulting in ecological environmental pollution and other problems [1–5]. At the same time, China's roads are in a stage of rapid development. By the end of 2021, the total mileage of expressways had exceeded 161,000 km. The problem of shortages of aggregate, as the main material from which expressways are constructed, has become more and more serious. Meanwhile, steel slag and natural aggregates have similar properties and can be mixed for use as aggregates. A large number of domestic laboratories' research on steel slag asphalt mixtures and test sections of steel slag asphalt mixtures have demonstrated that the use of steel slag in road materials is the primary means of eliminating them [6–9].

The steel slag accumulated and produced in China has the property of volume expansion. The main factor affecting the expansion of steel slag is free calcium oxide (f-CaO), which undergoes a slow hydration reaction when coming into contact with water. The process of generating calcium hydroxide ($Ca(OH)_2$) leads to the expansion and cracking of steel slag. There is a close relationship between f-CaO and the expansion of steel slag, and f-CaO on the surface of steel slag is more likely to cause expansion and cracking than that inside [10,11]. Road researchers have focused on the treatment of steel slag, and the reuse of steel slag after treatment has become an industry trend [12–14]. The main effective treatment methods include carbonization, acid treatment, surface coating, etc. The carbonization of steel slag effectively reduces the volume expansion of steel slag, while at the same time improving its hydration activity [15,16]. The acid treatment effectively reduces the expansion of steel slag. Commonly used acids include glacial acetic acid, oxalic acid, phosphoric acid, formic acid, etc. The main treatment methods include chemical dry modification, chelating treatment, etc., which can reduce the expansion of steel slag, and improve the activity of steel slag powder [17–19]. Treated steel slag is mainly applied in asphalt mixture. Coating resin, cement slurry, waterproofing agent, and other materials, when applied on the surface of steel slag, can hinder the contact between steel slag and water, thus inhibiting the expansion of the steel slag [20,21]. These methods are not able to effectively eliminate f-CaO in steel slag, and only inhibit the hydration of f-CaO. During the subsequent application process, the coating layer may fall off, resulting in the subsequent expansion of the steel slag. The long-term application of treated steel slag has not been verified and may lead to potential safety hazards. The immersion stability of steel slag asphalt mixture under the action of water is worth considering.

To solve the problem of the volume expansion of steel slag caused by water, and to evaluate the immersion stability of treated steel slag and its asphalt mixture, an oxalic acid solution was used to treat steel slag aggregate. This paper analyzes the mechanism of influence of oxalic acid and water erosion in the performance of steel slag and its asphalt mixture, evaluates the applicability of steel slag and its asphalt mixture following treatment with oxalic acid, and provides engineering suggestions and theoretical guidance for the practical application of steel slag in asphalt mixture.

2. Materials and Methods

2.1. Materials

BOF steel slag aggregate, limestone aggregate, and SBS(I-D) polymer modified asphalt were used as raw materials. The raw material used for treatment was a self-prepared oxalic acid solution; oxalic acid was produced by Tianjin Chemical Three Plant Co., Ltd. with the chemical formula $C_2H_2O_4{\cdot}2H_2O$ and a reagent content of not less than 99.6%, as a colorless transparent crystal or powder. The chemical composition of the selected steel slag, and its basic physical properties, are shown in Tables 1 and 2. Moreover, some properties of different steel slags and natural aggregates are compared. The main components of steel slag are calcium oxide, iron oxide, and silica, accounting for more than 10%, while mechanical properties such as Los Angeles abrasion have quite different values, which is mainly a result of the different processes employed in different steel mills [22,23]: SSD is pristine steel slag, which was provided by Shandong Iron and Steel Co., Ltd., while EOF steel slag is energy-optimized furnace slag. The basic physical properties of limestone and the experimental results of SBS-modified asphalt are shown in Tables 3 and 4. The physical indexes of fine limestone aggregate and asphalt meet the specification requirements. However, the density of fine limestone aggregate is generally lower than that of steel slag—below 3 g/cm^3. This is mainly because there is a large amount of iron in steel slag, which is one of the main reasons for the high density of steel slag [2].

Table 1. Chemical composition of steel slag (Wt.%).

	CaO	Fe_2O_3	SiO_2	MgO	Al_2O_3	P_2O_5	MnO
BOF Steel Slag	43.49	28.36	12.99	6.75	2.33	2.17	1.34
SSD [22]	37.57	27.75	16.35	6.95	3.21	2.69	3.56
EOF Steel Slag [23]	35.28	26.91	16.69	9.27	6.2	1.43	1.88

Table 2. Physical properties of steel slag.

Tested Parameters	Unit	BOF Steel Slag	EOF Steel Slag [23]	Natural Coarse Aggregate [23]	Technical Indicators
Crush Value	%	17.8	8	6.55	≤26
Los Angeles Wear (C)	%	16.9	1.16	8.27	≤28
Bibulous Rate	%	1.91	3	0.8	≤2.0
Gross Volume Relative Density	g/cm^3	3.465	2.86	2.75	—
Apparent Relative Density	g/cm^3	3.586	3.02	2.886	≥2.60
Adhesiveness	level	5	-	-	≥5

Table 3. Physical properties of limestone.

Tested Parameters	Unit	Test Value	Technical Indicators
Gross Volume Relative Density	g/cm^3	2.656	—
Apparent Relative Density	g/cm^3	2.724	≥2.45
Angularity (Flow Time Method)	s	38.9	—
Sand Equivalent	%	73	≥50

Table 4. Test results of SBS polymer-modified asphalt.

		Unit	Test Results	Technical Indicators
	Penetration (25 °C, 5 s, 100 g)	0.1 mm	53	40~60
	Penetration Index	-	0.151	≥0
	Softening Point (Global Method)	°C	79.5	≥60
	Ductility (5 °C)	cm	29	≥20
	Flash Point	°C	294.8	≥230
	Ignition Point	°C	304.8	-
	Solubility	%	99.76	≥99
	Elastic Recovery (25 °C)	%	82	≥75
	Rotational Viscosity of Brinell (135 °C)	Pa·s	1.73	≤3
	Density (15 °C)	g/cm^3	1.037	-
	Relative Density (25 °C)	-	1.038	-
	Storage Stability Segregation, 48 h Softening Point Difference	°C	2	≤2.5
Film Heating Test (163 °C, 5 h)	**Quality Change**	%	−0.209	±1.0
	Penetration Ratio (25 °C)	%	78	≥65
	Ductility (5 cm/min, 10 °C)	cm	21	≥15

2.2. Treatment Method

Steel slag is mainly used as coarse and fine aggregate in asphalt mixtures; therefore, 0–16 mm steel slag was used for the treatment. Different concentrations of oxalic acid solutions were prepared, and different treatment times were adopted for the preliminary steel slag treatment. The best treatment concentration was determined on the basis of the water absorption rate and the f-CaO content of treated steel slag, and the best treatment time was determined on the basis of the change in pH value. In the treatment process, the steel slag to be treated was put into the prepared 0.4 mol/L oxalic acid solution at 25 °C according to a solid–liquid ratio of 1:5. To avoid the CO_2 and water in the air influencing the treatment process, the treatment process was sealed after stirring for 3 min, and the treatment time was 84 h. Treated steel slag can be used after being dried at 105 °C. The specific treatment process is as follows (Figure 1).

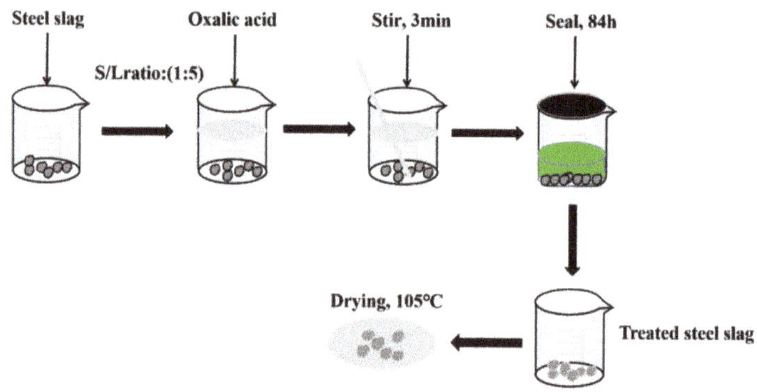

Figure 1. Treatment process of steel slag.

2.3. Experimental Method

According to the Test Method of Aggregate of Highway Engineering (JTG E42-2005), the basic physical properties of steel slag and limestone were tested. The basic index of SBS-modified asphalt, the road performance index of the steel slag asphalt mixture, and the mix design of steel slag asphalt mixture were adopted with reference to the Standard Test Methods of Bitumen and Bituminous Mixtures for Highway Engineering (JTG E20-2011). There are three index samples in each group tested by steel slag and its asphalt mixture, and the indexes of the samples need to meet the specifications. The experimental results are the average of the results of the three samples. The water stability of the mixture is tested by a Universal testing machine (UTM-30), including the Marshall stability test and freeze-thaw splitting test. The sample in the Marshall stability test needs to be kept in a water bath at a constant temperature of 60 °C for 30–40 min and 48 h, respectively. After the heat preservation is completed, it is put into a fixture for testing, and the loading rate is 50 mm/min. For the samples in the freeze–thaw test, the molded samples are divided into two groups. One group is stored at room temperature, and the other group is kept in a vacuum for 16 h. After taking out, the samples are kept in a water bath at 60 °C for 24 h. Finally, the samples of both groups are kept in a water bath at 25 °C for 2 h, and then taken out for testing. The loading rate is 50 mm/min. The rutting stability of the mixture is tested by a rutting tester (SYD-0719B). The formed sample is kept in a 60 °C incubator for 5 h, and then put into a test bench for testing. The test will be stopped when the test time is 1 h or the maximum deformation is 25 mm.

The mineral composition and phase composition of steel slag have a great influence on the properties of steel slag. To ensure the average of the selected steel slag, selected the original steel slag by the quartering method and ground it to below 0.075 mm. The material was tested by X-ray diffraction (XRD), the diffraction pattern was analyzed, the mineral composition of the steel slag was obtained, the equipment used in this experiment is Ultima IV, with the parameters of X-ray Cu Kα Radiation Diffraction Meter, wavelength of 0.15406 nm, voltage of 40 KV, current of 40 mA, testing speed of 5°/min, and testing angle of 10–90. X-ray fluorescence (XRF) is used to test the elements and substance composition in steel slag, and the equipment model is PANalytical Axios. Moreover, the microstructure of the steel slag is observed using a scanning electron microscope (SEM), its equipment model is Hitachi Regulus8100. Ethylene glycol differential thermal analysis—EDTA-TG—is an effective current method that is accurate and simple for measuring the content of f-CaO in steel slag. After grinding the steel slag selected by the quartering method to below 0.075 mm, the free total calcium in the steel slag is determined by titration with ethylene glycol, and the free total calcium content obtained using this method includes the content of calcium hydroxide. The content of calcium hydroxide in steel slag is determined by TG, and its equipment model is TA. The content of f-CaO is obtained by subtracting the content of calcium hydroxide from the free total calcium [24]. The 10-day soaking expansion rate of steel slag was defined with reference to the Test Method for Stability of Steel Slag (GB/T 24175-2009). The specific process is as follows: the steel slag mixture is mixed with the optimal water content, compacted by a standard heavy-duty compactor, and then put into the volume expansion rate mold, as shown in Figure 2. The initial dial gauge is read, and the mold is soaked in a 90 °C water bath for 6 h to completely immerse the sample. Before heating, the daily dial indicator reading of the sample height is recorded for 10 days. According to Formula (1) [25],

$$\gamma = \frac{d_{10} - d_0}{120} \times 100 \quad (1)$$

where γ is the water immersion expansion rate, %; 120 is the original height of the test piece, mm; d_{10} is the final reading of the dial indicator, mm; d_0 is the initial reading of the dial indicator, mm.

Figure 2. Steel slag volume expansion test setup. 1—Lower perforated plate; 2—steel slag; 3—surcharge board; 4—dial gauge.

The application process of steel slag and its asphalt mixture is greatly affected by water, which also directly causes steel slag expansion. To intuitively evaluate the changes in the performance of steel slag and its asphalt mixture under the action of water, the water erosion method is adopted [10,11]. Water erosion process: the steel slag aggregate and the formed mixture specimen are placed in a normal-temperature water bath box, and the steel slag aggregate and the asphalt mixture need to be completely submerged by the water. After soaking for a long setting time, they are taken out for performance testing. The adhesion test [26] was used to evaluate the adhesion between aggregate and asphalt. Two representative groups of limestone were taken, combined with untreated steel slag and treated steel slag, and put in asphalt for 45 s. Then, they were cooled in the air for 15 min, and placed in boiling water for 3 min and 30 min; their adhesion was evaluated using Formula (2):

$$S = \frac{m_1 - m_2 - m_3}{m_1 - m_2} \quad (2)$$

where S is the anti-stripping percentage of asphalt, m_1 is the original mass of asphalt aggregate, m_2 is the original mass of aggregate, and m_3 is the loose loss mass of the aggregate surface.

2.4. Mix Proportion Design of Steel Slag Mixture

At present, there are many specifications of asphalt mixture, and AC asphalt mixture is widely used in pavement. AC-13c is the main type in most experiments [12]. AC-13c asphalt mixture is a continuous gradation of fine particles, with a nominal maximum particle size of 13 mm. Therefore, this paper uses the AC-13c Marshall preparation method to prepare steel slag asphalt mixture [27].

As oxalic acid cannot directly affect the particle size of steel slag, the gradation curves of steel slag before and after treatment show little difference. However, in order to better standardize the experimental results, designed two gradations under the condition of meeting the lower limit of gradation. The specific gradation curves are shown in Figure 3. In addition, the percentage passing through the graded mesh quality is shown in Table 5. The gross bulk density, stability, flow value, and volume index of the two kinds of asphalt mixtures prepared with five asphalt–aggregates ratios (4.3%, 4.7%, 5.2%, 5.7%, and 6.2%) were tested. In addition, the best asphalt–aggregate ratios for steel slag asphalt mixtures before and after treatment were determined to be 5.2% (mass fraction) and 4.8% (mass fraction), respectively. The indexes of the Marshall specimen of the steel slag asphalt mixture under the best asphalt–aggregate ratio are shown in Table 6.

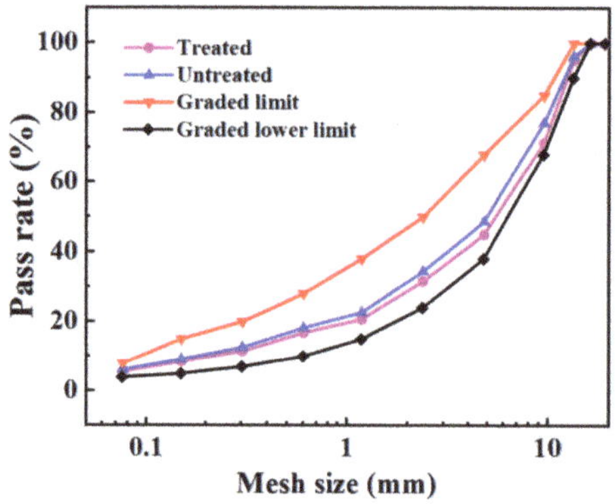

Figure 3. Asphalt mixture grading curve.

Table 5. Mass passing percentage of graded sieve.

Test Item	Percentage of Mass Passing the Following Sieve (mm)/%									
	16	**13.2**	**9.5**	**4.75**	**2.36**	**1.18**	**0.6**	**0.3**	**0.15**	**0.075**
AC-13	100.0	94.7	68.6	41.6	28.4	18.5	15	10.4	7.8	5.1

Table 6. Marshall test results of steel slag asphalt mixture at optimum asphalt dosage.

Optimum Oilstone Ratio	Gross Volume Relative Density (g/cm^3)	Void Fraction (%)	Voids in Mineral Aggregate (%)	Effective Asphalt Saturation (%)	Stability (KN)	Flow Value (mm)
Untreated (5.2%)	3.0	4.3	17.6	75.6	20.23	2.7
Treated (4.8%)	2.96	3.4	14.1	75.7	22.62	3.1

This section mainly discusses the gradation design and the best asphalt–aggregate ratio of steel slag asphalt mixture before and after treatment. The gradation curve of steel slag asphalt mixture before and after treatment has little difference. It is worth noting that the best asphalt–aggregate ratio of steel slag asphalt mixture after treatment has decreased by 0.4%. All calculated data are the average value from three tests.

3. Results and Discussion

3.1. Basic Properties of Steel Slag Aggregate

3.1.1. Expansion of Steel Slag

The main factors causing the volume expansion of steel slag are f-CaO, free magnesium oxide (f-MgO), and RO phase. f-CaO, the f-CaO caused volume expansion corresponding to 91.7% [28,29]. Because the determination methods of f-MgO are not uniform, the current determination methods are not able to achieve rapid and accurate results, and there is no specific method for the determination of RO phase, the f-CaO content of steel slag aggregate and the soaking expansion rate are the main indexes for judging the volume stability of steel slag [2]. Table 7 and Figure 4 show the changes in f-CaO content, expansion rate, and $Ca(OH)_2$ content of steel slag before and after treatment, respectively. After treatment with oxalic acid, the content of f-CaO decreased from 4.551% to 2.225%, and the expansion rate decreased from 3.59% to 1.69%. The mass loss of $Ca(OH)_2$ at 400~550 °C was calculated by TG analysis. The mass loss of water in steel slag at 400–550 °C can be expressed as the loss of calcium hydroxide [28]. After the steel slag is treated with oxalic acid, the content of calcium hydroxide is increased by 127%, which indicates that oxalic acid is able not only to consume f-CaO, but also promote its hydration. The soaking expansion rate of steel slag can intuitively indicate its volume stability. Formula (3) is used to fit the 10-day soaking expansion rate of steel slag [30]. In the early stage of steel slag soaking, the expansion rate increased greatly, because, under the conditions of the 90 °C water bath, f-CaO hydrates to form $Ca(OH)_2$, causing volume expansion. With the passage of time, the hydration reaction weakened, the content of f-CaO decreased, and the growth trend of expansion rate decreased [31]. In the process of treating steel slag with oxalic acid, it reacts with some basic oxide f-CaO, effectively reducing the content of f-CaO in the steel slag, reducing the expansion rate and expansion speed of the steel slag, and effectively reducing the expansion of the steel slag.

$$Y = a\left(1 - e^{(-bx)}\right) \tag{3}$$

where Y is the expansion rate of steel slag, %; a and b are constants; e = 2.71828183 . . . ; and x is time, day.

Table 7. Comparison of f-CaO content (Wt.%).

	Free Calcium	$Ca(OH)_2$	f-CaO
Untreated	4.852	0.31	4.551
Treated	2.848	0.623	2.225

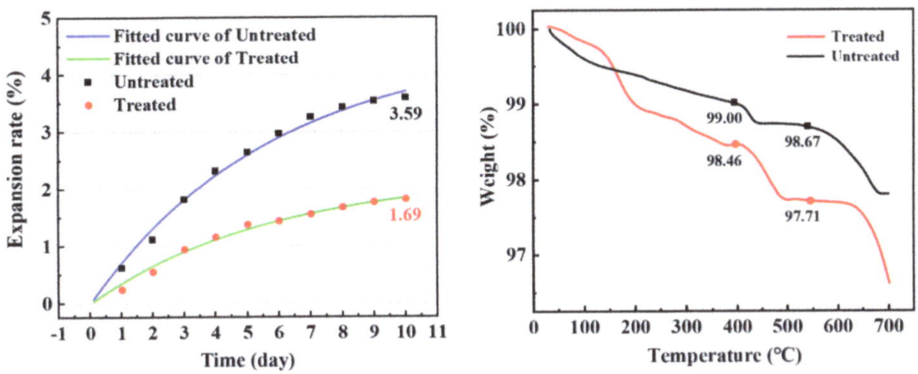

Figure 4. Comparison of steel slag expansion rate and $Ca(OH)_2$ mass loss.

3.1.2. Basicity and Adhesion

The adhesive force between aggregate and asphalt is one of the main factors determining the performance of the mixture. Because steel slag is a basic aggregate, there are a lot of metal cations such as Ca^{2+}, Mg^{2+}, Al^{3+}, Fe^{3+}, and Mn^{2+} on the surface of steel slag. In the process of its combination with asphalt, it can react with some substances in asphalt, such as asphaltene groups, to generate bituminous acid salt, and form an adsorption layer on the surface of steel slag, which has high chemical strength. Therefore, steel slag and asphalt have good adhesive force [13,32]. The asphalt mixture prepared using it has good performance. It is worth noting that in the process of treating steel slag with oxalic acid, the basicity of the steel slag may be decreased due to acid–base reaction, and the adhesive force may be reduced, such that there is a reduction in the performance of the mixture.

Therefore, by comparing the basicity of steel slag and limestone before and after treatment, and the adhesion with asphalt, the adhesion between the steel slag and asphalt after oxalic acid treatment was tested. Table 8 presents a comparison of the chemical composition and basicity between the steel slag and the limestone before and after treatment. After treatment, the content of calcium oxide (CaO) and iron oxide (Fe_2O_3) in the steel slag decreases, while the content of silicon dioxide (SiO_2) changes little, and the content of other metal oxides decreases slightly. After treatment, the basicity of steel slag decreases slightly, but it is still higher than that of limestone. Figure 5 shows the results of the adhesion test. Compared with untreated steel slag, there was a decrease in the adhesion of treated steel slag at 3 min and 30 min, and the asphalt peeling rate increased by 33.6% and 8%, respectively. However, the peeling rate of treated steel slag at 3 min and 30 min was lower than that of limestone, and its adhesion was still better than that of the limestone aggregate. It has been proved that oxalic acid is able to leach calcium, iron, and other metal ions from steel slag [33,34], and can be combined with it to form oxalic acid salt. The surface basicity of steel slag is reduced, the reaction with acid groups in asphalt is weakened, and the adhesive force is reduced, resulting in a decrease in its adhesion.

Table 8. Comparison of chemical composition and basicity of steel slag with limestone (Wt.%).

	CaO	Fe_2O_3	SiO_2	MgO	Al_2O_3	MnO	Basicity
Untreated	43.49	28.36	12.99	6.75	2.33	1.34	3.35
Treated	40.26	25.81	12.54	6.64	2.23	1.19	3.21
Limestone	36.27	4.64	13.29	1.62	4.59	1.74	2.73

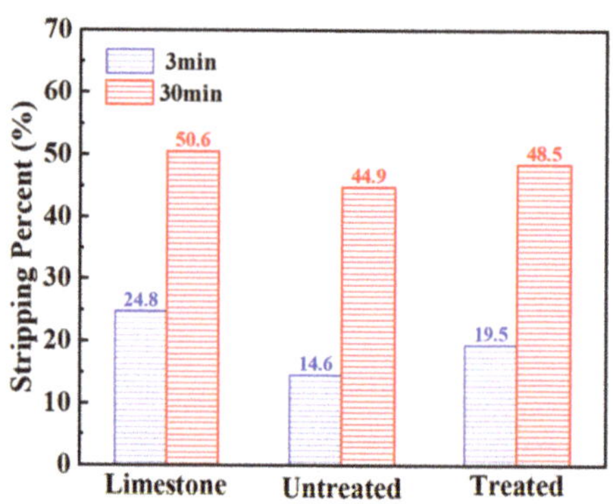

Figure 5. Boiling water stripping rate of aggregate.

3.1.3. Crush Resistance

At present, most countries use abrasion to express the anti-crushing ability of aggregate, but the importance of crush value in the asphalt mixture cannot be ignored. Therefore, in this section, the abrasion value and the crush value are used to jointly characterize the anti-crushing ability of the aggregate. Figure 6 shows the changes in abrasion value and crush value of steel slag aggregate with different water erosion times. After treatment with oxalic acid, the crush resistance of steel slag is improved, and the abrasion value and crush value are reduced by 27.7% and 7.9%, respectively. With increasing water erosion time, the crush resistance of steel slag before and after treatment showed a decreasing trend, but the abrasion and crush value of treated steel slag gradually became stable. After 10 days of water erosion, the abrasion and crush values of untreated steel slag increased by 2.4% and 3.1%, respectively, while those of treated steel slag increased by 2.1% and 3.3%, respectively. The overall decrease was small and the impact on the crush resistance of steel slag was low. Figure 7 shows the pH change in steel slag eroded by water. Under the long-term erosion of steel slag aggregate and water, silicate minerals and CaO in the steel slag aggregate undergo a hydration reaction with water, resulting in hydrated calcium silicate (C-S-H) and $Ca(OH)_2$. During the erosion process, the pH value changes obviously, and CO_2 in the air reacts with $Ca(OH)_2$ to generate calcium carbonate ($CaCO_3$), resulting in a white coating on the surface of the aqueous solution [35], as shown in Figure 8. Additionally, with increasing water erosion time, the amount of white coating gradually increases, and the pH value gradually increases. In the process of hydration to form C-S-H and $Ca(OH)_2$, the volume of steel slag aggregate expands, micro-cracks appear on the surface and pores of steel slag, and water immerses into the cracks, providing further hydration. However, due to the slow hydration process, only the surface can be hydrated, resulting in a slight decrease in the strength of the steel slag aggregate [36]. Under conditions of water erosion, the pH value of steel slag treated with oxalic acid changes slightly, the surface of the aqueous solution is clear, and its pH value exhibits a stable trend. In the process of treating steel slag with oxalic acid, most silicate minerals and f-CaO on the surface of steel slag are acidified to generate oxalic acid salt, reducing the hydration activity of the steel slag surface. With increasing water erosion time, the hydration between steel slag and water decreases, and the steel slag and water solution remain in a relatively stable state.

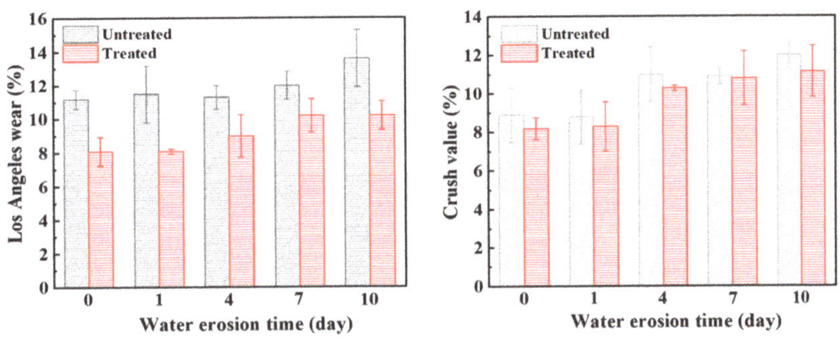

Figure 6. The Los Angeles wear and crush value of steel slag after different water erosion times.

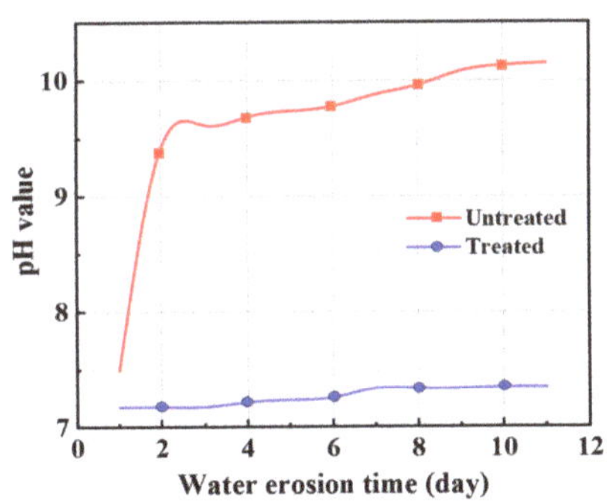

Figure 7. Comparison of pH values during the steel slag water erosion process.

Figure 8. Water erosion process of untreated steel slag: (a) untreated; (b) treated.

3.2. *Water Stability of Steel Slag Asphalt Mixture*

3.2.1. Marshall Stability

Table 9 shows the change in the stability of the steel slag asphalt mixture before and after the treatment with oxalic acid under different water erosion times, and Figure 9 shows the change in residual stability. Generally speaking, the residual stability of steel slag treated with oxalic acid is slightly better than that of untreated steel slag under the action of water erosion at different times. It is worth noting that the residual stability of steel slag before and after the treatment is mostly higher than 100%, which indicates that the strength of the mixture is improved after heat preservation for 48 h at 60 °C. The residual stability of the untreated steel slag increased gradually while undergoing erosion for durations of 0–4 days, but decreased for durations of 4–10 days of erosion, with an obvious decrease of 15.7%. After treatment, the residual stability of steel slag can be maintained at about 105%, being almost unaffected by water erosion. Steel slag with f-CaO content higher than 3% undergoes expansion, and the f-CaO content of the steel slag selected in this study was as high as 4.852%. Long-term immersion causes water to infiltrate into the asphalt, coming into contact with the steel slag. During the research process, it is necessary to keep the temperature in the water bath at 60 °C for 0.5 h or 48 h, causing the asphalt film on the surface of the steel slag to peel off, and the f-CaO on the surface is more likely to react with the water in order to generate $Ca(OH)_2$, resulting in the expansion of the steel slag [36], thereby filling the internal pores of the mixture, and improving the compactness and the resistance to deformation of the mixture. However, after being eroded by water for more than 4 days, most of the f-CaO on the surface of untreated steel slag continues to hydrate, and the hydration activity of the steel slag aggregate improves. After a 60 °C water bath, the hydration reaction is intensified, and the internal expansion of the steel slag aggregate is too great [31], which leads to the slow destruction of its structure, thus reducing its residual stability. Compared with untreated steel slag, oxalic acid is able to eliminate most silicate minerals and f-CaO on the surface of the steel slag and inhibit the hydration reaction, resulting in an insignificant change in the overall residual stability.

Table 9. Test results of stability.

	Soaking Time	0.5 h Marshall Stability (KN)	48 h Marshall Stability (KN)		Soaking Time	0.5 h Marshall Stability (KN)	48 h Marshall Stability (KN)
Untreated	0 days	20.11	21.11	Treated	0 days	21.81	22.9
	1 day	21.71	23.1		1 day	20.83	22.23
	4 days	19.42	21.72		4 days	19.86	20.98
	7 days	20.94	21.06		7 days	19.68	20.67
	10 days	22.99	21.66		10 days	20.72	22.16

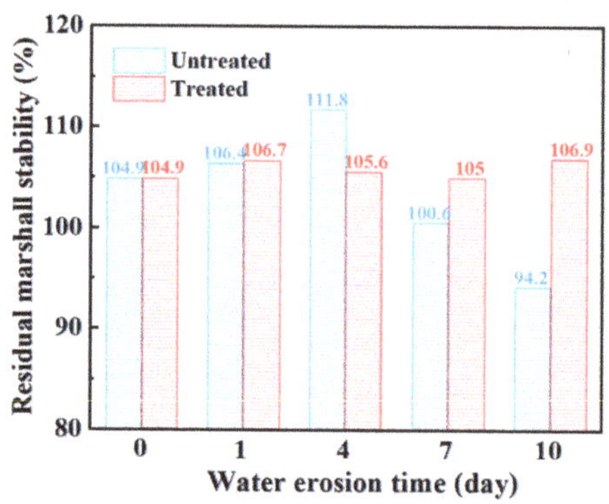

Figure 9. Marshall residual stability of steel slag asphalt mixture with different water erosion times.

3.2.2. Freeze–Thaw Splitting

Table 10 and Figure 10, respectively, show the changes in the freeze–thaw splitting strength and strength ratio of steel slag asphalt mixture before and after the treatment under different water erosion conditions. Under early water erosion conditions, the unfrozen and freeze–thaw splitting strength of the treated steel slag is higher than that of untreated steel slag. However, with increasing water erosion time, the splitting strength of treated steel slag gradually becomes lower than that of untreated steel slag. By observing the freeze–thaw splitting strength ratio of steel slag asphalt mixture before and after treatment, it is found that the freeze–thaw splitting strength ratio of treated steel slag is lower than that of untreated steel slag, and after 4 days of water erosion, there is an increase in the decreasing trend of the freeze–thaw splitting strength ratio. Generally speaking, the freeze–thaw splitting strength ratio of steel slag asphalt mixture before and after treatment meets 75% of the specification requirements. With increasing water erosion time, there was a decrease in the freeze–thaw splitting strength and splitting strength ratio of the steel slag asphalt mixture. Compared to the Marshall stability test, the freeze–thaw splitting test conditions are more stringent, making them better able to reflect the water damage resistance of the mixture. With increasing water erosion time, the emulsification between water and asphalt intensifies, and the asphalt layer peels off, leading to the overall structure of the mixture becoming loose. However, the vacuumizing process is able to fill most of the empty pores of the mixture, and through vacuumizing and freezing, the expansion effect is weakened, the asphalt shrinks and becomes brittle, the brittleness of the mixture is increased, and the adhesive force between the asphalt and the aggregate decreases [37]. However, the treated steel slag surface precipitates and transforms, resulting in calcium oxalate, which improves the strength of the steel slag [38,39], and under the condition of water erosion, the expansion effect of hydration reaction of steel slag in the water bath is inhibited, so the ratio of freeze–thaw cracking strength and splitting strength of treated steel slag decreases rapidly.

Table 10. Test results of freeze–thaw splitting strength.

	Water Erosion Time	Freeze–Thaw Cycle Splitting Strength (MPa)	Unfreeze–Thaw Cycle Splitting Strength (Mpa)		Water Erosion Time	Freeze–Thaw Cycle Splitting Strength (Mpa)	Unfreeze–Thaw Cycle Splitting Strength (Mpa)
	0 days	2.11	2.20		0 days	2.37	2.51
	1 d	2.05	2.14		1 day	2.16	2.3
Untreated	4 days	1.73	1.83	**Treated**	4 days	1.71	1.83
	7 days	1.77	1.88		7 days	1.62	1.77
	10 days	1.71	1.83		10 days	1.52	1.69

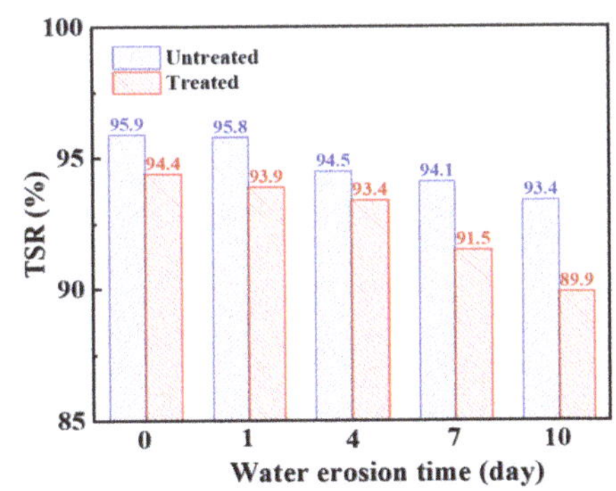

Figure 10. Freeze–thaw splitting strength ratio of steel slag asphalt mixture with different water erosion times.

3.3. Rutting Stability and Bending Crack Resistance of Steel Slag Asphalt Mixture

Under the same conditions with respect to gradation, aggregate, and asphalt type, the rutting stability of the mixture is mainly related to its strength and compactness. The higher the strength and compactness of the mixture, the higher its dynamic stability and the better its rutting stability [18,40]. Figure 11 shows the changes in dynamic stability and maximum bending strain of the steel slag mixture before and after treatment for different water erosion times. The dynamic stability of the steel slag before and after treatment decreased by 28.4% and 37%, respectively, after 10 days of water erosion, and the maximum bending strain decreased by 26.7% and 43.3%, respectively. It can be clearly seen that the rutting stability of treated steel slag is effectively improved, and the dynamic stability of the steel slag gradually decreases with increasing water erosion time. In the process of water erosion of the mixture, water emulsifies the asphalt, and the adhesion gradually decreases. Oxalic acid reduces the volume expansion of steel slag, but it can also reduce adhesion. Although most of the f-CaO on the surface of the steel slag is eliminated, under the condition of long-term immersion, the asphalt film keeps falling, the compactness decreases, and the bonding degree decreases [41]; however, the strength of the mixture will still decrease. This is the main reason for which the rutting stability of the steel slag mixture decreases with increasing water erosion time. At present, the influencing factors of the bending crack resistance of the mixture are not clear, and most researchers think that the bending stability of the mixture is related to its adhesion [42,43]. With increasing water erosion time, the bending crack resistance of the mixture gradually decreases. In the process of erosion for 0–1 day, the maximum failure strain of treated steel slag is higher than that of untreated steel slag. After 4 days of water erosion, it can be observed that the maximum failure strain of the treated steel slag is lower than that of the untreated steel slag, which corresponds to the freeze–thaw splitting strength reported in the previous section. The results show that the performance of the steel slag asphalt mixture treated with oxalic acid will be greatly reduced after a short duration of water erosion and in a low-temperature experiment. The authors believe that the decrease in the adhesion between asphalt and steel slag is mainly due to water erosion. The asphalt shrinks and becomes brittle, aggregate pores increase, and finally, its bending crack resistance decreases [40]. After the treatment, the metal ions on the surface of steel slag combined with asphalt decrease, resulting in its adhesion with asphalt decreasing rapidly with water erosion.

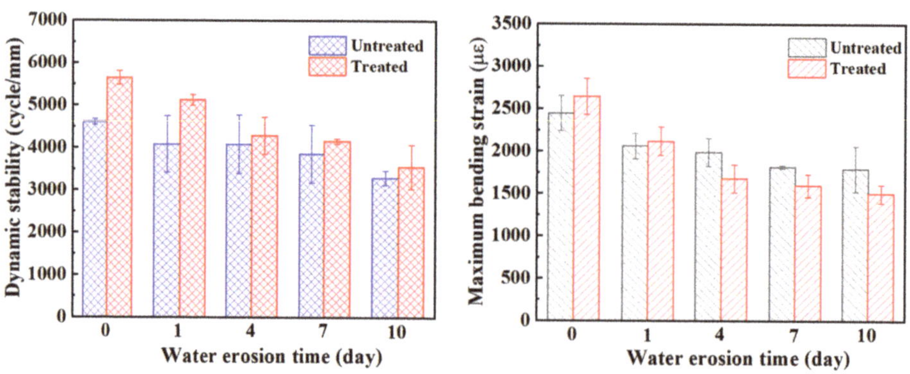

Figure 11. Dynamic stability and maximum bending strain of steel slag asphalt mixture under different water erosion times.

3.4. Mechanism of Action of Oxalic Acid

Figure 12 presents a comparison diagram of the mineral composition of steel slag before and after treatment with oxalic acid. Steel slag is a hydraulic mineral with complex mineral composition, mainly including tricalcium silicate (Ca_3SiO_5), dicalcium silicate (Ca_2SiO_4), calcium ferrite ($Ca_2Fe_2O_5$), and RO. After the steel slag is treated with oxalic acid, the diffraction peak of calcium oxalate (CaC_2O_4) appears. Some of the Ca_2SiO_4 and Ca_3SiO_5 diffraction peaks disappear and decrease. The diffraction peak of CaO disappears and changes to $Ca(OH)_2$. The presence of oxalic acid breaks the acid–base balance of the original system of steel slag, and some silicate minerals on the surface of the steel slag, such as Ca_2SiO_4 and Ca_3SiO_5, react with oxalic acid to form the CaC_2O_4 complex. During the treatment of steel slag, the hydration speed of f–CaO accelerates, and $Ca(OH)_2$ is generated. Under the closed treatment condition, CO_2 in the air fails to participate in the reaction, so there is $Ca(OH)_2$ after the treatment, which is consistent with the quality loss analysis results $Ca(OH)_2$ described above. Figure 13 presents a comparison of the SEM results of the surface of the steel slag. The surface of the untreated steel slag is complex and diverse, consisting of mostly spherical weak particles ($CaCO_3$), accompanied by a large number of loose network structures [44]. A large number of square blocks appear on the surface of the treated steel slag, which are judged to be $Ca_2C_2O_4$ crystals [45,46], accompanied by a small amount of flake $Ca(OH)_2$ [47]. Steel slag has a porous structure. During long-term treatment with oxalic acid, oxalic acid can acidify the surface of most steel slag, and penetrates into the pores of steel slag, forming CaC_2O_4 complexes on the surface and in the pores. The weak particles and reticular structure on the surface of the untreated steel slag are replaced and filled by the generated CaC_2O_4, resulting in precipitation transformation. The surface compactness of the steel slag increases and the strength of the steel slag increases slightly [48]. Oxalic acid can also promote the hydration reaction of f–CaO and react with some f–CaO in steel slag, which effectively reduces the expansion of the steel slag. The reaction principle is shown in Equations (4)–(6).

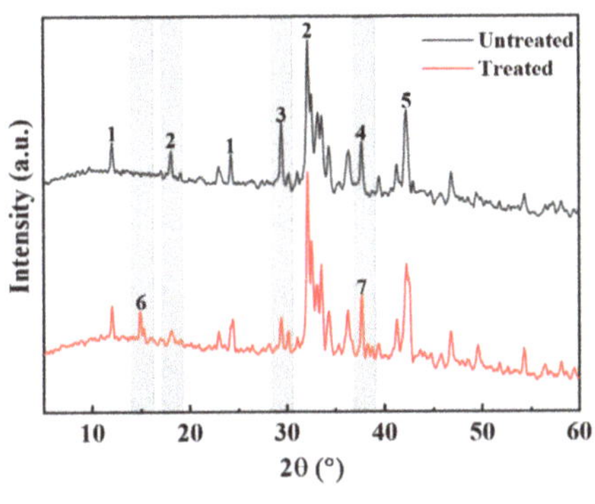

Figure 12. Comparison of mineral composition of steel slag (1—$Ca_2Fe_2O_5$; 2—Ca_2SiO_4; 3—Ca_3SiO_4; 4—CaO; 5—RO; 6—CaC_2O_4; 7—$Ca(OH)_2$).

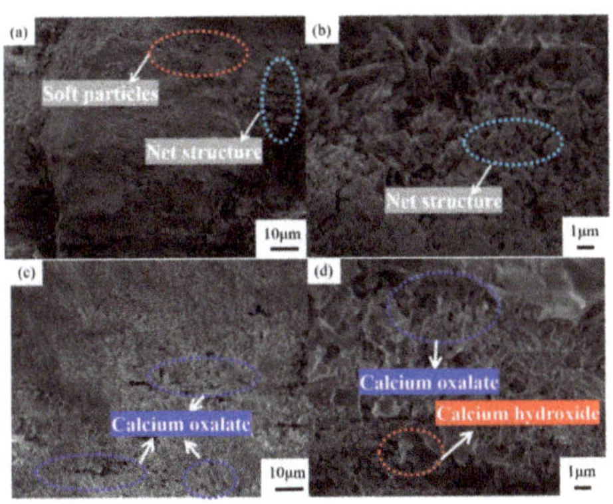

Figure 13. Microstructure of steel slag surface: (**a**,**b**) untreated, (**c**,**d**) treated.

During the treatment of steel slag with oxalic acid, a large number of calcium ions dissolve out and react with oxalic acid to generate $Ca_2C_2O_4$:

$$Ca^{2+}+H_2C_2O_4= CaC_2O_4+2H^+ \tag{4}$$

Oxalic acid can also promote the hydration of f-CaO:

$$\text{f-CaO} + H_2O = Ca(OH)_2 \tag{5}$$

$$Ca(OH)_2 + H_2C_2O_4 = CaC_2O_4 + 2H_2O \tag{6}$$

In Ding et al.'s research [19], oxalic acid was used to treat steel slag fine aggregate, and the treated steel slag was used as aggregate in concrete. In the process of oxalic acid treatment, calcium oxalate generated by the reaction between oxalic acid and f-CaO and calcite in cement precipitated on the concrete surface, forming a dense protective layer, which improved the compactness of the concrete and reduced the expansion of the concrete samples, reducing its linear expansion from 4.7% to 1.7%. The compressive and flexural strength of the final concrete samples were improved. In Huo et al.'s research [17,18], they used formic acid and glacial acetic acid to treat steel slag powder by dry modification. Because formic acid and glacial acetic acid can react with mineral components on the surface of steel slag to generate calcium formate and calcium acetate, the surface of steel slag will be etched during the reaction process, resulting in improvements with respect to the specific surface area and roughness of the treated steel slag powder, an increase in the number of pores on the surface of the steel slag powder, and easier access of water into the steel slag powder. The early hydration speed of the formed cement paste is faster, resulting in more hydration products. However, part of the f-CaO in the steel slag can react with formic acid and glacial acetic acid, consuming part of the f-CaO. Due to the compactness of the hydration products, the volume stability of the cement paste is finally improved. The final result is that formic acid and glacial acetic acid are able to improve the hydration activity of steel slag powder so as to improve the volume stability. In this study, oxalic acid was used to treat steel slag, and the treatment principle was similar to that of Yung, Huo, and others; however, we obtained a more mature conclusion. Firstly, after treatment with oxalic acid, a large number of square calcium oxalate crystals are formed on the surface and pores of the steel slag, and the compactness of the steel slag surface is improved, providing a certain strength, which contributes to improving the road performance of the steel slag asphalt mixture. However, acid can accelerate the hydration of most substances in steel slag and consume part of the f-CaO in the steel slag. After acid treatment, the hydration activity of steel slag is greatly reduced, and the hydration degree of f-CaO in steel slag is reduced when it meets water again, finally leading to a significant decrease in the expansion rate of the steel slag.

3.5. Mechanism of Influence of Water Erosion on the Phase Composition of Steel Slag

Figure 14 shows the changes in the mineral composition of the steel slag before and after treatment after water erosion for 0 days, 5 days and 10 days, respectively. The intensity of most mineral diffraction peaks of steel slag decreases, while some peaks, such as $Ca_2Fe_2O_5$, Ca_2SiO4, and $Ca_2C_2O_4$, decrease obviously, and are almost invisible in the XRD diffraction patterns. Most minerals on the surface of the steel slag can hydrate to form hydrated calcium silicate (C-S-H) during the process of water erosion, especially Ca_2SiO_4 and Ca_3SiO_4, which are invisible in the XRD diffraction patterns because of their amorphous forms [49–51]. Table 11 shows the changes in the chemical elements in the steel slag before and after the treatment with water erosion. Under the condition of water erosion, most of the elements in steel slag are decreased, as, for example in the cases of the Ca, Fe, Si, and other elements. Figure 15 shows the changes in the main chemical elements. The chemical elements in the steel slag after the treatment exhibit no obvious changes with water erosion, and the overall chemical elements tend to be stable compared with untreated steel slag, while the calcium element content decreases slightly, which is consistent with the analysis of mineral composition above. Figure 16 shows the change in the micro-morphology of steel slag before and after treatment with water erosion at different times. Water erosion can accelerate the hydration of calcium silicate and f-CaO on the surface of steel slag, resulting in floc C-S-H and flake $Ca(OH)_2$ [52–54]. With increasing water erosion time, flake $Ca(OH)_2$ and floc C-S-H on the surface of untreated steel slag become more and more obvious, and at the same time, a few microcracks appear. However, the surface of the treated steel slag is mostly massive $Ca_2C_2O_4$, and water erosion can accelerate the shedding degree of $Ca_2C_2O_4$. Moreover, floc C-S-H was not found on the surface of steel slag, which was due to the existence of a large amount of $Ca_2C_2O_4$ blocking the C-S-H. With increasing shedding degree of $Ca_2C_2O_4$, a small amount of fashioned $Ca(OH)_2$ begins to appear on the surface of the steel slag, which indicates that the f-CaO on the surface of the treated steel slag has been basically eliminated, which is in agreement with the previous experimental results.

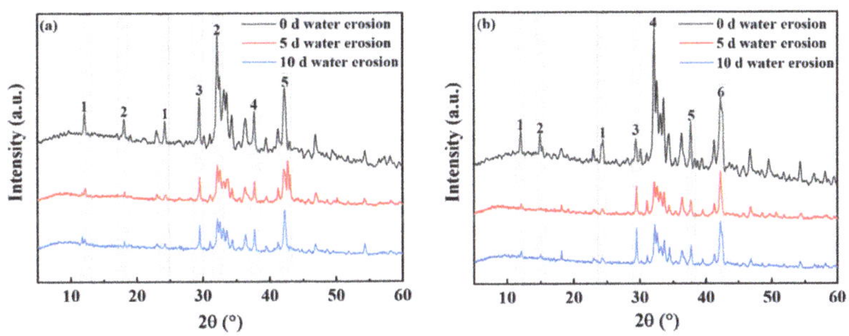

Figure 14. Mineral composition changes of untreated and treated steel slag under different water erosion times. (**a**) Untreated (1—$Ca_2Fe_2O_5$; 2—Ca_2SiO_4; 3—Ca_3SiO_4; 4—CaO; 5—RO). (**b**) Treated (1—$Ca_2Fe_2O_5$; 2—$Ca_2C_2O_4$; 3—Ca_3SiO_4; 4—Ca_2SiO_4; 5—$Ca(OH)_2$; 6—RO).

Table 11. Changes in chemical elements under different water erosion times (Wt.%).

	Ca	Fe	Si	Mg	Al	P	Mn
Untreated (0 days)	28.13	17.30	5.38	3.65	1.14	0.83	0.87
5 days	27.29	17.04	5.07	3.50	1.11	0.83	0.82
10 days	21.16	16.45	4.28	3.42	1.10	0.82	0.81
Treated (0 days)	26.30	15.95	5.28	3.63	1.1	0.82	0.79
5 days	23.84	14.83	5.26	3.58	1.09	0.8	0.7
10 days	22.34	14.26	5.24	3.57	1.08	0.79	0.69

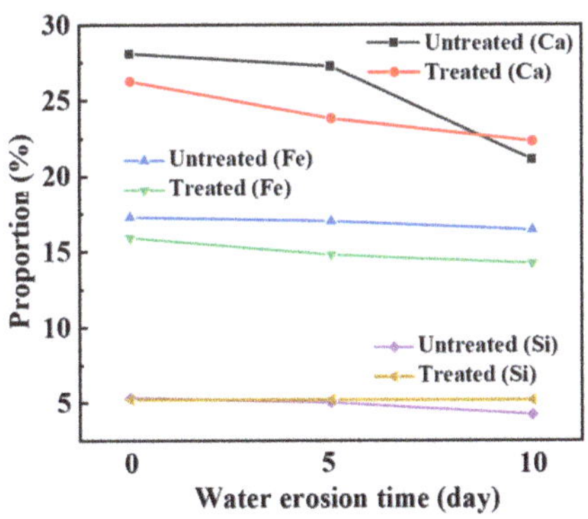

Figure 15. Influence of water erosion on main chemical elements in steel slag.

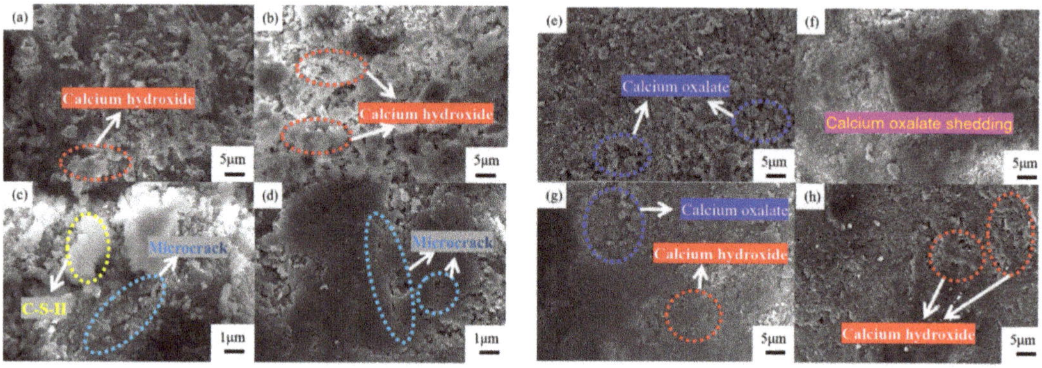

Figure 16. Changes in surface micro-morphology of steel slag under different water erosion times: untreated after (**a**) 1 day, (**b**) 4 days, (**c**) 7 days and (**d**) 10 days; and treated after (**e**) 1 day, (**f**) 4 days, (**g**) 7 days and (**h**) 10 days.

Most of the $Ca_2C_2O_4$ present on the surface of the treated steel slag is free in aqueous solution due to the action of water erosion. With increasing water erosion time, the degree of $Ca_2C_2O_4$ that is free and falling off increases, resulting in a slight decrease in the strength of the steel slag. At the same time, the shedding of the asphalt layer and the lack of basic metal content on the surface of steel slag have a great influence on adhesion, which leads directly to the obvious downward trend of the adhesion in the mixture with the action of water erosion, and the downward trend of the performance of the mixture [55,56].

4. Conclusions

In this study, a self-prepared oxalic acid solution was used to treat steel slag with the aim of inhibiting its expansion, and the steel slag and its asphalt mixture were subjected to water erosion at different times before and after treatment. Then, the basic properties of the steel slag and its asphalt mixture were tested by means of comparison. The mechanisms of influence of oxalic acid and water erosion on the mixed properties of steel slag and its asphalt mixture were discussed. The potential of steel slag treated with oxalic acid in an asphalt mixture was evaluated. The following conclusions were drawn:

(1) Oxalic acid is able to promote the hydration reaction of f-CaO on the surface of steel slag, eliminate some f-CaO in steel slag, and inhibit the volume expansion of steel slag. However, due to the leaching of alkaline ions, the adhesion decreases. After 7 days of water erosion, the crush resistance of treated steel slag tends to be stable. Water erosion has little effect on the crush resistance of steel slag.

(2) After oxalic acid treatment, the performance of the steel slag asphalt mixture is improved, but in the process of water erosion, its performance decreases obviously, especially with respect to rutting stability and bending crack resistance. The performance of the steel slag asphalt mixture after treatment is greatly reduced after being subjected to low temperature, and its low-temperature performance needs further consideration.

(3) Water erosion is able to accelerate the hydration of calcium silicate and f-CaO on the surface of steel slag, leading to the slight expansion of the steel slag surface, decreased adhesion, and a decline in the performance of the steel slag asphalt mixture. Oxalic acid can react with calcium silicate and f-CaO on the surface of steel slag to produce $Ca_2C_2O_4$. It can promote the hydration of f-CaO, filling the surface and most pores of the steel slag, effectively inhibiting the expansion effect of f-CaO hydration on the surface of steel slag after water erosion, and improving the strength of the steel slag.

(4) The surface of steel slag treated with oxalic acid mostly consists of the $Ca_2C_2O_4$ complex. With increasing water erosion time, the shedding degree of $Ca_2C_2O_4$ increases.

Most basic metal elements on the surface of steel slag are leached out, and the adhesion between steel slag and asphalt decreases more obviously, which finally leads to a decline in the performance of the asphalt mixture. After a long duration of water erosion, the properties of steel slag treated with oxalic acid and its asphalt mixture are similar to those of untreated steel slag, indicating that it is feasible to apply steel slag treated with oxalic acid in asphalt mixtures.

(5) This work solves the problem that steel slag is difficult to apply due to volume expansion, and the treated steel slag can be widely used in asphalt mixtures. This provides a solution to the excessive accumulation and output of steel slag at present, in China, and solves the environmental problems caused by steel slag and the lack of natural aggregates in China. It is worth noting that whether the performance of steel slag asphalt mixture changes under the long-term action of water erosion and how the performance changes under actual application conditions are the factors that determine whether the steel slag asphalt mixture can ultimately be applied. In future research, it is worth our continued consideration.

Author Contributions: Conceptualization, X.H. and F.Y.; methodology, X.H. and H.H.; validation, X.H., F.Y. and R.G.; formal analysis, F.Y.; investigation, X.H.; resources, F.Y.; data curation, R.G.; writing—original draft preparation, X.H.; writing—review and editing, X.H.; visualization, H.H.; supervision, R.G.; project administration, X.H.; funding acquisition, F.Y. All authors have read and agreed to the published version of the manuscript.

Funding: This research was supported by Yunnan Major Scientific and Technological Projects (grant no. 202102AG050043).

Institutional Review Board Statement: Not applicable.

Informed Consent Statement: Not applicable.

Data Availability Statement: Not applicable.

Conflicts of Interest: The authors declare no conflict of interest.

References

1. Huang, S.Y.; Huo, B.B.; Chen, C.; Zhang, Y.M. The Influence of Metakaolin on the Hydration of Steam-cured Steel Slag Blended Cement. *Mater. Rep.* **2022**, *36*, 73–78.
2. He, L.; Zhan, C.Y.; Lv, S.T.; James, G.; Gao, J.; Karol, K.J.; Jan, V.; Xie, J.; Lidija, R.; Lling, T.Q. Application Status of Steel Slag Asphalt Mixture. *J. Traffic Transp. Eng.* **2020**, *20*, 15–33.
3. Cui, P.; Wu, S.; Xiao, Y.; Yang, C.; Wang, F. Enhancement Mechanism of Skid Resistance in Preventive maintenance of Asphalt Pavement by Steel slag Asedb on Micro-surfacing. *Constr. Build. Mater.* **2020**, *239*, 117870. [CrossRef]
4. Loncnar, M.; Mladenovic, A.; Zalar, S.V.; Zupancic, M.; Vander, S.H.A. Leaching and Geochemical Modelling of an Electric Arc Furnace (EAF) and Ladle Slag Heap. *Toxics* **2022**, *10*, 10. [CrossRef]
5. Cui, P.D.; Wu, S.P.; Xiao, Y.; Hu, R.; Yang, T.Y. Environmental performance and functional analysis of chip seals with recycled basic oxygen furnace slag as aggregate. *J. Hazard. Mater.* **2021**, *405*, 124441. [CrossRef]
6. Wu, S.P.; Cui, P.D.; Xie, J.; LIU, Q.T.; Pang, L. Expansive Inhibition method of Steel Slag Aggregate and Volume Stability of Mixture: A Review. *China J. Highw. Transp.* **2021**, *34*, 166–179.
7. Kambole, C.; Paigge Green, P.; Kupolati, W.K.; Ndambuki, J.M.; Adeboje, A.O. Basic Oxygen Furnace Slag for Road Pavements: A Review of Material Characteristics and Performance for Effective Utilisation in Southern Africa. *Constr. Build. Mater.* **2017**, *148*, 618–631. [CrossRef]
8. Zpa, B.; Pm, A.; Mrk, A.; Mb, A.; Sh, B.; Ldp, A. Urban Mining for Asphalt Pavements: A Review—ScienceDirect. *J. Clean. Prod.* **2021**, *280*, 124916.
9. Li, C.; Chen, Z.W.; Xie, J.; Wu, S.P.; Xiao, Y. A Technological and Applicational Review on Steel Slag Asphalt Mixture. *Mater. Rep.* **2017**, *31*, 86–95, discussion 122.
10. Mo, L.T.; Lin, S.; Meng, X.Y.; Qu, L.C.; Chang, W.W.; Xiao, Y. Volume Expansion Characteristics and Crack Simulation of Steel Slag. *China J. Highw. Transp.* **2021**, *34*, 180–189.
11. Wang, C. Study on the Surface Modification Technology of Steel Slag and the Performance of Modified Steel Slag Asphalt Mixture. Master's Thesis, Kunming University of Science and Technology, Kunming, China, 2018. (In Chinese)
12. Xu, H.Q.; Wu, S.P.; Li, H.C.; Zhao, Y.C.; Lv, Y. Study on Recycling of Steel Slags Used as Coarse and Fine Aggregates in Induction Healing Asphalt Concretes. *Materials* **2020**, *13*, 889. [CrossRef]

13. Liu, J.Z.; Yu, B.; Hong, Q.Z. Molecular dynamics simulation of distribution and adhesion of asphalt components on steel sla. *Constr. Build. Mater.* **2020**, *255*, 119332. [CrossRef]
14. Zhou, M.K.; Xu, C.; Xiao, C. Studies on the Volumetric Stability and Mechanical Properties of Cement-Fly-Ash-Stabilized Steel Slag. *Materials* **2021**, *14*, 495. [CrossRef]
15. Zeng, H.M.; Liu, Z.C.; Wang, F.Z. Effect of Accelerated Carbonation Curing on Mechanical Property and Microstructure of High Volume Steel Slag Mortar. *J. Chin. Ceram. Soc.* **2020**, *48*, 1801–1807.
16. Yao, H.S.; Chen, S.J.; Chen, D.W.; Hamdy, M.N.; Hou, G.H. Study on Soundness of Steel Slag Block under Accelerated Carbonation. *Bull. Chin. Ceram. Soc.* **2020**, *39*, 187–193.
17. Huo, B.B.; Li, B.L.; Chen, C.; Zhang, Y.M. Surface Etching and Early Age Hydration Mechanisms of Steel Slag Powder with Formic Acid. *Constr. Build. Mater.* **2021**, *280*, 122500. [CrossRef]
18. Huo, B.B.; Li, B.L.; Chen, C.; Zhang, Y.M. Mechanism and Properties of Glacial Acetic Acid Modified Steel Slag Powder via Dry Chemical Modification Method. *J. Chin. Ceram. Soc.* **2021**, *49*, 948–954.
19. Ding, Y.C.; Cheng, T.W.; Liu, P.C.; Lee, W.H. Study on the Treatment of BOF Slag to Replace Fine Aggregate in Concrete. *Constr. Build. Mater.* **2017**, *146*, 644–651. [CrossRef]
20. Chen, Z.W.; Gong, Z.L.; Jiao, Y.Y.; Wang, Y.; Shi, K.; Wu, J.C. Moisture Stability Improvement of Asphalt Mixture Considering the Surface Characteristics of Steel Slag Coarse Aggregate. *Constr. Build. Mater.* **2020**, *251*, 118987. [CrossRef]
21. Ma, L.L.; Xu, D.B.; Wang, S.Y.; Gu, X.Y. Expansion Inhibition of Steel Slag in Asphalt Mixture by a Surface Water Isolation Structure. *Road Mater. Pavement Des.* **2019**, *21*, 2215–2229. [CrossRef]
22. Li, J.J.; Zhao, S.W.; Song, X.Q.; Ni, W.; Mao, S.L.; Du, H.H.; Zhu, S.T.; Jiang, F.X.; Zeng, H.; Deng, X.J.; et al. Carbonation Curing on Magnetically Separated Steel Slag for the Preparation of Artificial Reefs. *Materials* **2022**, *15*, 2055. [CrossRef]
23. Masilamani, A.; Ramalingam, M.; Kathirvel, P.; Murali, G.; Vatin, N.L. Mechanical, Physico-Chemical and Morphological Characterization of Energy Optimised Furnace (EOF) Steel Slag as Coarse Aggregate in Concrete. *Materials* **2022**, *15*, 3079. [CrossRef]
24. Du, J.; Liu, J.X.; Li, M. Determination of Free Calcium Oxide in Steel Slags by Combination of Glycol-EDTA Titrimetry and Thermogravimetry-differential Thermoanalysis (TG-DTA). *Phys. Test. Chem. Anal. (Part B Chem. Anal.)* **2013**, *49*, 961–964.
25. *GB/T 24175-2009*; Test Method for Stability of Steel Slag. China Standards Press: Beijing, China, 2009.
26. Chen, Z.W.; Xie, J.; Xiao, Y.; Chen, J.Y.; Wu, S.P. Characteristics of Bonding Behavior between Basic Oxygen Furnace Slag and Asphalt Binder. *Constr. Build. Mater.* **2014**, *64*, 60–66. [CrossRef]
27. *JTG E20-2011*; Standard Test Methods of Bitumen and Bituminous Mixtures for Highway Engineering. China Standards Press: Beijing, China, 2011.
28. Wang, G.; Wang, Y.; Gao, Z. Use of Steel Slag as A Granular Material: Volume Expansion Prediction and Usability Criteria. *J. Hazard. Mater.* **2010**, *184*, 555–560. [CrossRef]
29. Bernard, E.; Dipayan, J. Forces of Hydration that can Cause Havoc in Concrete. *Concr. Int.* **2003**, *25*, 51–57.
30. Dong, C.L. Expansion Inhibition Mechanism of Steel Slag with Hydrophobic Surface and Properties of Its Asphalt Mixture. Master's Thesis, Southeast University, Nanjing, China, 2019. (In Chinese)
31. Choi, S.Y.; Yang, E.I. An Experimental Study on Alkali Silica Reaction of Concrete Specimen Using Steel Slag as Aggregate. *Appl. Sci.* **2020**, *10*, 6699. [CrossRef]
32. Gao, Z.X.; Shen, A.Q.; Zhai, C.W.; Guo, Y.C.; Yu, P. Determination of Volumetric Parameters and Impacting Mechanism of Water Stability for Steel Slag Asphalt Mixture. *J. Traffic Transp. Eng.* **2018**, *18*, 1–10.
33. Dorian, P.G.; Rubén, V.M.; Margarita, H.E. Solar Photoassisted Advanced Oxidation of Synthetic Phenolic Wastewaters Using Ferrioxalate Complexes. *Sol. Energy* **2008**, *83*, 306–315.
34. Ambikadevi, V.R.; Lalithambika, M. Effect of Organic Acids on Ferric Iron Removal From Iron-stained Kaolinite. *Appl. Clay. Sci.* **2000**, *16*, 133–145. [CrossRef]
35. Mikhail, S.A.; Turcotte, A.M. Thermal Behaviour of Basic Oxygen Furnace Waste Slag. *Thermochim. Acta.* **1995**, *263*, 87–94. [CrossRef]
36. Guo, R.X.; Wang, C.; Niu, Z.L.; Yan, F.; Xia, H.T.; Liu, X.Y. Effect of Water Erosion on the Dynamic Modulus of Steel Slag Asphalt Mixture. *Bull. Chin. Ceram. Soc.* **2018**, *37*, 166–172.
37. Li, J.R.; Tang, B.M.; Liu, R.Q.; Li, C.; Chen, F. Influence of Incineration Fly Ash Granulated Particles on Water Stability of Asphalt Mixture. *J. Build. Mater.* **2021**, *24*, 1048–1053.
38. Zhang, H.N.; Xu, A.J.; Cui, J.; He, D.F.; Tian, N.Y. Overview and Trend of Steel Slag Recycling and Reuse. *Steelmaking* **2012**, *28*, 74–77.
39. Deng, K.; Wu, K.; Shi, H.S. Status and Progress of Resource Utilization of Steel Slag in Cement and Concrete. *Res. Appl. Build. Mater.* **2015**, *1*, 7–10, 13.
40. Chen, W.; Wei, J.C.; Xu, X.Z.; Zhang, X.M.; Han, W.Y.; Yan, X.P.; Hu, G.L.; Lu, Z.Z. Study on the Optimum Steel Slag Content of SMA-13 Asphalt Mixes Based on Road Performance. *Coatings* **2021**, *11*, 1436. [CrossRef]
41. Zhang, H.T.; Gong, M.Y.; Yang, B.; Yang, H.S. Comparative Study on Road Performance of Different Composite Modified Asphalt Mixtures. *J. Highw. Transp. Res. Dev.* **2019**, *36*, 1–6.
42. Xie, J.G.; Yang, C.X.; Fang, X.Q.; He, Y.F.; Xu, H.D. Cooling Attribute and Road Performance of Infrared Radiation Asphalt Mixture. *J. Nanjing Univ. Aeronaut. Astronaut.* **2011**, *43*, 105–109.

43. Wu, Z.M.; Gao, P.W.; Chen, D.F.; Zhang, H.B.; Chen, X.; Lin, Y.C. Effect of Chlorine Deicers on Performance of Low Temperature Crack Resistance of Asphalt Mixture. *Highw. Eng.* **2012**, *37*, 26–30.
44. Gu, H.X.; Wu, Q.S.; Wu, Y.; Min, Z.A. Strength and Microstructure of Carbonated Cement- Steel Slag Composite Gelled Material. *J. Mater. Sci. Eng.* **2019**, *37*, 35–39.
45. Arvaniti, E.C.; Lioliou, M.G.; Paraskeva, C.A.; Payatakes, A.C.; Ostvold, T.; Koutsoukos, P.G. Calcium Oxalate Crystallization on Concrete Heterogeneities. *Chem. Eng. Res. Des.* **2010**, *88*, 1455–1460. [CrossRef]
46. Petrova, E.V.; Gvozddev, N.V.; Rashkovich, L.N. Growth and Dissolution of Calcium Oxalate Monohydrate Crystals. *J. Optoelectron. Adv. M.* **2004**, *6*, 261–268.
47. Lin, R.S.; Kim, T.; Wang, X.Y.; Du, W. Potential Application of MoS_2 Nanoflowers as Photocatalysts in Cement: Strength, Hydration, and Dye Degradation Properties. *J. Clean. Prod.* **2022**, *330*, 129947. [CrossRef]
48. Doherty, B.; Pamplona, M.; Selvaggi, R.; Miliani, C.; Matteini, M.; Sgamellotti, A.; Brunetti, B. Efficiency and resistance of the artificial oxalate protection treatment on marble against chemical weathering—ScienceDirect. *Appl. Surf. Sci.* **2007**, *253*, 4477. [CrossRef]
49. Lin, C.J.; Dai, W.J.; Li, Z.F.; Wang, Y.S. Study on the Inorganic Synthesis from Recycled Cement and Solid Waste Gypsum System: Application in Grouting Materials. *Constr. Build. Mater.* **2020**, *251*, 118930. [CrossRef]
50. Gruskovnjak, A.; Lothenbach, B.; Winnefeld, F.; Figi, R.; Ko, S.C.; Adler, M.; Mader, U. Hydration Mechanisms of Super Sulphated Slag Cement. *Cem. Concr. Res.* **2008**, *38*, 983–992. [CrossRef]
51. Huang, X.Q.; Zhao, X.R.; Bie, S.Q.; Yang, C.F. Hardening Performance of Phosphogypsum-Slag-Based Material. *Procedia Environ. Sci.* **2016**, *31*, 970–976. [CrossRef]
52. Lin, R.S.; Wang, X.Y.; Yi, H. Effects of Cement Types and Addition of Quartz and Limestone on the Normal and Carbonation Curing of Cement Paste. *Constr. Build. Mater.* **2021**, *305*, 124799. [CrossRef]
53. Lin, R.S.; Han, Y.; Wang, X.Y. Macro–Meso–Micro Experimental Studies of Calcined Clay Limestone Cement (LC3) Paste Subjected to Elevated Temperature. *Cem. Concr. Compos.* **2020**, *116*, 103871. [CrossRef]
54. Zhang, X.S.; Li, H.B.; Li, S.; Ding, Y.F.; Zhang, H.B.; Tong, Y.F.; Hua, S.D. Test and Microstructural Analysis of a Steel Slag Cement-Based Material Using the Response Surface Method. *Materials* **2022**, *15*, 3114. [CrossRef]
55. Podoll, R.T.; Becker, C.; Irwin, K.C. *Surface Analysis by Laser Ionization of the Asphalt Aggregate Bond*; March Progress Report SHRP-87-AIIR-07; SRI International: Menlo Park, CA, USA, 1991.
56. Tarrer, A.R.; Wagh, V. *The Effect of the Physical and Chemical Characteristics of the Aggregate on Bonding*; Strategic Highway Research Program; National Research Council: Washington, DC, USA, 1991.

materials

MDPI

Article

Study on the Stability of Bio-Oil Modified Prime Coat Oil Based on Molecular Dynamics

Shuang Shi [1], Lanqin Lin [1], Zhaoguang Hu [2], Linhao Gu [1,*] and Yanning Zhang [1]

1 School of Transportation, Southeast University, Southeast University Road #2, Nanjing 211189, China
2 China Road and Bridge Corporation, 88 Outer Andingmen Street, Beijing 100011, China
* Correspondence: gu_linhao@163.com

Abstract: To explore the effect of different emulsifier contents on the stability performance of biomass-emulsified asphalt, three types of emulsified asphalt with 1%, 3%, and 5% anionic emulsifiers were prepared and analyzed by molecular dynamics simulation and macroscopic experiments. Firstly, we used molecular simulation software (Material Studio, MS) to construct a model of biomass-emulsified asphalt with different emulsifier contents and analyzed the microscopic mechanism of the emulsifier to improve the stability of the emulsified asphalt by the radial distribution function, interaction energy, interfacial layer thickness, and solubility parameters of the emulsified asphalt system with different emulsifier contents. The results were validated by macro and micro tests including storage stability, particle size determination, and infrared spectroscopy. The results show that at low emulsifier contents, the emulsifier can reduce the interfacial tension between the oil–water interface and expand the transition region between the two phases (interfacial layer thickness), which will prevent interparticle agglomeration and reduce the emulsion particle size, thus reducing the settling rate and ensuring the stability of the emulsion. When the emulsifier content is further increased beyond the critical micelle concentration, the emulsifiers will agglomerate with each other and show larger peaks in the radial distribution function, and the phenomenon of emulsifier agglomeration will appear in the five-day storage stability test, resulting in a corresponding decrease in the proximity of the infrared absorption peak area ratio in the same wavelength band of the upper and lower layers of the biomass-emulsified asphalt, and the emulsion stability decreases instead.

Keywords: road engineering; stability; molecular dynamic; bio-oil asphalt/water emulsion; particle size; infrared spectrum analysis

Citation: Shi, S.; Lin, L.; Hu, Z.; Gu, L.; Zhang, Y. Study on the Stability of Bio-Oil Modified Prime Coat Oil Based on Molecular Dynamics. *Materials* **2022**, *15*, 6737. https://doi.org/10.3390/ma15196737

Academic Editors: Qiang Li, Jiaqing Wang, Kaijian Huang, Dongdong Ge and Fangyuan Gong

Received: 6 September 2022
Accepted: 26 September 2022
Published: 28 September 2022

Publisher's Note: MDPI stays neutral with regard to jurisdictional claims in published maps and institutional affiliations.

Copyright: © 2022 by the authors. Licensee MDPI, Basel, Switzerland. This article is an open access article distributed under the terms and conditions of the Creative Commons Attribution (CC BY) license (https://creativecommons.org/licenses/by/4.0/).

1. Introduction

As the country's economy continues to develop, the convenience and comfort of transportation have become key factors in determining the further improvement of the country's economy. In recent years, China's transportation system has also become more and more developed, road density is increasing, and the construction of highways is becoming more and more advanced. However, the material properties between the two structural layers of roads have large differences, and the layers are not in a completely continuous state; when subjected to a large horizontal shear stress, the surface layer will occur nudging, cracking, and other pavement deteriorations. Therefore, the current highway construction process often uses prime coat oil construction for interlayer disposal to achieve the following functions: First, the consolidation of loose aggregates on the surface of the grass-roots level improves the surface strength of the base layer level. Second, sealing the grass-roots level and playing a role in waterproofing. Third, increasing the adhesion between the base layer level and the surface layer ensures continuity between the base layer, so that the neutral surface is always located in the semi-rigid base layer, allowing the asphalt surface layer to withstand the larger layer bottom bending, tensile stress, and fatigue damage [1–4].

However, the prime coat oil needs to have a certain storage stability and sufficient penetration depth to ensure the use of the construction process and the pavement structure to play the above role. An asphalt/water emulsion is more stable after breaking the emulsion, will not volatilize, and has high safety and good economic and environmental benefits; furthermore, with the continuous optimization of the production process of asphalt/water emulsion and continued improvement of the production process, the application of an asphalt/water emulsion in the road structure has become more widespread, so in the prime coat oil, an asphalt/water emulsion has also begun to replace kerosene-diluted asphalt to optimize road performance [5,6].

In addition, with the implementation of the "double carbon" plan advocated by the state, the recycling of resources and the environmental protection of resources have put forward higher requirements, and the extraction of biomass materials from animal and plant remains, restaurant waste oil, used handicrafts, and other resources has gradually attracted the attention of the road engineering industry [7]. Researchers at home and abroad have modified asphalt with biomass materials, and the feasibility, self-healing, anti-aging, and anti-rutting properties of biomass asphalt have been fully investigated from multiple perspectives, and it has been found that biomass materials can indeed enhance some specific properties of asphalt materials [8–12]. Through preliminary research and experimental studies, it was found that adding biomass oil as a penetrating agent to asphalt/water emulsions can also effectively enhance the penetration depth of prime coat oils, but there is a lack of corresponding micro-level studies around how much emulsifier dosage can make the biomass prime coat oil achieve better stability overall. If the stability of biomass prime coat oils can be investigated from a microscopic point of view with different emulsifier contents, a positive impact on the optimization of prime coat oils and their application in asphalt pavements will be achieved. For microscopic-level research, molecular dynamics has simulation characteristics at the nanoscale, which can simulate the physical motion of system atoms or molecules and then effectively calculate the intermolecular interactions and system energy, and the method can track the dynamic evolution process of complex systems over time and reveal the adsorption properties and action mechanism of molecules at interfaces in minute time scales [13]. Fan et al. [14] used molecular simulation software to construct a water/emulsifier/asphalt system for molecular dynamics simulation and studied the effect of the interaction between asphalt and emulsifier on asphalt–water interface properties by parameters such as interface formation energy, interfacial layer thickness, and diffusion coefficient. He et al. [15] conducted a comprehensive analysis of the molecular dynamics of an asphalt system, constructed the general equilibrium steps of the molecular dynamics model of asphalt systems, proposing to verify the validity of the model with density, glass transition temperature, viscosity, etc., and summarized the progress of the molecular dynamics approach to study the behavior mechanism of asphalt nano aggregation, self-healing, and modification. Zhang et al. [16] analyzed the "self-healing" mechanism of graphene-microencapsulated asphalt by infrared spectroscopy, fluorescence microscopy, and molecular dynamics simulation. Yu et al. [17] added bio-based epoxy soybean oil to cold mix epoxy asphalt and found that the difference between the cohesive energy density and solubility of asphalt and epoxy monomer decreased, and the interaction energy of the whole system increased by Fourier variation infrared spectroscopy and molecular dynamics simulation. Kapyelata et al. [18] utilized the simulation technique to study the carbon nanotube/recycled polyethylene asphalt interaction and found that carbon nanotubes can promote the distribution of asphaltene fractions, thus improving the storage stability of the recycled polyethylene-modified asphalt. Based on the above study, this paper will investigate the effect of emulsifier content on the stability of a biomass asphalt/water emulsion based on the molecular dynamic (MD) simulation technique combined with macro and micro tests. At present, most studies on the stability of asphalt/water emulsions based on molecular dynamics focus on how the emulsifier reduces the interfacial tension between the asphalt and water phases, and the commonly used characterization parameters are interaction energy, interfacial layer thickness, and compatibility parameters. In this paper, the effect of emulsifier dosage on the stability of a biomass anionic asphalt/water emulsion will be investigated from three perspectives, namely, interaction energy, interfacial layer thickness, and radial distribution function, using MS, and the optimal emulsifier dosage will be determined and validated by macro and microscopic tests of storage stability, particle size analysis, and IR spectroscopy.

2. Test Material and Microscopic Model

2.1. Test Materials and Instruments

2.1.1. Instruments and Sample Preparation

In this paper, three kinds of macro or micro tests were involved, namely, infrared spectroscopy, laser particle size measurement, and storage stability experiments. For the accuracy of the test results, three parallel tests were conducted for each test, and the average of the three test results was taken as the final result.

For the infrared tests, we used a Nicolet iS10 Fourier infrared spectrometer, and the minimum resolution was 0.5 cm^{-1}, the experimental parameters were set as 64 scans and 4 cm^{-1} resolutions, and then omic software was used to perform baseline correction and smoothing on the obtained IR spectral curves. Finally, the inflection point tangent was used to measure the infrared characteristic peak area of the pre-processed IR spectral map.

For particle size data acquisition, an LS200 laser particle sizer was used in this paper. The instrument has an analysis range of 0.1 μm to 750 μm and can derive parameters such as refractive index as well as shading ratio for various media. The test parameters were set as follows: the background sampling time was 10 s, the single sampling time was also set to 10 s, and the result type was volume (V).

Storage stability experiments were performed by injecting 250 mL of emulsified bitumen solution into stability test tubes and leaving them on a smooth platform for 5 days. After five days, the emulsified bitumen solution was exported through the upper and lower branch tubes and evaporated. Finally, the storage stability of the asphalt/water emulsion was determined through the upper and lower branches of the asphalt/water emulsion solid content difference.

2.1.2. Materials

In this study, Korea SK-70# asphalt was chosen as the base asphalt; its physical properties are shown in Table 1. All indicators met the relevant technical regulations in China.

Table 1. Conventional indexes of base asphalt.

	25 °C Penetration (0.1 mm)	25 °C Ductility (cm)	Softening Point (°C)	60 °C Brinell Viscosity (Pa·s)
Test results	63.7	>100	47	203
Index requirements	60–80	>100	>46	>160
Test method	GB/T0606-2011	GB/T0605-2011	GB/T0606-2011	GB/T0625-2011

In addition, this paper used self-developed anionic emulsifier to prepare an asphalt/water emulsion, the main component was which was sodium dodecyl diphenyl ether desulphonated. According to the existing basic test results, 3% of the emulsifier content of the asphalt/water emulsion samples prepared more uniform particles and the emulsification effect was better, so in the test group, 1%, 3%, and 5% were set as the emulsification content.

Biomass oil was produced independently, and its main components were soft fatty acid, linolenic acid, oleic acid, and stearic acid, and the specific gravity ratio of the components was 14:33:37:15. The best dosing was studied through indoor tests and was determined as asphalt:biomass oil = 10:1.

2.2. Molecular Model of Biomass Asphalt/Water Emulsion

2.2.1. Determination of Asphalt/Water Emulsion Composition and Model Proportioning Calculation

In this paper, the average molecular structure model of asphalt consisting of twelve components proposed by Li et al. based on SARA four-component analysis was chosen as the molecular model of matrix asphalt [19–21], and the ratio of the twelve components was adjusted according to the four-component ratio of SK-70# matrix asphalt [22] (Table 2), and the final molecular composition of each component's molecular composition is shown in Table 3. Then the emulsifier molecules, water molecules, and biomass oil molecules were assigned according to 50% solids content and 10:1 asphalt/biomass oil, and the asphalt/water emulsions with 1%, 3%, and 5% emulsifier content were established, respectively, with the specific molecular parameters shown in Table 4.

Table 2. Proportion of four components of SK-70 base asphalt (%) [22].

Types of Asphalt	Asphaltene	Saturate	Aromatics	Resin
SK-70#	12.44	26.62	44.62	16.32

Table 3. Model parameters of 12-component base asphalt.

Components		Molecular Formula	Number of Molecules	Model Component(%)	Actual Component (%)	Proportion Error (%)
Asphaltene	Phenol	$C_{42}H_{54}O$	1	13.28	12.44	0.84
	Pyrrole	$C_{66}H_{81}N$	1			
	Thiophene	$C_{51}H_{62}S$	1			
Aromatics	Quinolinohopa-ne	$C_{40}H_{59}N$	2	26.93	26.62	0.31
	Thioisorenicrat anc	$C_{40}H_{60}S$	1			
	Benzobisbenz-othiophene	$C_{18}H_{10}S_2$	3			
	Pyridinohopan-e	$C_{36}H_{57}N$	2			
	Trimethylbenz-eneoxane	$C_{29}H_{50}O$	2			
Resin	PHPN	$C_{35}H_{44}$	14	44.17	44.62	0.45
	DOCHN	$C_{30}H_{46}$	2			
Saturate	Squalane	$C_{30}H_{62}$	4	15.91	16.32	0.41
	Hopane	$C_{35}H_{62}$	2			

Table 4. Molecular parameters of other components of bio-oil asphalt/water emulsion.

Components		Molecular Formula	Molecular Weight (g/mol)	Number of Molecules	Ratio to Asphalt (%)
Emulsifier	1%	$C_{24}H_{32}O_7Na_2S_2$	542	1	2.7
	3%			3	6.5
	5%			5	10.3
H_2O	1%	H_2O	18	904	97.3
	3%			867	94.1
	5%			826	89.6
Biomass of oil	Hexadecenoic acid	$C_{16}H_{32}O_2$	256.4	1	9.1
	Linolenic acid	$C_{18}H_{32}O_2$	280.5	2	
	Oleic acid	$C_{18}H_{34}O_2$	282.5	2	
	Stearic acid	$C_{18}H_{36}O_2$	284.5	1	

2.2.2. Molecular Modeling of Asphalt/Water Emulsion

In this study, the kinetic calculation of molecular systems was carried out on the Material studio 2021 computational simulation platform, and the COMPASS II force field environment was selected, which can classify organic and inorganic molecular systems. Firstly, the molecular models were constructed under the Visualizer module of Material studio, according to Figure 1, and the bond angles and bond lengths were adjusted using the Clean function. Then, the Amorphous Cell module was used to encapsulate the molecular models by the number of ratios in Tables 3 and 4, and finally the Geometry Optimize function in the Forcite module was used to optimize each box to achieve the most stable configuration (lowest molecular energy).

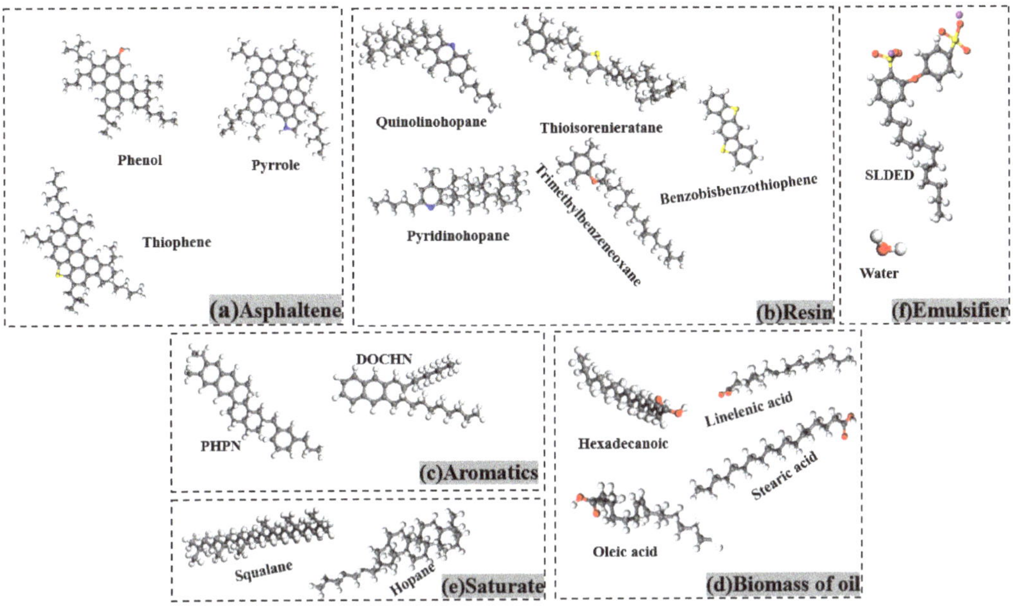

Figure 1. Composition diagram of bio-oil asphalt/water emulsion model.

In addition, due to the relatively complex configuration of biomass asphalt, it was necessary to anneal it to eliminate the unreasonable internal structure, and finally to perform two separate kinetic calculations of different systems to make it closer to the actual material properties and ensure the accuracy of the simulation. To make sure that the emulsifier could better achieve the effect of reducing the interfacial tension between water and biomass asphalt, the build layer's function was used to assemble the three components in the order of water, emulsifier, and biomass asphalt, and in the last, a vacuum layer of 30 Å was reserved to prevent the effect of periodic boundary conditions.

3. Emulsifier Dosing on the Stability of the Biomass Asphalt/Water Emulsion

3.1. Radial Distribution Function

Before conducting the molecular dynamics simulations, this paper first analyzed the effect of emulsifiers on matrix asphalt roughly by infrared spectroscopy; the test results are shown in Figure 2.

The common characteristic peaks of each control group in the figure were attributed as follows: the anti-symmetric stretching vibration of methylene (–CH_2–) near wave number 2948 cm^{-1} and the symmetric stretching vibration of –CH_2– near wave number 2849 cm^{-1}, indicating that the alkyl chains in the alkane compounds in the control group were arranged in an orderly manner at this time. The alkane methylene (–CH_3) asymmetric variable angle vibration was near wave number 1455 cm^{-1}, the alkane methylene symmetric variable angle vibration peak was near wave number 1375 cm^{-1}, and wave numbers 803 cm^{-1}, 745 cm^{-1}, and 727 cm^{-1} were the range of wave numbers in the fingerprint region, which is the benzene ring substitution region, and the characteristic peaks appearing in this range were caused by the C–H bond on the benzene ring swinging out of plane vibrations. In addition to the same characteristic peaks mentioned above, it was seen that the matrix asphalt had a double-peak structure at wave numbers 1536 cm^{-1} and 1574 cm^{-1} for the stretching vibration of the double bond and the bending vibration of H–O, H–N. However, with the addition of the emulsifier and the increase of dosage, it was seen that the double peaks gradually evolved into single peaks and move towards the vicinity of wave number 1593 cm^{-1}. The reason for this situation may be caused by the induction of the C=C double bond of the benzene ring in the asphalt with the O atom of the emulsifier molecule, resulting in the high spectral band of the splitting to about 1593 cm^{-1}. The evaporated residues of the biomass asphalt/water emulsion around wave numbers 1151 cm^{-1} and 1124 cm^{-1} showed wave peaks at these two locations, and the peak intensity became progressively larger with increasing emulsifier content as a result of SO3 symmetric stretching vibrations of aromatic sulfonates in more emulsifiers. Furthermore, near the wave number 628 cm^{-1}, with the increase of emulsifier dosage, there was a wave peak with a gradual increase of signal intensity, which is part of the C–S stretching vibration. Simulation results also proved that 3% emulsifier content was near the critical micelle concentration of the anionic emulsifier used in this paper, which had good stability performance and could guide the amount of emulsifier.

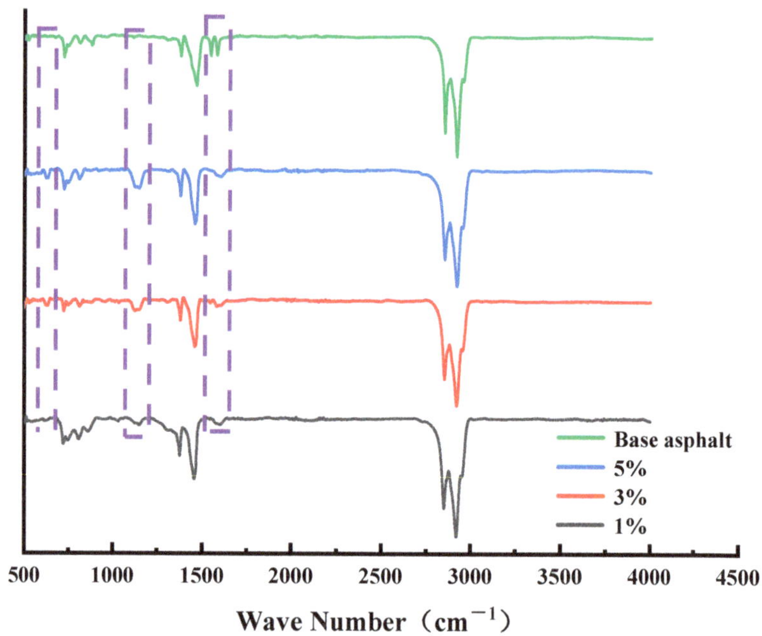

Figure 2. Infrared spectra of biomass asphalt/water emulsion with different emulsifier contents (The purple dashed line represents the main characteristic peak changes occurring in the emulsified asphalt relative to the matrix asphalt).

This paper first used the molecular simulation method to analyze the radial distribution function of the biomass asphalt/water emulsion; the radial distribution function characterizes the probability of occurrence of other particles around a particle and thus can be used to reflect the stacking of the polymer model. In this paper, the radial distribution function was calculated for three emulsifier contents of a biomass asphalt/water emulsion, and the results are shown in Figure 3.

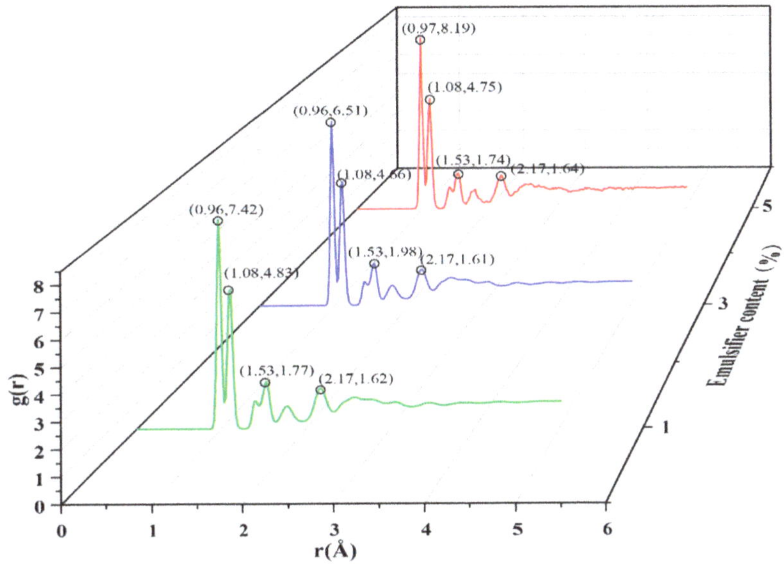

Figure 3. Radial distribution function of asphalt/water emulsion with different emulsifier contents.

As seen in Figure 3, the overall trend of the radial distribution function corresponding to the three emulsifier contents was the same, all forming four more obvious peaks, with the first and second peaks being significantly larger than the peaks of the other two peaks. The first peak position of these contents of emulsifier was located at 0.96 Å, and the peak size showed that 1% > 5% > 3%, while for the other peaks, respectively, they appeared at 1.08 Å, 1.53 Å, and 2.17 Å, and the values were basically similar. From the results of the radial distribution function, it can be seen that the agglomeration phenomenon of biomass asphalt/water emulsion appeared to be weakened and then enhanced during the increase of emulsifier addition, which is because the emulsifier has hydrophilic chain segments at one end and hydrophobic chain segments at the other end with good surface activity, and its adsorption on the two-phase interfacial film is enhanced with the increase of emulsifier dosage, and the emulsifier spontaneously generates a water interface between the two phases, which makes the surface-free energy function of the emulsion as low as possible to give the emulsion high stability. However, as the emulsifier concentration increases to a certain value, the emulsifier molecules start to cluster to form aggregates, with the lipophilic chain segments clustering together and the hydrophilic chain segments reaching outward to the aqueous phase to form the so-called micelles, and the lowest emulsifier concentration that can form micelles is the critical micelle concentration, referred to as the CMC value [23–25]. As shown in Figure 4, the biomass asphalt and water molecules of the biomass asphalt/water emulsion after molecular dynamics simulations were hidden, and through observation of the distribution and structure of the emulsifier, it found that with the increase of the emulsifier dosage, at 5% emulsifier dosage, after kinetic calculations, a part of the emulsifier molecules started to approach each other between the lipophilic groups, and the emulsifier molecules started to cluster. This conclusion is similar to Quan Xiujie's radial distribution function simulation results for emulsified asphalt with different emulsifiers. The peak value of emulsified asphalt in the first stage will increase first and then decrease with the increase of emulsifier content [26].

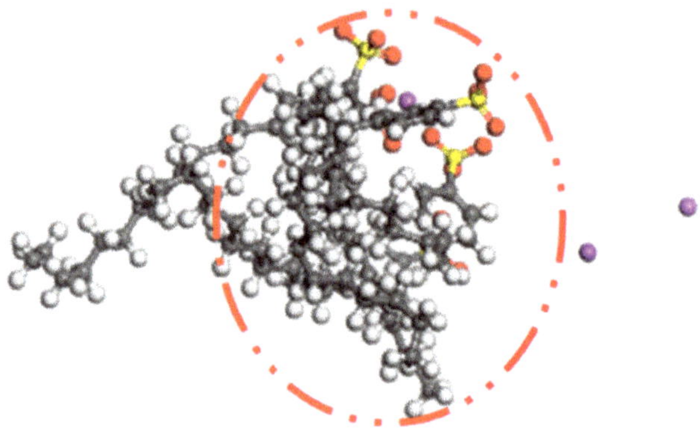

Figure 4. Agglomeration of emulsifier with 5% emulsifier content.

3.2. Interaction Energy

During molecular dynamics simulations, the intermolecular interaction energy is often used to quantitatively characterize the strength of intermolecular interactions and thus predict the compatibility between materials of the blended system. In this paper, the total intermolecular interaction energy, van der Waals interaction energy, and electrostatic interaction energy were used as evaluation indicators to evaluate the stability of the system [27–31].

The energy values of the biomass asphalt/water phase, biomass asphalt, and the water phase after kinetic simulation were each calculated by the Energy function in the Forcite module and calculated according to the interaction energy calculation equation; the results are shown in Figure 5.

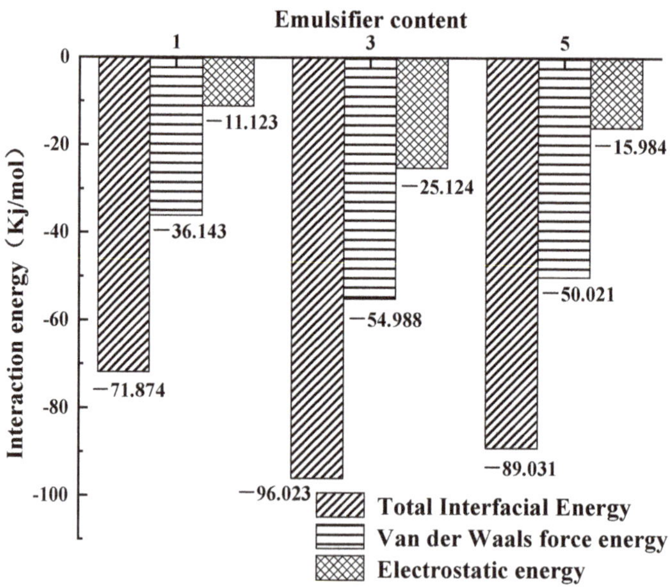

Figure 5. Interaction energy of asphalt/water emulsion with different emulsifier contents.

It can be clearly seen that the total interfacial action energy, van der Waals force action energy, and electrostatic force action energy of the biomass asphalt/water emulsion at the three emulsifier contents were all negative, indicating that the emulsifier played a role in reducing the energy at the interface between the biomass asphalt phase and the water phase at all three emulsifier dosages, thus reducing the interfacial tension between them. However, under the same simulation conditions, different emulsifier contents showed different action energies, again indicating that the emulsifier content had some influence on the stability at the interface. With the increase of the emulsifier content, all three action energies showed a trend of first increasing and then decreasing, i.e., the stability of biomass asphalt/water emulsion would become stable at the beginning due to the reduction of interfacial tension by the emulsifier, but when the amount of emulsifier further increased beyond the so-called critical micelle concentration, the whole system would become unstable. This is because the force between the hydrophilic group polar head of the emulsifier molecule and the water molecule was mainly van der Waals force, while the anionic emulsifier molecule itself had a negative charge, resulting in a strong electrostatic force of mutual attraction between it and the positively charged asphalt molecule, thus having the ability to strengthen the interaction between the oil and water phases. However, as mentioned above, when the emulsifier content was too high, the emulsifier flocculation phenomenon occurred, and then this part of the emulsifier could not play or completely play its role, the interaction between the oil and water phase weakening instead.

In addition, as the emulsifier strengthened the interaction between the oil and water phases, the oil and water phases penetrated each other, the thickness and volume of the interface layer formed at the interface increased, the activation of the emulsifier on the interface became stronger, and the stability of the asphalt/water emulsion was better [27]. Therefore, this paper further analyzed the stability situation with the help of molecular dynamics simulations to calculate the relative concentration distribution curves of biomass asphalt, water, and emulsifier to characterize the interfacial layer thickness. To quantitatively describe the effect of different emulsifier contents on the oil/water interfacial layer, the thickness of the oil/water interfacial layer was defined using the "10–90%" thickness principle in this paper [30]. The thickness range of the interfacial layer is roughly marked with yellow squares in Figure 6, and the black arrows point to the enlarged view of the interfacial layer, and in addition, the simulation and calculation results are shown in Figure 6. According to the change of temperature and emulsifier content, Fan Weiyu [14] also used molecular dynamics simulation to launch a simulation containing emulsified asphalt's asphalt/water interfacial energy and diffusion coefficient in the corresponding research. Finally, he also drew the conclusion, which is the same as the one in this paper, that the emulsifier can abate the repulsion between the asphalt/water interface, but this effect will increase with the increase of emulsifier content and then begin a weakening trend, that is, there is too much emulsifier, but it cannot make asphalt and water increase their mutual solubilization, which means a poor emulsifying effect.

As seen from the figure, water molecules and biomass asphalt molecules in the simulation both moved to the middle of the system, and the distribution area of the water phase and the oil phase appeared to have a certain degree of overlap, representing a certain degree of mutual solubility between the water phase and the oil phase. The emulsifier monolayer film was distributed in the transition region of the asphalt–water interface, because the hydrophilic group of the emulsifier had a strong van der Waals force effect with the water molecules, causing some water molecules to be adsorbed in the surrounding polar base, thus expanding the transition region of the water molecule interface. Furthermore, the emulsifier had a strong negative charge, which extended to the oil phase, and the two had a strong electrostatic adsorption effect, so that oil molecules and the lipophilic base also had a certain amount of mutual solubility, thus increasing the transition region of the oil phase, and furthermore, there was a certain amount of mutual solubility between oil and water, causing the tension of the asphalt–water interface to be reduced. Therefore, the wider the thickness of the interfacial layer, the larger the transition area between the water phase and the oil phase, the weaker the interfacial repulsion between oil and water, and the lower the interfacial tension between oil and water. The thickness of the interfacial layer increased at first with the amount of emulsifier from 1% to 3%. However, when the amount was further increased, the thickness of the interfacial layer decreased. This is because when the emulsifier content reached the critical micelle concentration, with continued increases in the emulsifier content, the emulsifier would agglomerate, and due to the stronger hydrogen bonding of water molecules to emulsifier molecules compared to asphalt molecules as Figure 7 shows, the adsorption at the interface would no longer increase, but the emulsifier would be concentrated in the water phase, and thus the interface layer thickness would tend to increase in flatness or due to the strong van der Waals forces between the emulsifier molecules, so that the original adsorption at the interface part of the emulsifier molecules occurred in the water phase, thus reducing the thickness of the interface layer. The present results also demonstrated again that 3% emulsifier content gave the best stability of the biomass asphalt/water emulsion among the three emulsifier dosages.

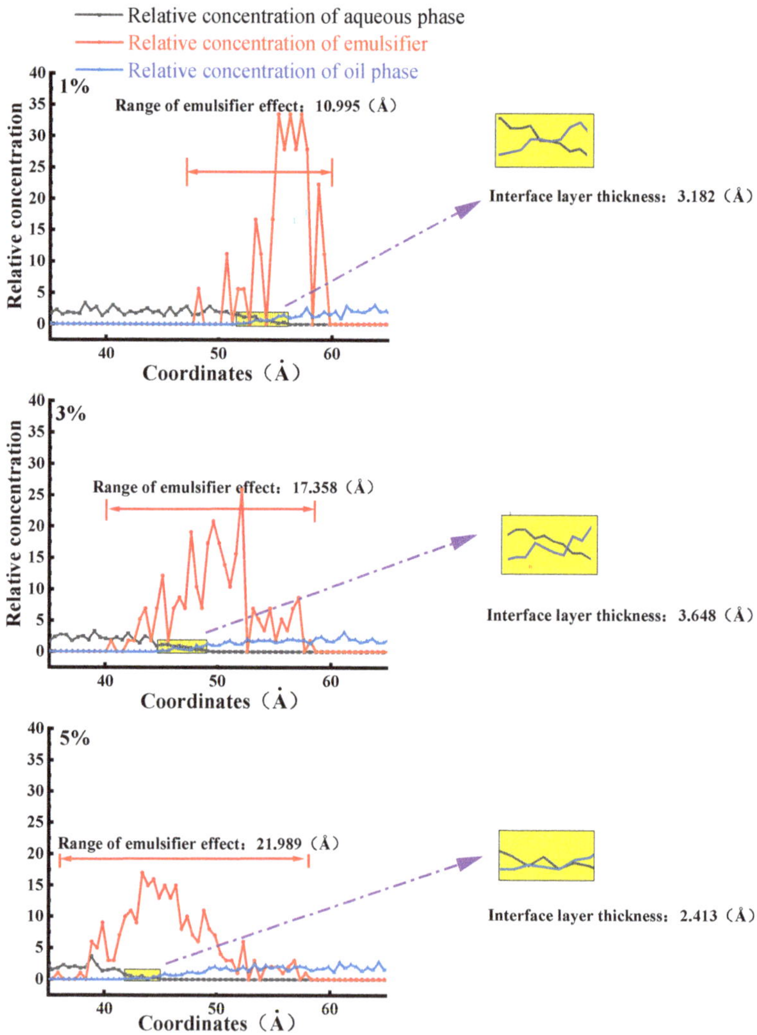

Figure 6. Concentration distribution of each component of bio-oil asphalt/water emulsion along the z-axis.

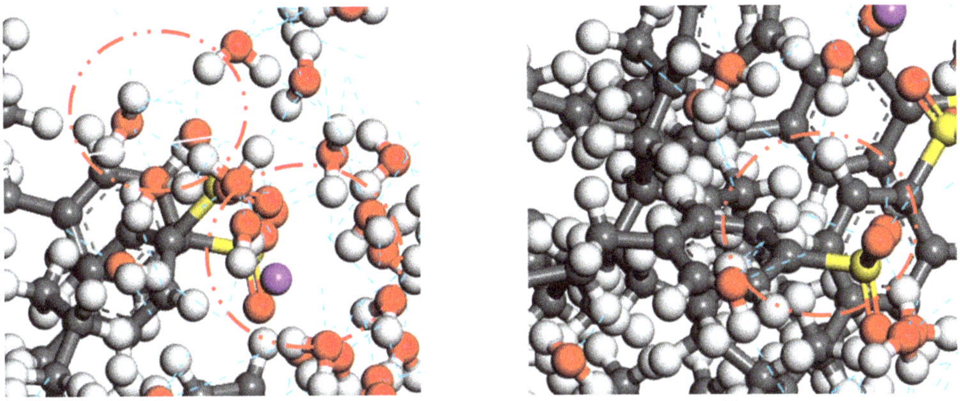

Figure 7. Hydrogen bonds between molecules of bio-oil asphalt/water emulsion. (**a**) Hydrogen bond between water and emulsifier. (**b**) Hydrogen bond between bio-oil asphalt and emulsifier.

3.3. Particle Size Measurement

According to Stokes' law, as shown in Equation (1):

$$v = \frac{2gr^2(\rho_1 - \rho_2)}{9\eta} \quad (1)$$

where v is the settling velocity of asphalt/water emulsion particles; g is the acceleration of gravity; r is the radius of asphalt/water emulsion particles; ρ_1 is the relative density of asphalt; ρ_2 is the relative density of the water phase; and η is the viscosity of the water phase.

Asphalt/water emulsion particles will settle under the action of gravity, and their settling rate is proportional to the square of the particle size, which means that the particle size of an asphalt/water emulsion has the greatest influence on its storage stability, so the stability performance of biomass asphalt/water emulsions can be evaluated by the particle size test. In this paper, the particle sizes of three biomass asphalt/water emulsions were tested by an LS-200 laser particle size meter, and the results are shown in Table 5.

Table 5. Particle size of bio-oil asphalt/water emulsion with different emulsifier contents.

Material	**Bio-Oil Asphalt/Water Emulsion**		
Emulsifier Content	**1%**	**3%**	**5%**
D_{10}	1.712	1.579	1.647
D_{20}	2.013	1.793	1.899
D_{30}	2.310	1.941	2.193
D_{40}	2.439	2.197	2.407
D_{50}	2.648	2.315	2.591
D_{60}	2.901	2.487	2.797
D_{70}	3.201	2.664	3.134
D_{80}	3.798	3.012	3.287
D_{90}	4.312	3.510	3.978
Volumetric mean particle size	2.917	2.416	2.736
Span of particle size	0.982	0.834	0.899

Different opinions have been put forward by domestic and foreign scholars as to which particle size index is a better guide for the performance of asphalt/water emulsions, mainly focusing on D_{50}, volume mean particle size, and particle size span (span) [32–37], while the definition of particle size span is shown in Equation (2):

$$Span = \frac{D_{90} - D_{10}}{D_{50}} \tag{2}$$

In this paper, these three parameters were also selected as evaluation indexes for the particle size of the asphalt/water emulsion, and the results are shown in Figure 8.

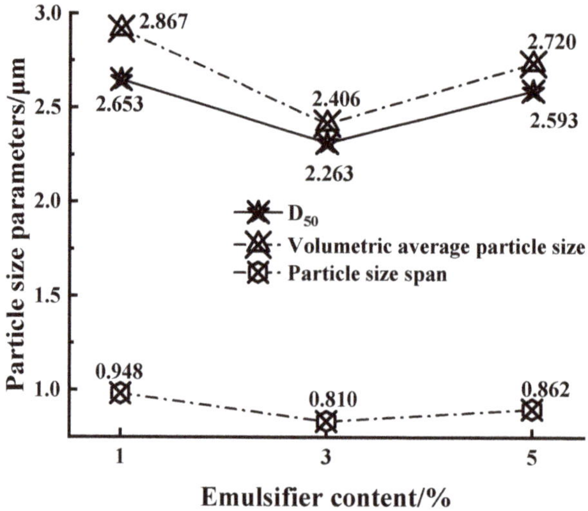

Figure 8. Particle size indexes of bio-oil asphalt/water emulsion with different emulsifier contents.

Based on previous studies, it was mostly believed that the average particle size of asphalt/water emulsion particles should be less than 2–4 μm to achieve better stability performance and permeability [32,33,37]; as shown in Figure 8, the prepared asphalt/water emulsion met the requirements in terms of particle size. However, in contrast, in terms of the D50 and the volume average particle size, it was found that with the increase of the amount of emulsifier, the two particle size indicators showed a trend of firstly decreasing and then increasing, which is consistent with the results of the interaction energy simulation, where the thickness of the interfacial layer, with an increasing content of emulsifier, firstly decreased and then increased, and there could be different degrees of thickness of the interfacial film between oil and water to prevent particle aggregation forming larger asphalt/water emulsion particles. In addition, because the emulsifier content was lower, the ionization of the emulsifier molecules on the surface of the particles with a charge, the electrostatic repulsion of the same charge to prevent asphalt particles, and, within a certain range, to increase the amount of emulsifier, could make the asphalt particles increase the charge strength, resulting in electrostatic repulsion also becoming stronger, and the storage stability of the asphalt/water emulsion becoming better. However, when the amount of emulsifier exceeded the critical micelle concentration, the particle size of the asphalt/water emulsion increased with the amount of emulsifier and became larger, because when the concentration of emulsifier reached the critical micelle concentration, the asphalt–water interfacial tension no longer decreased and tended to stabilize, and the emulsifier in addition to the asphalt surface aggregated into a film, the excess emulsifier molecules formed their own lipophilic base inward, and the hydrophilic base formed outward multi-molecular aggregates, that is, micelles or micelles. The formation of the average particle size of the emulsion increased. The particle size span reflected the uniformity of particle size distribution of the asphalt/water emulsion, and the sample with 3% emulsifier dosage also had the best uniformity at the same time. From the particle size perspective, this indicates that the biomass asphalt/water emulsion with 3% emulsifier dosage had better stability.

3.4. Storage Stability

The storage stability test mainly consisted of the sample closed stability test tube being placed on the test tube clamp at room temperature for 5 days and nights; a resting process; daily observation of the emulsion to determine whether there was delamination, precipitation, or discoloration, etc.; and finally, analyzing the upper and lower branches of the asphalt/water emulsion solid content difference to determine the storage stability of the asphalt/water emulsion, and the three different emulsifier contents of asphalt/water emulsion storage stability, as shown in Figure 9.

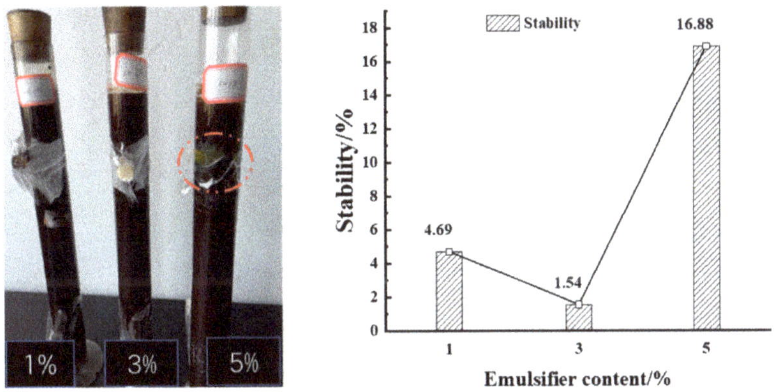

(a) (b)

Figure 9. Effect of emulsifier dosing on the storage stability performance of biomass asphalt/water emulsion. (**a**) Flocculation of Asphalt/water emulsion. (**b**) Stability.

As can be seen from Figure 9, when the emulsifier content was relatively small, the prepared asphalt/water emulsion had good storage stability and could meet the values required by the specification. With the increase of the amount of emulsifier, the storage stability of the biomass asphalt/water emulsion showed a trend of decreasing and then increasing, which was similar to the emulsifier agglomeration obtained from the simulation results of the radial distribution function, namely, the peak of the first peak of the test group with 5% emulsifier content in the simulation was significantly higher than the other two groups, which was also reflected in the stability experiments, as shown in Figure 9a, At 5% emulsifier content, the emulsifier showed a significant agglomeration into clusters. Therefore, the stability test tube test also proved that a certain degree of increase in the emulsifier content could improve the storage stability of asphalt/water emulsions, but when the emulsifier content was further increased beyond the critical micelle concentration, the asphalt/water emulsion appeared to agglomerate and flocculate, thus affecting the stability and related properties. However, the accuracy of the stability test tube test was affected by the placement and evaporation of the test tube and the lack of analysis of the changes in the composition of the asphalt/water emulsion after modification and the differences in the composition of the upper and lower branches after dissociation. Figure 10 shows the infrared spectra of the evaporated residues of the asphalt/water emulsion in the upper and lower branches after the 5 d stability test tube with different emulsifiers.

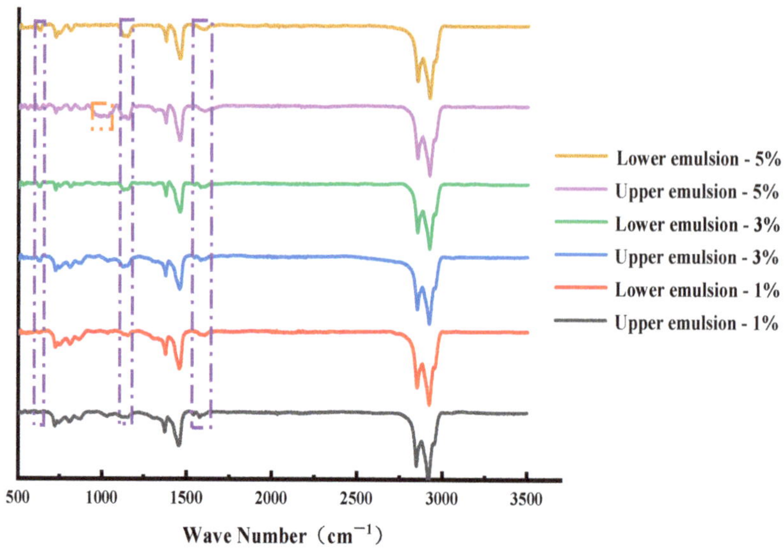

Figure 10. Infrared spectra of bio-oil asphalt/water emulsion at different layers with different emulsifier contents (The purple dashed line represents the main characteristic peak changes occurring in the emulsified asphalt relative to the matrix asphalt).

The basic wave peaks on the wave spectrum are described in Section 2.1.1 except that the evaporated residue of the upper biomass asphalt/water emulsion with 5% emulsifier content showed significant wave peaks at wave numbers 999 cm^{-1} and 1028 cm^{-1} that other controls did not have, as C–S vibrations between the S atom of the emulsifier and the C atom of the aromatic ring, which corresponded to the storage stability test where the upper 5% asphalt/water emulsion solution showed. The emulsifier agglomeration phenomenon corresponded to the phenomenon that when the emulsifier is mixed too much, the excess emulsifier will be aggregated in the water phase and thus suspended in the upper part of the biomass asphalt/water emulsion, resulting in a large gap between the upper and lower solution components and a decrease in storage stability. On the contrary, when the emulsifier content was 1% and 3%, the IR spectra of the upper and lower solutions did not show much difference, which proved that the prepared asphalt/water emulsion had good storage stability [38].

To further compare the effect of different emulsifier admixtures on the storage stability of biomass asphalt/water emulsions, the difference in the ratio of the absorption peak area corresponding to the emulsifier of the evaporated residue of the upper and lower emulsions of the storage stability test tube to the absorption peak area corresponding to the matrix asphalt was used for evaluation. The absorption peaks corresponding to the matrix asphalt were selected as wave numbers 2948 cm^{-1}, 2849 cm^{-1}, 1455 cm^{-1}, 1375 cm^{-1}, etc., while the emulsifiers were selected as the most significant wave peaks 1151 cm^{-1}, 1124 cm^{-1}, 628 cm^{-1}, etc. The characteristic peak area values and the different results are shown in Table 6 and Figure 11 [39,40].

As shown in Figure 11, the ratio of the characteristic peak area of the emulsifier to the characteristic peak area of the matrix asphalt in the upper and lower branches of the stability test tube tended to decrease and then increase with the increase of the emulsifier, which is similar to the results obtained from the storage stability test.

Table 6. Infrared characteristic peak area of evaporation residue of bio-oil asphalt/water emulsion with different emulsifier contents.

Control Group \ Wave Number (cm^{-1})	2948	2849	1455	1375	1151	1124	628
Upper 1%	693.693	431.286	779.685	137.132	26.114	16.332	6.988
Lower 1%	713.879	459.798	779.638	154.987	25.369	13.874	5.746
Upper 3%	681.125	402.146	720.987	120.126	30.999	32.007	53.365
Lower 3%	696.348	396.984	738.969	118.741	32.941	32.104	55.001
Upper 5%	642.153	431.123	790.258	132.110	86.212	64.988	51.214
Lower 5%	649.336	440.369	805.112	140.136	48.764	34.589	73.951

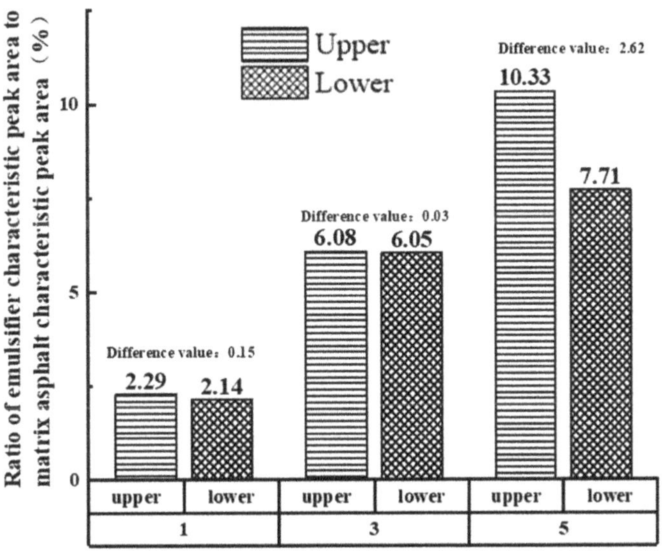

Figure 11. Proportion of emulsifier characteristic peak area in upper and lower branch tubes of storage-stabilized tubes.

4. Conclusions

1. With the increase of emulsifier dosing, the peak of the radial distribution function of biomass asphalt/water emulsions tends to decrease and then increase, which is consistent with the storage stability experimental results; however, when the emulsifier content is too large, it will lead to the phenomenon of emulsifier agglomeration.

2. Emulsifiers can promote the mutual solubility of the asphalt–water phase, increasing the thickness of the interfacial layer between the two components, and the interfacial layer can prevent the aggregation of particles, thus ensuring smaller emulsion particles and enhancing the stability of the emulsion, but when the content of emulsifier is too high, the emulsifier forms micelles, resulting in an increase in the average particle size of the emulsion, and the emulsifier does not give full play to the role of enhancing the interaction of the asphalt–water phase, making the thickness of the interfacial layer between the two components smaller, and thus reducing stability.

3. Infrared spectroscopy analysis found that the infrared characteristic peak area ratio of emulsifier to matrix asphalt of the emulsion evaporation residue of the upper and lower layers of biomass asphalt/water emulsions with different emulsifier contents showed a trend of firstly increasing and then decreasing with the increase of emulsifier content,

which is consistent with the stability performance trend derived earlier, indicating that the infrared absorption peak area ratio can evaluate the stability performance of biomass asphalt/water emulsions.

4. When biomass asphalt/water emulsion has the best emulsifier dosage, the emulsion can achieve the best stability performance, and finding the best emulsifier dosage of the corresponding emulsion can have high value for engineering applications, but the impact of emulsifier content on other properties of biomass asphalt/water emulsion is not explored in this paper; follow-up research should focus on the analysis of emulsifier content on the permeability of biomass asphalt/water emulsions and other road performance measures.

Author Contributions: Conceptualization, S.S.; methodology, L.L.; software, L.G.; validation, Y.Z.; formal analysis, S.S. and L.G.; investigation, S.S.; resources, Z.H.; data curation, S.S.; writing—original draft preparation, S.S.; writing—review and editing, Z.H.; visualization, Y.Z.; supervision, L.G.; project administration, Z.H.; funding acquisition, S.S. and L.G. All authors have read and agreed to the published version of the manuscript.

Funding: This research was funded by the fellowship of China Postdoctoral Science Foundation (2021M690613), the National Key Research and Development Project (2020YFB1600102), the National Natural Science Foundation of China (52108404) and the Natural Science Foundation of Jiangsu Province (BK20210251), and Fundamental Research Funds for the Central Universities (2242022R20062). In addition, the authors would like to thank Advanced Analysis & Testing Center of Southeast University for the assistance in experiments. The APC was funded by the Fundamental Research Funds for the Central Universities (2242022R20062).

Institutional Review Board Statement: Not applicable.

Informed Consent Statement: Not applicable.

Data Availability Statement: All data, models, and code generated or used during the study appear in the submitted article.

Conflicts of Interest: The authors declare that there is no conflict of interest related to the publication of this paper.

References

1. Zhang, X.; Gao, J.; Kong, X.; Pei, C. Reasonable selection and application of permeable layer materials for Semi-rigid base asphalt pavement. *J. Beijing Inst. Civ. Eng. Archit.* **2005**, *21*, 17–20.
2. Wang, K. Research on Permeable formation of semi-rigid base. *J. Highw. Transp. Sci. Technol.* **2000**, *17*, 23–26.
3. Cui, Z. Application of SBR modified emulsified asphalt permeable layer on Beijing-Zhang Expressway. *Natl. Def. Transp. Eng. Technol.* **2003**, *1*, 64–66.
4. Shu, Z.; Zhang, B. Application and research progress of permeable oil. *West. Leather* **2019**, *41*, 7–8.
5. Nuria, Q.; Camila, B.; Luisa, C. Storage stability of bimodal emulsions vs. monomodal emulsions. *Appl. Sci.* **2017**, *7*, 1267.
6. Ouyang, J.; Sun, Y.; Zarei, S. Fabrication of solvent-free asphalt emulsion prime with high penetrative ability. *Constr. Build. Mater.* **2020**, *230*, 117020. [CrossRef]
7. Qu, X.; Ding, H.; Wang, C.; Liu, Y.; Wang, H. Research on micro properties of bio-oil modified asphalt based on molecular dynamics simulation technique. *Mater. Rep.* **2022**, *19*, 1–11.
8. Li, N.; Zhu, Z.; Li, P. Study on applicability of biomass oil to replace asphalt for road use. *Renew. Energy Resour.* **2022**, *40*, 448–454.
9. Wang, N.; Chen, F.; Ma, T.; Luan, Y.; Zhu, J. Compaction Performance of Cold Recycled Asphalt Mixture Using SmartRock Sensor. *Autom. Constr.* **2022**, *140*, 104377. [CrossRef]
10. Zhang, J.; Zhang, H. Laboratory evaluation of waste cooking oil-based bio-oil modified asphalt binder. *J. Beijing Univ. Technol.* **2018**, *44*, 904–909.
11. Shi, K.; Fu, Z.; Song, R.M.; Liu, F.; Ma, F.; Dai, J. Waste chicken fat oil as a bio-oil regenerator to restore the performance of aged asphalt: Rheological properties and regeneration mechanism. *Road Mater. Pavement Des.* **2021**, 1–25. [CrossRef]
12. Girimath, S.; Singh, D. Effects of bio-oil on performance characteristics of base and recycled asphalt pavement binders. *Constr. Build. Mater.* **2019**, *227*, 116684. [CrossRef]
13. Zhao, Y.; Yang, Z.; Wang, J.; Li, J.; Zheng, Y. Advances in molecular dynamics simulation of self-healing behavior of asphalt binders. *Chem. Prog.* **2022**, 1–17.
14. Fan, W.; Zhao, P.; Kang, J.; Nan, G. Application of molecular simulation technology to asphalt/water emulsion study. *J. China Univ. Pet. (Nat. Sci. Ed.)* **2014**, *38*, 179–185.

15. He, L.; Li, G.; Zheng, Y.; Alexiadis, A.; Valentin, J.; Kowalski, K.J. Research progress and prospect of molecular dynamics of asphalt systems. *Mater. Rep.* **2020**, *34*, 19083–19093.
16. Zhang, E.H.; Liu, S.; Shan, L.Y.; Wang, Y.Y.; Tan, Y.Q. Study on the dual mechanism healing mechanism of graphene microencapsulated asphalt. *Chin. J. Highw.* **2022**, *35*, 91–99.
17. Yu, X.; Wang, J.; Si, J.; Mei, J.; Ding, G.; Li, J. Research on compatibility mechanism of biobased cold-mixed epoxy asphalt binder-sciencedirect. *Constr. Build. Mater.* **2020**, *250*, 118868. [CrossRef]
18. Yu, C.; Hu, K.; Yang, Q.; Wang, D.; Zhang, W.; Chen, G.; Kapyelata, C. Analysis of the storage stability property of carbon nanotube/recycled polyethylene-modified asphalt using molecular dynamics simulations. *Polymers* **2021**, *13*, 1658. [CrossRef]
19. Li, D.D.; Greenfield, M.L. Chemical compositions of improved model asphalt systems for molecular simulations. *Fuel* **2014**, *115*, 347–356. [CrossRef]
20. Xu, G.J.; Wang, H. Study of cohesion and adhesion properties of asphalt concrete with molecular dynamics simulation. *Comput. Mater. Sci.* **2016**, *112*, 161–169. [CrossRef]
21. Fardin, K.; Rajesh, K. Molecular simulations of asphalt rheology: Application of time–temperature superposition principle. *J. Rheol.* **2018**, *62*, 941–954.
22. Chen, H.; He, M.; Ji, X.; Huang, Y. Gray correlation analysis of asphalt performance and four fractions. *J. Chang. Univ. (Nat. Sci. Ed.)* **2014**, *34*, 1–6.
23. Wu, M.; Xu, G.; Luan Yi Zhu, Y.; Ma, T.; Zhang, W. Molecular dynamics simulation on cohesion and adhesion properties of the emulsified cold recycled mixtures. *Constr. Build. Mater.* **2022**, *333*, 127403. [CrossRef]
24. Li, Y.; Sang, S.; Wang, K.; Zhang, S.; Chen, L.; Zhang, Y.; Tan, J. Preparation and properties of nonionic waterborne epoxy resin emulsifier. *Paint. Coat. Ind.* **2021**, *51*, 25–32.
25. Wang, J.; Han, X.; Li, T.; Yu, J. Mechanism and Application of Emulsifiers for Stabilizing Emulsions: A Review. *Food Sci.* **2020**, *41*, 303–310.
26. Quan, X. *Effect of Hydrophilic Groups on the Stability and Demulsification Process of Dodecyl Anionic Emulsion Asphalt*; Chongqing Jiaotong University: Chongqing, China, 2021.
27. Su, M.; Zhang, H.; Zhang, Y.; Zhang, Z. Molecular Dynamics Simulation of Compatibility and Mechanical Properties of Sbs and Asphalt. *J. Chang. Univ. (Nat. Sci. Ed.)* **2017**, *37*, 24–32.
28. Wang, L.; Zhang, L.; Liu, Y. Compatibility of rubber powder and asphalt in rubber powder modified asphalt by molecular dynamics. *J. Build. Mater.* **2018**, *21*, 689–694.
29. Li, C.; Fan, S.Y.; Xu, T. Method for evaluating compatibility between sbs modifier and asphalt matrix using molecular dynamics models. *J. Mater. Civ. Eng.* **2021**, *33*, 04021207. [CrossRef]
30. Rivera, J.L.; Mccabe, C.; Cummings, P. Molecular simulations of liquid-liquid interfacial properties: Water-n-alkane and water-methanol-n-alkane systems. *Phys. Rev. E* **2003**, *67*, 011603. [CrossRef]
31. Tang, F.; Ma, T.; Zhang, J.; Guan, Y.; Chen, L. Integrating Three-Dimensional Road Design and Pavement Structure Analysis Based on BIM. *Autom. Constr.* **2020**, *113*, 103152. [CrossRef]
32. Wu, J.; Du, B.; Li, H. Performance evaluation method of prime coat oil based on average particle size of asphalt/water emulsion. *Highway* **2016**, *61*, 265–269.
33. He, L.; Wen, X.; Hou, Y.; Deng, W. Influence factors of particle size and distribution of anionic asphalt/water emulsion. *J. Chongqing Jiao Tong Univ. (Nat. Sci. Ed.)* **2021**, *40*, 99–104.
34. Sun, Z.; Ma, S.; Zhang, J.; Fuan, S. Microstructure characterization and performance investigation of a rapid hardening asphalt/water emulsion seal mixture. *Road Mater. Pavement Des.* **2021**, *22*, 2767–2782. [CrossRef]
35. Liu, B.J.; Hou, W.L. Influence of storage conditions on the stability of asphalt emulsion. *Pet. Sci. Technol.* **2017**, *35*, 1217–1223. [CrossRef]
36. Kiihnl, L.; Braham, A.F. Developing a particle size specification for asphalt emulsion. *Constr. Build. Mater.* **2021**, *293*, 123414. [CrossRef]
37. Zhang, H. Influence of asphalt/water emulsion particle size distribution on asphalt properties. *Transp. World* **2016**, 110–111.
38. Wei, W.; Guo, P.; Tang, B. Review of the research on diffusion efficiency of virgin-aged asphalt in recycled asphalt mixture. *Mater. Rep.* **2017**, *31*, 109–114.
39. Zhou, Y.; Li, Q.; Tong, Y.; Qu, F.; Wen, G.; Wu, C. Effect of regenerant on macroproperties and microstructure of sbs modified asphalt. *Highway* **2022**, *67*, 302–309.
40. Chen, J.; Li, X.; Han, L.; Shen, Y.; Fang, X. Study on modification mechanism of asphalt by waste tire rubber powder. *Shanxi Archit.* **2021**, *47*, 110–111+196.

materials

Article

Analysis of the Influence of Waste Seashell as Modified Materials on Asphalt Pavement Performance

Guopeng Fan [1], Honglin Liu [1,2,*], Chaochao Liu [1], Yanhua Xue [1], Zihao Ju [1], Sha Ding [3], Yuling Zhang [1] and Yuanbo Li [1]

1 National Engineering Research Center of Highway Maintenance Technology, Changsha University of Science & Technology, Changsha 410114, China
2 Henan Transport Investment Group Co., Ltd., Zhengzhou 450016, China
3 Wuhan Hanyang Municipal Construction Group Co., Ltd., Wuhan 430050, China
* Correspondence: hlliu@hncj.edu.cn

Abstract: An increasing amount of waste seashells in China has caused serious environmental pollution and resource waste. This paper aims to solve these problems by using waste seashells as modified materials to prepare high-performance modified asphalt. In this study, seashell powder (SP) and stratum corneum-exfoliated seashell powder (SCESP) were adopted to prepare 10%, 20% and 30% of seashell powder-modified asphalt (SPMA) and stratum corneum-exfoliated seashell powder-modified asphalt (SCESPMA) by the high-speed shear apparatus, respectively. The appearance and composition of two kinds of SPs were observed and determined by the scanning electron microscope (SEM). The types of functional groups, temperature frequency characteristics, low temperature performance and adhesion of SPMA were tested by the Fourier-transform infrared (FTIR) spectrometer, dynamic shear rheometer (DSR), bending beam rheometer (BBR) and contact angle meter. The results show that the SP and SCESP are rough and porous, and their main component is $CaCO_3$, which is physically miscible to asphalt. When the loading frequency ranges from 0.1 Hz to 10 Hz, the complex shear modulus (G*) and phase angle (δ) of SPMA and SCESPMA increase and decrease, respectively. At the same load frequency, SCESPMA has a larger G* and a smaller δ than SPMA. At the same temperature, SCESPMA has a larger rutting factor (G*/sin δ) and better high-temperature deformation resistance than SPMA. SP and SCESP reduce the low-temperature cracking resistance of asphalt, of which SCESP has a more adverse effect on the low-temperature performance of asphalt than SP. When SP and SCESP are mixed with asphalt, the cohesion work (*Waa*), adhesion work (*Was*) and comprehensive evaluation parameters of water stability (*ER1*, *ER2* and *ER3*) of asphalt are improved. It is shown that both SP and SCESP have good water damage resistance, of which SCESP has better water damage resistance than SP. These research results have important reference value for the application of waste biological materials in asphalt pavement.

Keywords: asphalt pavement; seashell powder; modified asphalt; high- and low-temperature performance; water stability

Citation: Fan, G.; Liu, H.; Liu, C.; Xue, Y.; Ju, Z.; Ding, S.; Zhang, Y.; Li, Y. Analysis of the Influence of Waste Seashell as Modified Materials on Asphalt Pavement Performance. *Materials* **2022**, *15*, 6788. https://doi.org/10.3390/ma15196788

Academic Editor: Simon Hesp

Received: 16 August 2022
Accepted: 24 September 2022
Published: 30 September 2022

Publisher's Note: MDPI stays neutral with regard to jurisdictional claims in published maps and institutional affiliations.

Copyright: © 2022 by the authors. Licensee MDPI, Basel, Switzerland. This article is an open access article distributed under the terms and conditions of the Creative Commons Attribution (CC BY) license (https://creativecommons.org/licenses/by/4.0/).

1. Introduction

The repeated load of vehicles can cause damage to the asphalt pavement. In order to improve the road performance of asphalt pavement, road workers have developed a variety of modifiers [1,2] to enhance the service quality and service life of asphalt pavement [3,4]. With the continuous improvement of environmental protection requirements, the concept of sustainable development has been introduced into road engineering, and more and more bio-based modified materials have been applied to asphalt pavement.

Bio-based modified materials have a series of advantages, such as regeneration, environmental protection and low price. Different bio-based modified materials have various effects on the road performance of asphalt [5]. At present, bio-based modified materials are mainly divided into bio-oil- and animal–plant-modified materials. They can modify some pavement performance of asphalt to obtain modified asphalt that is more beneficial to engineering application value.

The direct use of animals and plants after simple processing as modified asphalt materials has been gradually developed in recent years, and many scholars have conducted a series of researches in this field. Coconut seashell, coconut seashell fiber and palm seashell powders can increase the softening point temperature and rutting factor of asphalt [6–8]. Adding lignin and oil palm seashell powder to asphalt can significantly enhance the high-temperature stability of asphalt [9,10]. The addition of peanut seashell powder to asphalt can strengthen the rutting resistance of asphalt, and weaken its low-temperature performance and fatigue performance. When the content is 10%, the peanut seashell powder has better high-temperature storage stability [11]. Sargassum can promote high-temperature performance of asphalt. It has the best high-temperature stability when the content is 2.5%, and has good compatibility with asphalt [12]. Adding waste leather to 70 # asphalt improves its high-temperature performance, creep recovery ability and anti-aging performance [13]. Waste crayfish seashell powder can improve the high-temperature performance and creep resistance of asphalt [14]. Adding fish scales into neat asphalt can improve the adhesion, viscoelasticity, temperature sensitivity and permanent deformation resistance of asphalt [15]. The fish scale powder treated with silane coupling agent has little effect on the low-temperature and fatigue properties of asphalt, but it can change the viscoelastic components of asphalt, thereby improving the high-temperature stability of asphalt [16]. The addition of eggshell to asphalt mixture can reduce the amount of mineral powder and asphalt. At the same time, the road performance and mechanical property tests of modified asphalt are carried out. It is found that the high-temperature stability of asphalt is improved by the addition of eggshell, which can be applied to asphalt pavement engineering [17]. Some scholars have found that the eggshell powder has a rough, porous and loose structure, which improves the strange hardness and thermal stability. Its interaction with asphalt is a physical miscibility process [18]. When eggshell powder is mixed into asphalt, it makes asphalt hard and increases moisture susceptibility [19]. Laurel scallop, goby scallop and pectin scallop powders can significantly improve the elasticity of asphalt, reduce its low-temperature crack resistance, and keep the types of functional groups inside it unchanged [20]. The medium and low content of oyster powder can improve the low-temperature, rutting and fatigue properties of asphalt. This material can be used as a substitute for filler in asphalt mixtures [21]. The snail seashell powder can be well mixed with asphalt, which can reduce the void of asphalt mixture, thus improving the stability of asphalt mixture, but will reduce the permeability and rotary viscosity of asphalt [22]. Combined with the existing research, it is found that animal-modified materials can improve the high-temperature stability of asphalt to varying degrees. Some animal-modified materials can improve the fatigue performance of asphalt, but have negative effects on the low-temperature performance of asphalt to varying degrees. At the same time, most of these modified materials are physical modifications of asphalt.

China is rich in freshwater shellfish and aquatic products. Some seashell materials are currently used in handicraft manufacturing, decoration materials, agriculture and animal husbandry. These fields have high technical parameter requirements for seashell materials, and not all seashell products can meet these requirements. Therefore, a large number of seashells are directly dumped as garbage every year, resulting in great waste and environmental pollution. However, a large number of stone, asphalt and other materials are required in the construction process of asphalt pavement. If waste seashells can be used in asphalt pavement, the waste seashells pollution will be effectively reduced. At present, relevant scholars have been looking for countermeasures to reduce the environmental pollution caused by waste seashells. Existing researches have studied some waster seashells as modified asphalt materials. There is a lack of research on the road performance of Hyriopsis cumingii seashell powder-modified asphalt. Therefore, it is very meaningful to apply the Hyriopsis cumingii seashell powder to modify the asphalt, thus improving the road performance of asphalt pavement.

2. Materials Selection and Sample Preparation

2.1. Materials

2.1.1. Aggregate and Asphalt

The limestone was used for the test and its surface energy was shown in Table 1. The A-70 # asphalt produced by SK Company in Korea was selected. According to the requirements of relevant specifications, the technical indexes of asphalt were tested [23,24], and its relevant properties are shown in Table 2.

Table 1. Surface energy parameters of limestone.

Aggregate	γ_s/mJ·m^{-2}	γ_s^d/mJ·m^{-2}	γ_s^p/mJ·m^{-2}
Limestone	48.76	43.66	5.10

Table 2. Test results of A-70 # neat asphalt.

Items		Test Specification: JTG E20-2011		
		Measured Values	Technical Requirements	Reference Methods
Penetration (25 °C, 100 g, 5 s) (0.1 mm)		69.9	60~80	T0604-2011.
Penetration index (PI)		0.201	−1.5~+1.0	T0604-2011.
Length (5 cm/min, 10°C) (cm)		33	≥15	T0605-2011.
Length (5 cm/min, 15°C) (cm)		>100	≥100	T0605-2011.
Softening point (Global method) (°C)		49.5	≥46	T0606-2011.
Wax content (distillation) (%)		0.5	<2.2	T0615-2011.
Flash point (°C)		335	≥260	T0611-2011.
Solubility (%)		99.87	≥99.5	T0607-2011.
Density (25 °C) (g/cm)^3)		1.037	Measured	T0603-2011.
60 °C Dynamic viscosity (Pa·s)		199.8	≥180	T0620-2000.
Rolling thin film oven test (163 °C, 85 min)	Mass loss (%)	−0.065	≤±0.8	T0609-2011.
	Residual penetration ratio (%)	81	≥61	T0604-2011.
	Residual ductility (10 °C)	7	≥6	T0605-2011.
	Residual ductility (15 °C)	28	≥15	T0605-2011.

2.1.2. Seashell and Seashell Powder

The seashell of Hyriopsis cumingii was used in the test. Previous studies have shown that the seashell of Hyriopsis cumingii is divided into three layers. From the outside to the inside, it can be divided into the stratum corneum, prismatic layer and nacre, of which the prismatic layer with many holes is a porous structure, as shown in Figure 1 [25].

There are two kinds of seashell powders used in the test, that is, seashell powder (SP) and stratum corneum-exfoliated seashell powder (SCESP). The production process of the two is as follows: put the seashell and seashell with exfoliated stratum corneum into a small-scale high-speed multifunctional pulverizer for grinding, as shown in Figure 2, and then sift the powders through a 0.15 mm sieve to obtain the required ones, as shown in Figures 3 and 4. The physical characteristics of shells and shell powder are shown in Table 3.

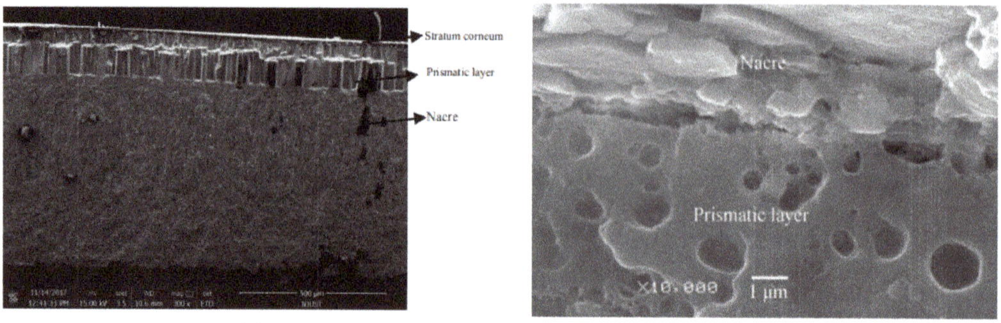

(**a**) (**b**)

Figure 1. Microstructure of Hyriopsis cumingii seashell. (**a**) Distribution of seashell structure layer; (**b**) Distribution of holes in the seashell.

Figure 2. Small high-speed multifunctional pulverizer.

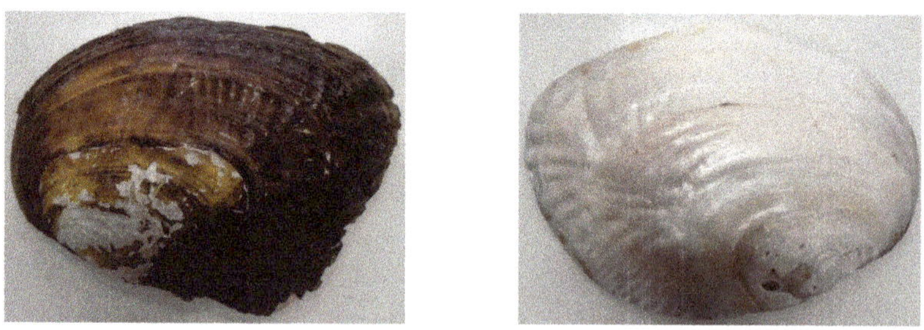

(**a**) (**b**)

Figure 3. Seashell and stratum corneum-exfoliated seashell. (**a**) Seashell; (**b**) Stratum corneum-exfoliated seashell.

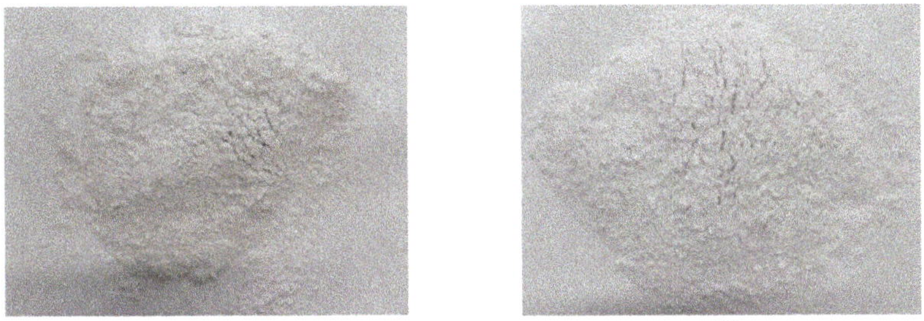

Figure 4. Seashell powder and stratum corneum-exfoliated seashell powder; (a) seashell powder; (b) Stratum corneum-exfoliated seashell powder.

Table 3. Physical characteristics of shell and shell powder.

Seashell Name	Seashell Structure	Types of Seashell Powder	Appearance	Particle Size of Seashell Powder
Hyriopsis cumingii	Stratum corneum Prismatic layer Nacre	Seashell powder	Light yellow powder	<0.15 mm
		Stratum corneum-exfoliated seashell powder	Milky white powder	

2.1.3. Reagent for Contact Angle Test

The distilled water, glycerol and formamide were used in the test, and their surface energy parameters are shown in Table 4.

Table 4. Surface energy parameters of reagents.

Reagents	γ_l/mJ·m^{-2}	γ_l^d/mJ·m^{-2}	γ_l^p/mJ·m^{-2}
Distilled water	72.8	21.8	51
Glycerol	64	34	30
Formamide	58	39	19

2.2. *Sample Preparation*

There have been studies on the use of biological materials to prepare modified asphalt. Generally, 5~30% of biological materials are used to prepare modified asphalt to study the road performance of modified asphalt [20,26]. The preparation of the seashell powder-modified asphalt sample in this study refers to the relevant existing research methods [20]. The sample preparation process is as follows: pour 2100 g neat asphalt into 7 sample cups evenly; heat the asphalt in the sample cup to 150 °C until it becomes complete flow; mix SP and SCESP into 6 sample cups, and stir the mixture for 30 min at 2500–3000 r/min under the high-speed shear instrument, as shown in Table 5. The seven modified asphalt samples were prepared as follows: neat asphalt (NA-70), 10% of SPMA (SPMA-10%), 20% of SPMA (SPMA-20%), 30% of SPMA (SPMA-30%), 10% of SCESPMA (SCESPMA-10%), 20% of SCESPMA (SCESPMA-20%), and 30% of SCESPMA (SCESPMA-30%).

Table 5. Contents of SP and SCESP in modified asphalt.

Sample No.	1	2	3	4	5	6	7
Content of SP (g)	0	10	20	30			
Content of SCESP (g)	-	-	-	-	10	20	30

2.3. *Test Method*

2.3.1. Micro Characteristic Test

In order to study the appearance and chemical composition of SP and SCESP, in this study, the tungsten filament scanning electron microscope (SEM) Evo10 of Zeiss company in Oberkhein, Germany was used to observe the appearance and morphology of SP and SCESP, and the Bruker X-ray energy dispersive spectrometer (EDS) equipped on the instrument was adopted to analyze their main elements and obtain their main chemical components.

2.3.2. Fourier-Transform Infrared (FTIR) Spectroscopy

The FTIR spectroscopy has important advantages in the analysis of organic compounds. It can be used for the analysis of organic compounds with asphalt and modified asphalt. In this study, the Bruker Tensor-27 FTIR spectrometer with scanning range of 500 cm^{-1}–4000 cm^{-1} and scanning times of 64 was used to analyze the functional groups of seven asphalt samples. The FTIR spectrometer was produced by Bruker Corporation in Billerica, USA. Four groups of parallel tests were conducted on each asphalt sample.

2.3.3. Dynamic Shear Rheometer (DSR) Test

For purpose of study the effects of SP and SCESP on the high-temperature rheological properties of asphalt, the viscoelastic properties of SPMA and SCESPMA under different load frequencies were also investigated. The dynamic shear rheometer MCR302 of Anton Paar company in Graz, Austria was used to carry out temperature-frequency scanning test [27]. The loading frequency of frequency scanning test ranges from 0.1 Hz to 10 Hz, and the test temperatures include 58 °C, 64 °C, 70 °C and 76 °C. The complex shear modulus (G^*) and phase angle (δ) of NA-70 and six modified asphalts were tested. The size of the sample is 25 mm in diameter and 1 mm in thickness. Four groups of parallel tests were carried out for each asphalt sample.

2.3.4. Bending Beam Rheometer (BBR) Test

Referring to the relevant requirements of AASHTO 2015, the creep stiffness (S) and creep rate (m) of NA-70 and six modified asphalts were tested with the TE-BBR tester of Cannon Company in State College, USA, and it was required that the beam specimen at 60 s have a $S \leq 300$ MPa and a $m \geq 0.3$. The test temperatures of −12 °C, −18 °C and −24 °C were selected, and four groups of parallel tests were performed.

2.3.5. Contact Angle Test

There is a close relationship between the water stability of asphalt and its adhesion and cohesion properties, and the adhesion and cohesion properties of asphalt can be obtained by contact angle test. As shown in Figure 5, the DSA100 contact angle meter of KRUSS Company in Hamburg, Germany was first used to measure the contact angle of asphalt sample and limestone. The distilled water, glycerol and formamide with known surface energy parameters were chosen as reagents. The interface diagram of reagents and limestone was shown in Figure 6. Five parallel tests were then carried out on each reagent and sample. If the data difference is large, the number of experiments will be supplemented. Finally, the average value of the five test results within the allowable error range was taken as the value of contact angle.

In order to evaluate the deformation resistance of asphalt and the adhesion between the asphalt and limestone, the cohesion and adhesion work indexes were introduced in this study. The calculation process is as follows.

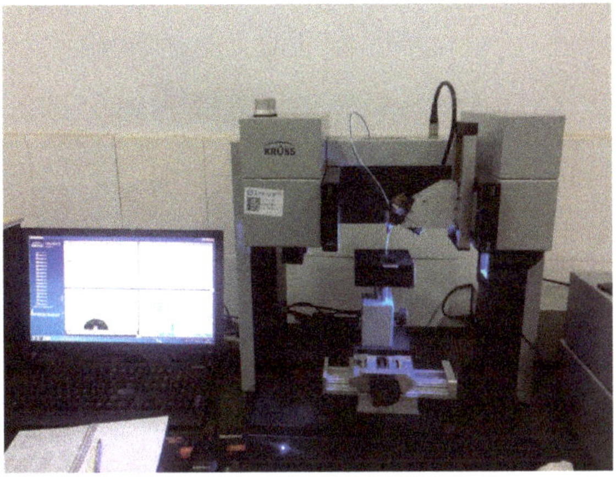

Figure 5. Contact angle meter.

Figure 6. Schematic diagram of interface between the reagent and limestone.

The surface energy is the work required to produce a new interface per unit area of material at certain temperatures, and consists of a dispersion component and a polarity component, which can be expressed as Equation (1):

$$\gamma = \gamma^d + \gamma^p \tag{1}$$

where: γ is the surface energy; γ^d is the dispersive component of the surface energy; and γ^p is the polar component.

The cohesion work W_{aa} refers to the deformation resistance of asphalt and modified asphalt slurry when relative displacement is generated under external force [28]. The greater the cohesion work, the stronger the deformation resistance. It can be expressed as Equation (2):

$$W_{aa} = 2\gamma_a \tag{2}$$

where: W_{aa} is the surface energy of asphalt or modified asphalt.

The adhesion characteristics between the asphalt (or modified asphalt) and limestone can be characterized by the adhesion work W_{as}. The greater the adhesion work, the better the stability of the two after combination. Its expression is as follows:

$$W_{as} = \gamma_a + \gamma_s - \gamma_{as} \tag{3}$$

where: W_{as} is the adhesion work between the asphalt (or modified asphalt) and limestone; γ_a is the surface energy of asphalt or modified asphalt; γ_s is the surface energy of limestone; and γ_{as} is the interface energy of asphalt (or modified asphalt) and limestone [29], which can be expressed as Equation (4).

$$\gamma_{as} = \gamma_a + \gamma_s - 2\sqrt{\gamma_a^d \gamma_s^d} - 2\sqrt{\gamma_a^p \gamma_s^p} \tag{4}$$

where: γ_a^d and γ_s^d are dispersion components of asphalt (or modified asphalt) and limestone surface, respectively; γ_a^p and γ_s^p are polar components of asphalt (or modified asphalt) and limestone surfaces, respectively.

The Young Equation is shown in Equation (5).

$$\gamma_s = \gamma_{as} + \gamma_a \cos\theta \tag{5}$$

The Young–Dupre Equation (6) can be obtained from Equations (1)–(5).

$$W_{as} = \gamma_a(1+\cos\theta) = 2\sqrt{\gamma_a^d\gamma_s^d} + 2\sqrt{\gamma_a^p\gamma_s^p} \tag{6}$$

The water damage resistance of asphalt mixture can be characterized by the spalling work (W_{asw}). The W_{asw} is characterized by the external work required for the stripping of asphalt (or modified asphalt) and limestone under the action of water [30], which can be expressed by Equation (7).

$$W_{asw} = W_{aw} + W_{sw} - W_{as} \tag{7}$$

where W_{aw} is the adhesion work between the asphalt (or modified asphalt) and distilled water, and W_{sw} is the adhesion work between the limestone and distilled water.

However, this index is not comprehensive. At present, scholars at home and abroad have found through research that the water damage resistance of asphalt and limestone is greatly related to the adhesion work, cohesion work and peeling work. Therefore, three water stability parameters *ER1*, *ER2* and *ER3* of asphalt and limestone were proposed [31]. The larger their values, the stronger the water damage resistance of asphalt and limestone, as shown in Equations (8)–(10).

$$ER1 = \left|\frac{W_{as}}{W_{asw}}\right| \tag{8}$$

$$ER2 = \left|\frac{W_{as} + W_{aa}}{W_{asw}}\right| \tag{9}$$

$$ER3 = \left|\frac{\min(W_{as}, W_{aa})}{W_{asw}}\right| \tag{10}$$

3. Results and Discussion

3.1. Microscopic Characteristics of Seashell Powder

The SEM has unique advantages in observing the micro morphology of materials. Therefore, this study used this equipment to observe the appearance and chemical composition of SP and SCESP. The effects of appearance morphology and chemical composition of the two particles on the performance of NA-70 were explored, and the test results are shown in Figure 7 and Table 6.

From Figure 7, it can be seen that the surface of SP and SCESP are rough with more folds, unevenness, internal pores and porosity. The SCESP has more internal holes and pores than the SP. It has been found that there are many holes in the prismatic layer of the seashell, which may be closely related to the existence of a large number of holes in SP and SCESP. There is more contact area between the SP (or SCESP) and NA-70 after high-speed shear stirring, which helps form a strong adsorption capacity between them.

Table 5 shows that the main elements of SP and SCESP are C, O, CA, Si, S, Al and Mg, and the main compounds are $CaCO_3$, SiO_2, Al_2O_3 and MgO; the content of $CaCO_3$ is the highest and the composition is similar to that of limestone, which indicates that when SP and SCESP are added to NA-70, $CaCO_3$ plays a major role in the change in road performance of NA-70.

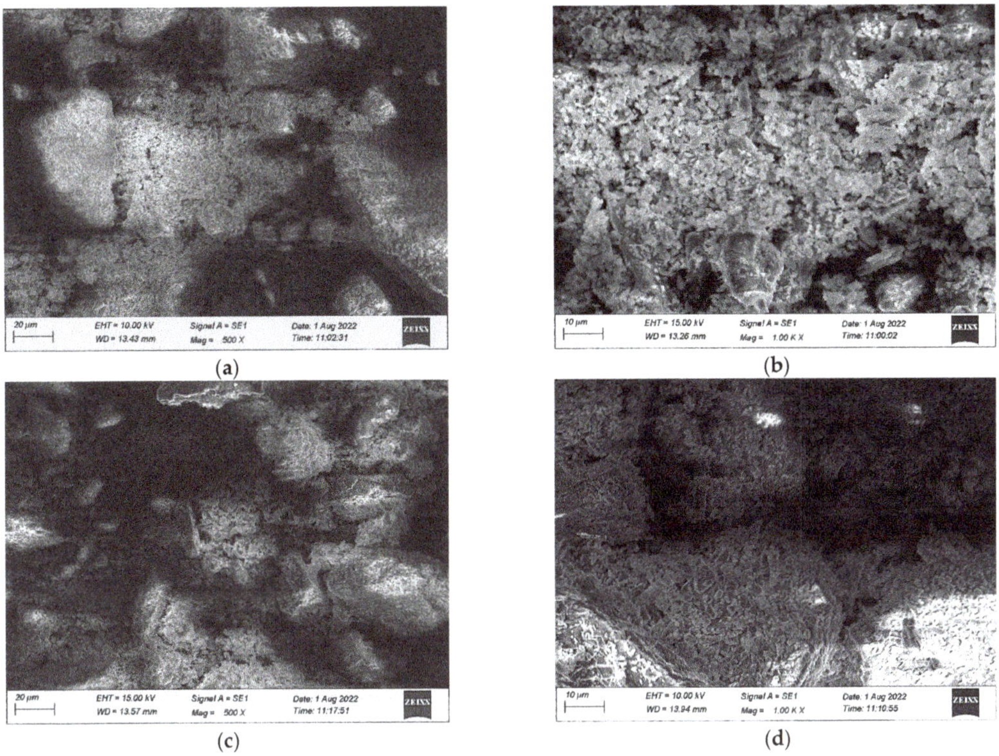

Figure 7. Microscopic images of SP and SCESP. (**a**) SCESP × 500; (**b**) SCESP × 1000; (**c**) SP × 500; (**d**) SP × 1000.

Table 6. Main components of SP and SCESP.

Sample Type	Main Components						
SP	Element	C	O	Ca	Si	S	Al
	Weight (%)	42.54	37.80	12.39	3.86	1.85	1.56
	Chemical compound	$CaCO_3$		SiO_2		Al_2O_3	
	Weight (%)	76.32		15.25		8.43	
SCESP	Element	C	O	Ca	Si	S	Mg
	Weight (%)	33.25	44.55	18.58	1.85	0.91	0.86
	Chemical compound	$CaCO_3$		SiO_2		MgO	
	Weight (%)	86.62		8.23		5.15	

3.2. Functional Group Analysis

There are many functional groups in asphalt [32]. In order to analyze the effects of SP and SCESP on the functional groups of NA-70, this study used the FTIR spectroscopy to analyze the functional groups of SPMA and SCESPMA, and the test results are shown in Figure 8.

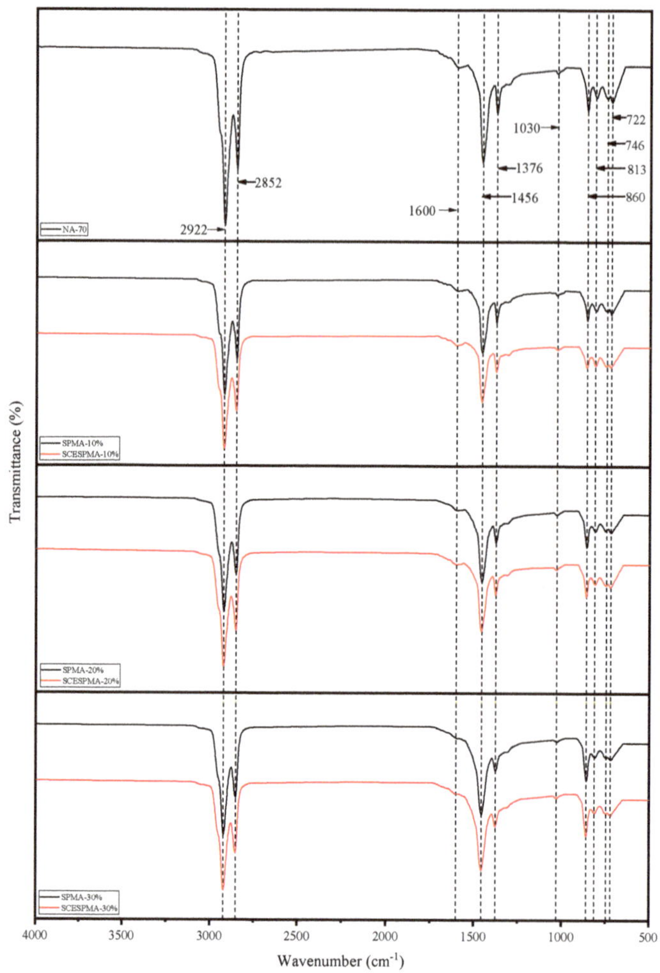

Figure 8. Functional groups of SPMA and SCESPMA.

It can be seen from Figure 8 that in the range of 4000–500 cm^{-1}, NA-70, SPMA and SCESPMA have the same absorption peak. There are characteristic peaks caused by stretching vibration near 2924 cm^{-1} and 2853 cm^{-1}, C=C stretching vibration characteristic peaks near 1600 cm^{-1}, aliphatic C-H bending deformation near 1456 cm^{-1} and 1376 cm^{-1}, sulfoxide group (S=O) absorption peaks near 1030 cm^{-1}, and C-H bond bending vibration near 860 cm^{-1}, 813 cm^{-1}, 746 cm^{-1} and 722 cm^{-1}. It is shown that no new chemical components are generated after the two are mixed, which is physically miscible. Therefore, the modification of SP and SCESP to NA-70 is physical.

3.3. Analysis of High-Temperature Rheological Properties

In order to study the effect of frequency on their viscoelasticity, the frequency sweep of SPMA and SCESPMA was carried out to obtain G* and δ. The curves with frequency are shown in Figures 9 and 10, respectively.

It can be seen from Figure 9 that the G* of SPMA and SCESPMA increases linearly with the increase in load frequency at 58 °C~88 °C. The specific rules are: SCESPMA-30% > SPMA-30% > SCESPMA-20% > SPMA-20% > SCESPMA-10% >SPMA 10% > NA-70.

When the load frequency ranges from 0.1 Hz to 10 Hz, the G* of SPMA and SCESPMA increases; when the temperature increases, the G* of SPMA and SCESPMA decreases. It is shown that the load frequency has an important influence on the viscoelasticity of SPMA and SCESPMA. The addition of SPMA and SCESPMA can improve the load resistance of asphalt. The SCESP has better deformation resistance, indicating that the stratum corneum of seashell has an adverse effect on the deformation resistance of asphalt. In addition, SPMA is more sensitive to temperature.

It can be seen from Figure 10 that at 58 °C~76 °C, the δ of NA-70, SPMA and SCESPMA gradually decreases with the increase in frequency. At 0.1 Hz~1.59 Hz, the value of δ decreases significantly, while at 1.59 Hz~10 Hz, the value of δ decreases slowly. The specific rules are: NA-70 > SCESPMA-10% > SCESPMA-20% > SPMA-10% > SCESPMA-30% > SPMA-20% > SPMA-30%. It is indicated that the temperature has a significant effect on the viscoelasticity of SPMA and SCESPMA; as the temperature increases, the effect is more significant. At low frequency, SPMA and SCESPMA show more viscous performance, while with the increase in frequency, the viscous performance decreases, and the elastic performance increases.

To analyze the high-temperature deformation resistance of SPMA and SCESPMA, the rutting factors (G* and sin δ) at 58 °C, 64 °C, 70 °C and 76 °C were measured at 1.59 Hz, and the results are shown in Figure 11.

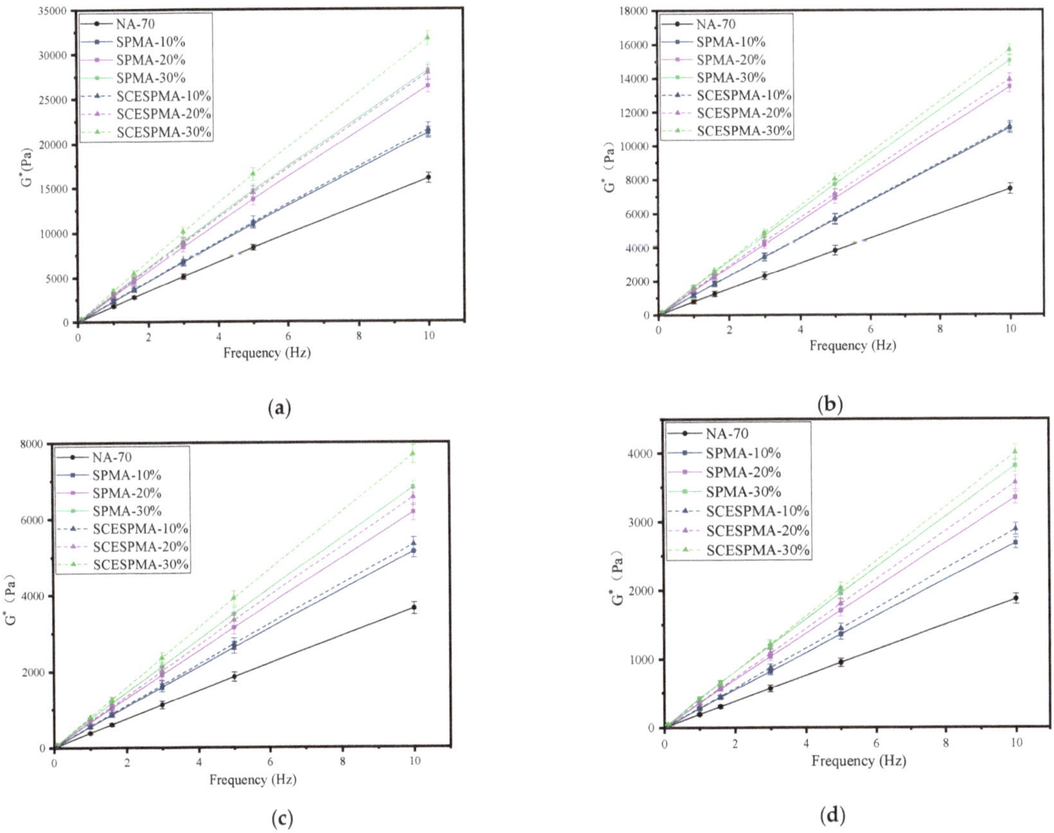

Figure 9. Relationship among G * and frequency of SPMA and SCESPMA at different temperatures. (**a**) 58 °C; (**b**) 64 °C; (**c**) 70 °C; (**d**) 76 °C.

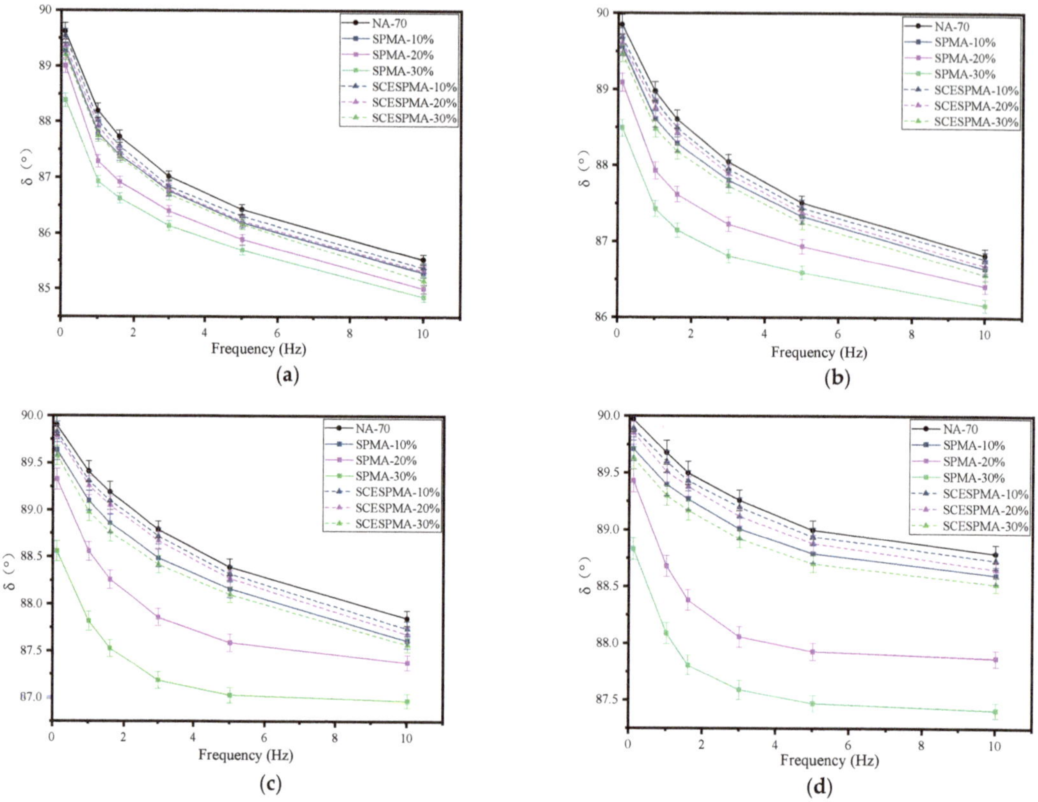

Figure 10. Relationship among δ and frequency of SPMA and SCESPMA at different temperatures. (**a**) 58 °C; (**b**) 64 °C; (**c**) 70 °C; (**d**) 76 °C.

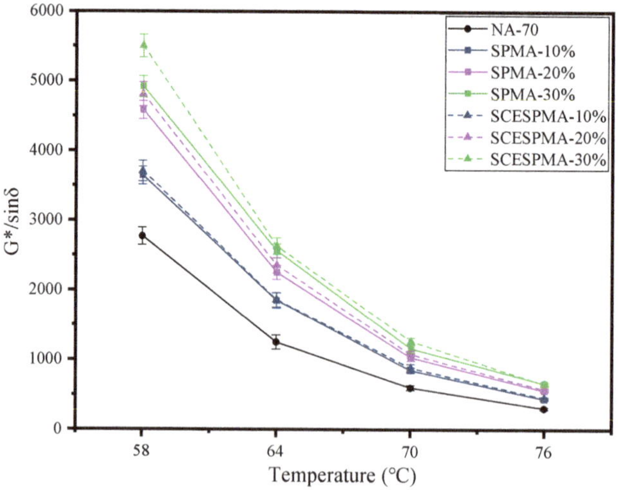

Figure 11. G* and sin δ of SPMA and SCESPMA at different temperatures.

From Figure 11, it can be seen that the G* and sin δ of NA-70, SPMA and SCESPMA gradually decrease from 58 °C to 76 °C, and their specific rules are: SCESPMA-30% > SPMA-30% > SCESPMA-20% > SPMA-20% > SCESPMA-10% > SPMA-10% > NA-70. It is shown that the temperature has significant effect on the G* and sin δ of SPMA and SCESPMA, which becomes less and less significant with the increase in temperature. SCESPMA has better high-temperature rutting resistance than SPMA. Previous studies have analyzed the impact of bio-based modified materials on the high-temperature performance of asphalt, and few have analyzed the impact of different components of these materials on the G*/sin δ of asphalt [11]. The effects of SP and SCESP on the G*/sin δ of asphalt were analyzed, respectively, in this study, and the potential of seashell materials to improve the high-temperature performance of asphalt was studied in detail.

The U.S. Strategic Highway Research Program (SHRP) uses the rutting factors G* and sin δ as rating indexes for the high-temperature performance of asphalt [33,34]. In this study, a continuous high-temperature performance gradation (PG) analysis was conducted for SPMA and SCESPMA with reference to ASTM D 7643-16, and the specific results are shown in Figure 12.

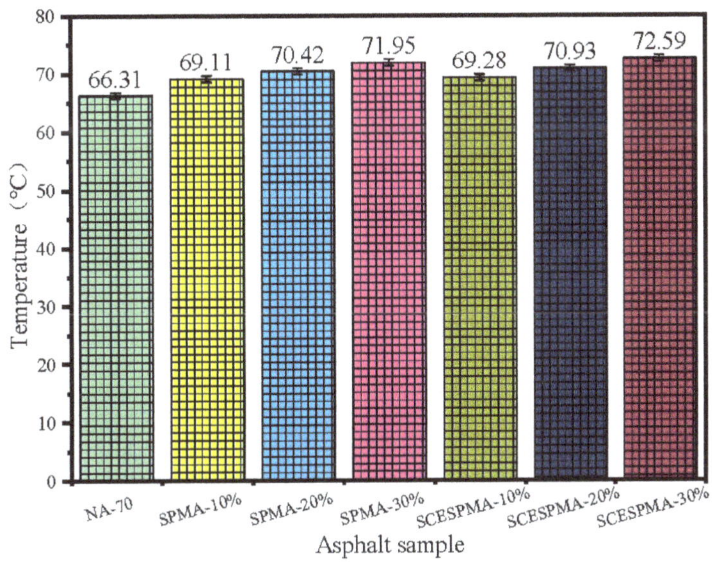

Figure 12. Continuous high-temperature PG analysis of SPMA and SCESPMA.

As shown in Figure 12, the temperature T_{cH} of continuous high-temperature PG of NA-70 increases as the contents of SP and SCESP increase. Compared with that of NA-70, the T_{cH} of SCESPMA-10%, SCESPMA-20% and SCESPMA-30% increases by 4.49%, 6.98% and 9.49%, respectively, and the T_{CH} of SPMA-10%, SPMA-20% and SPMA-30% increases by 4.22%, 6.21% and 8.52%, respectively.

It can be found that the temperature T_{CH} of SCESPMA is higher than that of SPMA at the same content. It is indicated that SCESP improves the high-temperature performance of NA-70, which may be related to the prismatic layer of seashell with more holes. The holes increases the contact surface between the seashell powder and NA-70, thus improving its high-temperature rutting resistance.

3.4. Analysis of Rheological Properties at Low Temperature

The evaluation parameters of low-temperature performance of asphalt are usually *S* and *m* [35,36]. In this study, the *S* and *m* values of NA-70 and SPMA and SCESPMA were tested, and the values of S and m are shown in Figure 13. To analyze the low-temperature cracking resistance of NA-70, SPMA and SCESPMA more comprehensively, this study used the comprehensive evaluation parameter of low-temperature performance $\lambda = S/m$ (the smaller the λ value, the better the low temperature performance), and the values of λ are shown in Figure 14.

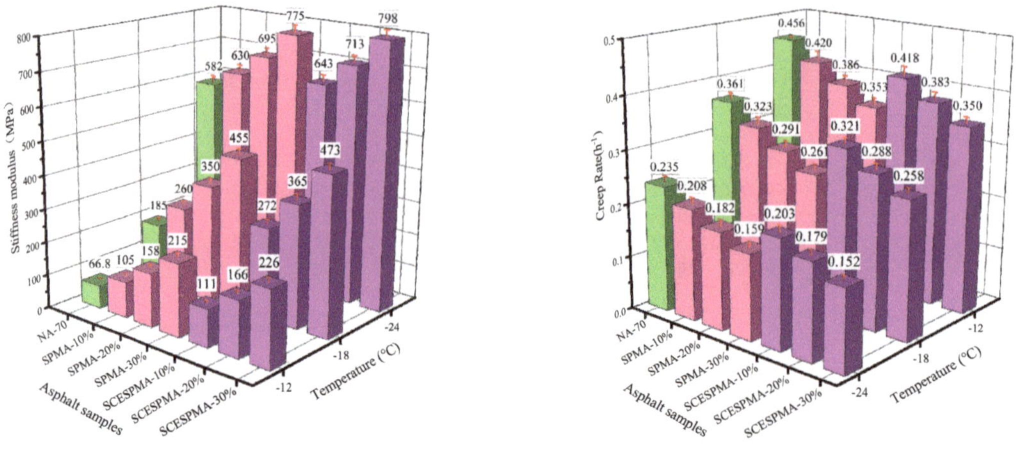

(a) (b)

Figure 13. S and m of SPMA and SCESPMA at different temperatures. (**a**) S; (**b**) m.

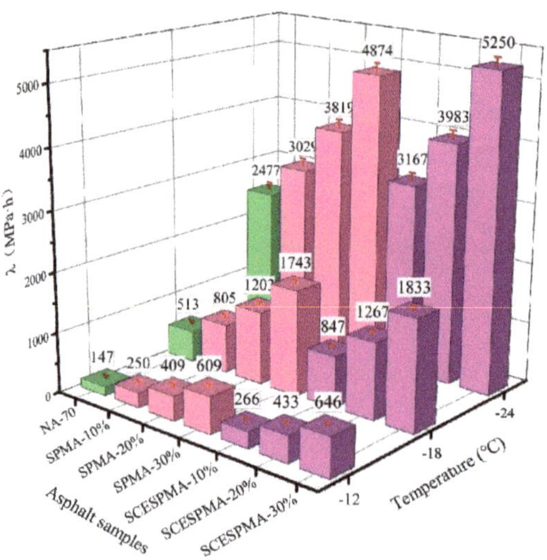

Figure 14. λ of SPMA and SCESPMA at different temperatures.

It can be observed from Figure 13a that the S value of SCESPMA are larger than that of SPMA at the same temperature; with the increase in SP and SCESPMA content, the S value of NA-70 increases. The results show that the low-temperature bending creep performance of NA-70 with SP and SCESPMA added is worse, and the low-temperature bending creep performance of SPMA is better than that of SCESPMA at the same content.

It is obvious from Figure 14 that the λ value of SPMA is smaller than that of SCESPMA at the same temperature, and the λ values of both SPMA and SCESPMA increase as the contents of SP and SCESP increase. It is indicated that SPMA has better low-temperature crack resistance than SCESPMA. Previous studies have found that adding other seashell materials to asphalt will generally reduce its low-temperature crack resistance [20,21], which is consistent with the results of this study. However, this study finds that the stratum corneum of seashell has better low-temperature crack resistance than other structural layers.

In this study, NA-70, SPMA and SCESPMA were analyzed by using the index T_{cL} of continuous low-temperature PG with reference to ASTM D 7643-16, and the values of T_{cL} are shown in Figure 15.

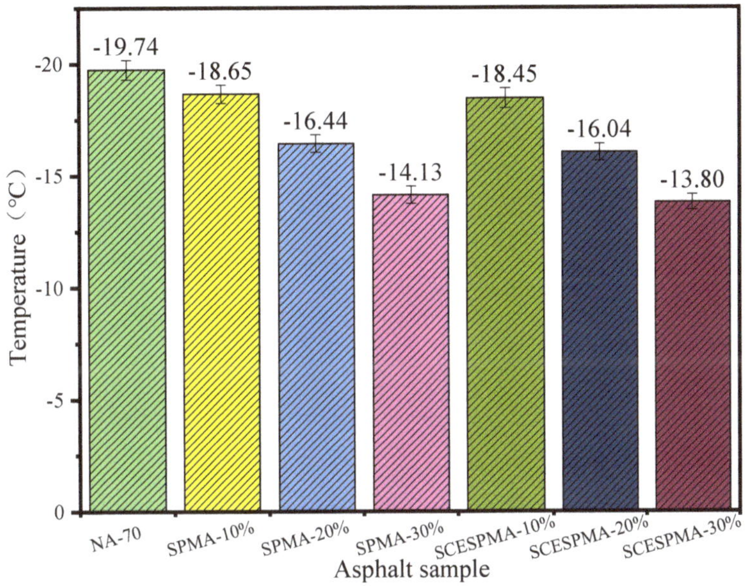

Figure 15. Continuous low-temperature PG analysis of SPMA and SCESPMA.

In Figure 15, it is shown that the temperature T_{cL} of continuous low-temperature PG of NA-70 increases with the increase in SP and SCESP contents; compared with that NA-70, the T_{cL} of SPMA increases by 5.52%, 16.72%, and 28.44%, respectively, and the T_{cL} of SCESPMA increases by 6.51%, 18.73% and 30.10%, respectively.

3.5. Analysis of Water Stability

To verify the reliability of contact angle test results, after obtaining the contact angles of NA-70, SPMA and SCESPMA with three reagents, a linear fitting was performed on the surface energy γ_l and $\gamma_l \cos\theta$ of different reagents, and the test results are shown in Figure 16.

According to Figure 16, γ_l and $\gamma_l \cos\theta$ of NA-70 and SPMA and SCESPMA show good linear correlation above 0.89, indicating that the contact angle test data are reliable.

Based on the data in Tables 1 and 3, W_{aa}, W_{as} and W_{asw} of NA-70, SPMA and SCESPMA were obtained by Equations (2), (3) and (7), and the test results are shown in Figure 17.

As shown in Figure 17, the specific rules of *Waa* and *Was* of SPMA and SCESPMA are: SCESPMA-30% > SPMA-30% > SCESPMA-20% > SPMA-20%> SCESPMA-10% > SPMA-10% > NA-70. It is found that compared with NA-70, SCESPMA improves *Waa* and *Was* by 9.91~27.70% and 5.44~15.04%, respectively, and SPMA enhances *Waa* and *Was* by 7.08~26.23% and 4.26~14.51%, respectively. The specific rules of *Wasw* of SPMA and SCESPMA are: NA-70 > SCESPMA-10% > SPMA-10% > SCESPMA-20%> SPMA-20%> SCESPMA-30% > SPMA-30%. It is shown that compared with NA-70, the *Wasw* of SCESPMA and SPMA is reduced by 0.38~1.83% and 0.72~2.1%, respectively.

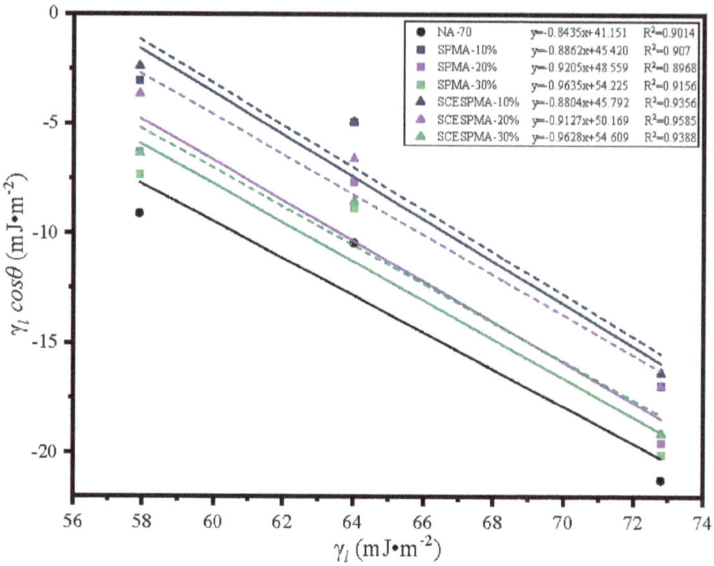

Figure 16. Relationship between γ_l and $\gamma_l cos\theta$.

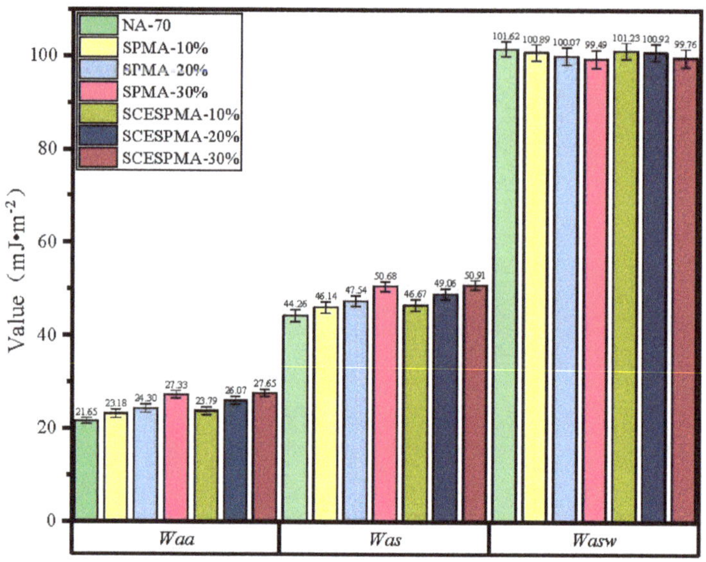

Figure 17. Adhesion properties of SPMA and SCESPMA.

The results show that both SP and SCESP improve the *Waa* and *Was* of NA-70 except for the W_{asw}. SCESPMA has better adhesion and cohesion than SPMA, indicating that the stratum corneum in the seashell is not conducive to improving the adhesion and cohesion of NA-70. The existing research on bio-based modified asphalt materials mainly focuses on its high- and low-temperature rheological properties, micro morphology and chemical composition [15,37], and rarely involves the adhesion and cohesion properties of seashell powder-modified asphalt. The results of this study show that the water stability of asphalt and limestone is not only related to W_{asw}, but also has a close relationship with W_{aa} and W_{as}, and the above three indexes should be considered comprehensively in this study. Hence, this study introduced the comprehensive evaluation parameters *ER1*, *ER2* and *ER3* for asphalt and limestone, so as to perform a comprehensive evaluation on the water stability of SPMA, SCESPMA and limestone, and the test results are shown in Figure 18.

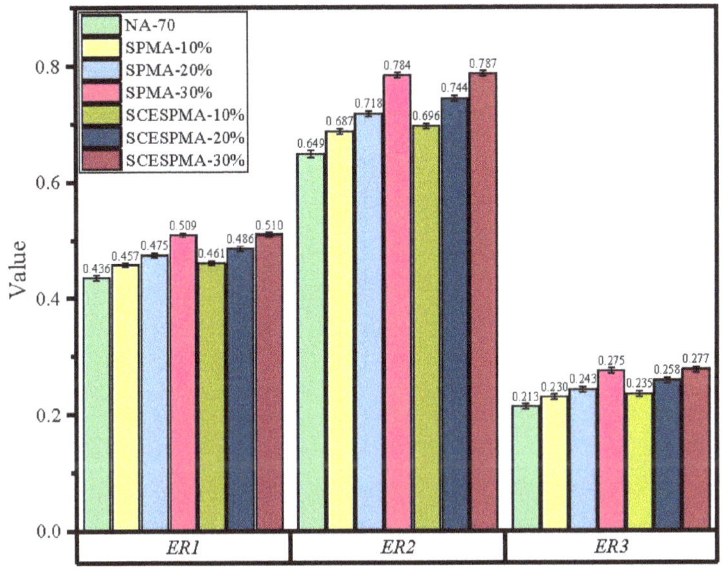

Figure 18. Comprehensive water stability evaluation parameters of SPMA and SCESPMA.

It can be seen from Figure 18 that the specific rules of *ER1*, *ER2* and *ER3* of SPMA and SCESPMA are: SCESPMA-30% > SPMA-30% > SCESPMA-20% > SPMA-20%> SCESPMA-10% > SPMA-10% > NA-70; as the contents of SP and SCESP increase, *ER1*, *ER2* and *ER3* increase; and at the same content, the *ER1*, *ER2* and *ER3* values of SCESPMA are larger than those of SPMA. It shows that SP and SCESP can improve the water damage resistance of NA-70 and limestone, and SCESP is more conducive to improving the water damage resistance of NA-70 than SP, which indicates that the stratum corneum in the seashell is not conducive to the water stability of asphalt and limestone.

4. Conclusions

In this study, the potential of SP and SCESP as asphalt modifiers was discovered, and the high-low temperature performance and water stability of SPMA and SCESPMA were studied. The main conclusions can be summarized as follows.

(1) The surface of SP and SCESP is rough with more pores and holes, and their main component is $CaCO_3$. SCESP has more pores and holes than SP, and has better adsorption effect during mixing with asphalt, thus giving the modified asphalt a better adhesion effect. From the perspective of the type and content of characteristic functional groups of modified asphalt. The modification of SP and SCESP to NA-70 is a physical process.

(2) Temperature–frequency tests show that SP and SCESP improve the high-temperature deformation resistance of NA-70. The G* of SPMA and SCESPMA is linear with the load frequency. At the same temperature, the G* of SCESPMA is more sensitive to the load frequency than that of SPMA, and SCESP has better high-temperature improvement effect on NA-70 than SP.

(3) BBR tests indicate that SP and SCESP increase the creep modulus (S) and reduce the creep rate (m) of NA-70, resulting in a decrease in the low-temperature crack

resistance of NA-70, and SCESP is more detrimental to the low temperature crack resistance of NA-70 than SP at the same content.

(4) The contact angle test shows that SP and SCESP enhance the cohesion work, adhesion work and water stability of NA-70; in addition, at the same content, SCESP strengthens the cohesion work, adhesion work and water stability of NA-70 better than SP.

Author Contributions: Data curation, Y.X.; Formal analysis, Z.J.; Investigation, Y.Z. and Y.L.; Project administration, Validation S.D.; Supervision, Writing—review & editing C.L.; Validation, Y.Z. and Y.L.; Writing—original draft, G.F.; Writing—review & editing, H.L. All authors have read and agreed to the published version of the manuscript.

Funding: This work was supported by Guopeng Fan's Postgraduate Scientific Research Innovation Project of Hunan Province (CX20190642), and the Research on green recycling technology of asphalt mixture (20192khx111).

Institutional Review Board Statement: Not applicable.

Informed Consent Statement: Not applicable.

Data Availability Statement: Data available on request due to restrictions e.g., privacy or ethical.

Acknowledgments: This work was supported by the Postgraduate Scientific Research Innovation Project of Hunan Province (CX20190642), and the Research on green recycling technology of asphalt mixture (20192khx111).

Conflicts of Interest: The authors declare there are no conflict of interest.

References

1. Lv, S.; Yuan, J.; Peng, X.; Cabrera, M.B.; Guo, S.; Luo, X.; Gao, J. Performance and optimization of bio-oil/Buton rock asphalt composite modified asphalt. *Constr. Build. Mater.* **2020**, *264*, 120235. [CrossRef]
2. Wang, P.; Tian, X.; Zhang, R.; Zhen, S. Effect of Waterborne Epoxy Resin on Properties of Modified Emulsified Asphalt and Its Microstructure. *J. Mater. Civ. Eng.* **2021**, *33*, 4021177. [CrossRef]
3. Chen, W.; Li, Y.; Chen, S.; Zheng, C. Properties and economics evaluation of utilization of oil shale waste as an alternative environmentally-friendly building materials in pavement engineering. *Constr. Build. Mater.* **2020**, *259*, 119698. [CrossRef]
4. Tang, J.; Fu, Y.; Ma, T.; Zheng, B.; Zhang, Y.; Huang, X. Investigation on low-temperature cracking characteristics of asphalt mixtures: A virtual thermal stress re-strained specimen test approach. *Constr. Build. Mater.* **2022**, *347*, 128541. [CrossRef]
5. Zhang, F.; Hu, C.; Zhang, Y. Research for SEBS/PPA compound-modified asphalt. *J. Appl. Polym. Sci.* **2018**, *135*, 46085. [CrossRef]
6. Buhari, R.; Zabidi, N.F.; Abdullah, M.E.; Abu Bakar, S.K.; Kamarudin, N.H.M. Rheological Behaviour of Coconut Shell Powder Modified Asphalt Binder. *J. Teknol.* **2016**, *78*, 17–22. [CrossRef]
7. Bethary, R.T.; Intari, D.E.; Fathonah, W.; Andika, S. The Evaluation of The use of Palm Shell Ash Waste to Polymer Modified Asphalt Mixture. *IOP Conf. Ser. Earth Environ. Sci.* **2021**, *830*, 012010. [CrossRef]
8. Jeffry, S.N.A.; Jaya, R.P.; Hassan, N.A.; Yaacob, H.; Satar, M.K.I.M. Mechanical performance of asphalt mixture containing nano-charcoal coconut shell ash. *Constr. Build. Mater.* **2018**, *173*, 40–48. [CrossRef]
9. Fu, X.; He, M.; Liu, Y. Research on Short-term Aging of Lignin Modified Asphalt. *E3S Web Conf.* **2021**, *233*, 01125. [CrossRef]
10. Ritonga, W.; Ritonga, N.Y.; Rangkuti, M.A.; Eddiyanto, E.; Prawijaya, S. The effect of oil palm shell ash to asphalt characteristics. *J. Phys. Conf. Ser.* **2019**, *1282*, 012031. [CrossRef]
11. Arabani, M.; Esmaaeli, N. Laboratory evaluation on effect of groundnut shell ash on performance parameters of asphalt bind-er and mixes. *Road Mater. Pavement Des.* **2020**, *21*, 1565–1587. [CrossRef]
12. Salazar-Cruz, B.; Zapien-Castillo, S.; Hernández-Zamora, G.; Rivera-Armenta, J. Investigation of the performance of asphalt binder modified by sargassum. *Constr. Build. Mater.* **2020**, *271*, 121876. [CrossRef]
13. Lv, S.; Guo, Y.; Xia, C.; Liu, C.; Hu, L.; Guo, S.; Wang, X.; Cabrera, M.B.; Li, M. Investigation on high-temperature resistance to permanent deformation of waste leather modified asphalt. *Constr. Build. Mater.* **2021**, *282*, 122541. [CrossRef]
14. Lv, S.; Xia, C.; Yang, Q.; Guo, S.; You, L.; Guo, Y.; Zheng, J. Improvements on high-temperature stability, rheology, and stiffness of asphalt binder modified with waste cray-fish shell powder. *J. Clean. Prod.* **2020**, *264*, 121745. [CrossRef]
15. Lv, S.; Hu, L.; Xia, C.; Cabrera, M.B.; Guo, Y.; Liu, C.; You, L. Recycling fish scale powder in improving the performance of asphalt: A sustainable utilization of fish scale waste in asphalt. *J. Clean. Prod.* **2020**, *288*, 125682. [CrossRef]
16. Lv, S.; Hu, L.; Xia, C.; Peng, X.; Cabrera, M.B.; Guo, S.; You, L. Surface-treated fish scale powder with silane coupling agent in asphalt for performance improvement: Conventional properties, rheology, and morphology. *J. Clean. Prod.* **2021**, *311*, 127772. [CrossRef]
17. Razzaq, A.K.; Yousif, R.A.; Tayh, S.A. Characterization of hot mix asphalt modified by egg shell powder. *Int. J. Eng. Res. Technol.* **2018**, *11*, 481–492.

18. Ji, G.; Wang, X.; Guo, Y.; Zhang, Y.; Yin, Q.; Luo, Y. Study on the Physical, Chemical and Nano-Microstructure Characteristics of Asphalt Mixed with Recycled Eggshell Waste. *Sustainability* **2021**, *13*, 11173. [CrossRef]
19. Huang, J.; Kumar, G.S.; Ren, J.; Sun, Y.; Li, Y.; Wang, C. Towards the potential usage of eggshell powder as bio-modifier for asphalt binder and mixture: Workability and mechanical properties. *Int. J. Pavement Eng.* **2021**, *23*, 3553–3565. [CrossRef]
20. Guo, Y.; Wang, X.; Ji, G.; Zhang, Y.; Su, H.; Luo, Y. Effect of Recycled Shell Waste as a Modifier on the High-and Low-Temperature Rheological Properties of As-phalt. *Sustainability* **2021**, *13*, 10271. [CrossRef]
21. Nciri, N.; Shin, T.; Lee, H.; Cho, N. Potential of Waste Oyster Shells as a Novel Biofiller for Hot-Mix Asphalt. *Appl. Sci.* **2018**, *8*, 415. [CrossRef]
22. Modupe, A.E.; Fadugba, O.G.; Busari, A.A.; Adeboje, A.O.; Aladegboye, O.J.; Alejolowo, O.O.; Chukwuma, C.G. Sustainability Assessment of the Engineering Properties of Asphalt Concrete Incorporating Pulverized Snail Shell Ash as Partial Replacement for Filler. *IOP Conf. Ser. Earth Environ. Sci.* **2021**, *665*, 012057. [CrossRef]
23. Chu, C.; Zhu, J.; Wang, Z.-A. Investigation of the Road Performance of AH-30 Bitumen and Its Mixture: Comparison with AH-70 and SBS-Modified Bitumen. *Adv. Mater. Sci. Eng.* **2021**, *2021*, 3124047. [CrossRef]
24. *JTG/T 5521–2019*; Technical Specifications for Highway Asphalt Pavement Recycling: Industry Recommended Standards of the People's Republic of China. China Communications Press: Beijing, China, 2019.
25. Shao, H.; Zhu, J.; Zhou, Q.; Lei, J.; Xu, J.; Wang, K. Characteristics of microstructure and size change of the shell of *Hyriopsis cumingii*. *Acta Mater. Compos.-Tate Sin.* **2019**, *36*, 2398–2406.
26. Geng, L.; Liu, Y.; Xu, Q.; Han, F.; Yu, X.; Qin, T. Development of bio-based stabilizers and their effects on the performance of SBS-modified asphalt. *Constr. Build. Mater.* **2020**, *271*, 121889. [CrossRef]
27. *AASHTO T 315-20*; Standard Method of Test for Determining the Rheological Properties of Asphalt Binder Using a Dynamic Shear Rheometer (DSR). American Association of State Highway and Transportation Officials: Washington, DC, USA, 2019.
28. Sun, Y.S.; Chen, X.L.; Han, Y.X.; Zhang, B. Research on Performance of the Modified Asphalt by Diatomite-Cellulose Composite. *Adv. Mater. Res.* **2010**, *158*, 211–218.
29. Zheng, Y.X.; Cai, Y.C.; Zhang, Y.M. Laboratory Study of Pavement Performance of Basalt Fiber-Modified Asphalt Mixture. In *Advanced Materials Research*; Trans Tech Publications: Zurich, Switzerland, 2011; pp. 175–179.
30. Yang, J.; Zhang, Z.; Shi, J.; Yang, X.; Fang, Y. Comparative analysis of thermal aging behavior and comprehensive performance of high viscosity asphalt (HVA) from cohesion, adhesion and rheology perspectives. *Constr. Build. Mater.* **2022**, *317*, 125982. [CrossRef]
31. Han, M.; Li, J.; Muhammad, Y.; Hou, D.; Zhang, F.; Yin, Y.; Duan, S. Effect of polystyrene grafted graphene nanoplatelets on the physical and chemical properties of asphalt binder. *Constr. Build. Mater.* **2018**, *174*, 108–119. [CrossRef]
32. Da Silva, A.M.; De Souza, K.R.; Jacobs, G.; Graham, U.M.; Davis, B.H.; Mattos, L.V.; Noronha, F.B. Steam and CO_2 reforming of ethanol over Rh/CeO_2 catalyst. *Appl. Catal. B Environ.* **2011**, *102*, 94–109. [CrossRef]
33. Yu, M.; Li, J.; Cui, X.; Guo, D.; Li, X. Antiageing Performance Evaluation of Recycled Engine Oil Bottom Used in Asphalt Rejuvenation. *Adv. Mater. Sci. Eng.* **2019**, *2019*, 2947170. [CrossRef]
34. Zhang, H.; Gong, M.; Huang, Y.; Miljković, M. Study of the high and low-temperature behavior of asphalt based on a performance grading system in Northeast China. *Constr. Build. Mater.* **2020**, *254*, 119046. [CrossRef]
35. Chen, Z.; Yi, J.; Chen, Z.; Feng, D. Properties of asphalt binder modified by corn stalk fiber. *Constr. Build. Mater.* **2019**, *212*, 225–235. [CrossRef]
36. Yang, Q.; Li, X. Mechanism and Effectiveness of a Silicone-Based Warm Mix Additive. *J. Mater. Civ. Eng.* **2019**, *31*, 04018336. [CrossRef]
37. Wang, X.; Guo, Y.; Ji, G.; Zhang, Y.; Zhao, J.; Su, H. Effect of Biowaste on the High- and Low-Temperature Rheological Properties of Asphalt Binders. *Adv. Civ. Eng.* **2021**, *2021*, 5516546. [CrossRef]

MDPI
St. Alban-Anlage 66
4052 Basel
Switzerland
Tel. +41 61 683 77 34
Fax +41 61 302 89 18
www.mdpi.com

Materials Editorial Office
E-mail: materials@mdpi.com
www.mdpi.com/journal/materials

www.ingramcontent.com/pod-product-compliance
Lightning Source LLC
Chambersburg PA
CBHW061836210926
51850CB00006B/902